Lecture Notes in Mathematics

Edited by A. Dold and B. Eckmann

Analysis
cations

nternational Conference
pril 11–15, 1983

Edited by A. Truman and D. Williams

Springer-Verlag
Berlin Heidelberg New York Tokyo 1984

Editors

Aubrey Truman
David Williams
Department of Mathematics and Computer Science
University College of Swansea
Singleton Park, Swansea SA2 8PP, Wales

AMS Subject Classification (1980): 60H05, 60H10

ISBN 3-540-13891-9 Springer-Verlag Berlin Heidelberg New York Tokyo
ISBN 0-387-13891-9 Springer-Verlag New York Heidelberg Berlin Tokyo

Printing and binding: Beltz Offsetdruck, Hemsbach / Bergstr.
2146/3140-543210

PREFACE

This volume contains a number of papers presented at the Workshop on Stochastic Analysis and its Applications, held in Swansea from 11 April to 15 April 1983, together with some more recent research papers by the Swansea school. The applications include such diverse topics as stochastic mechanics and the Titius-Bode law, non-standard Dirichlet forms and polymers, statistical mechanics, quantum stochastic processes, the applications of local-time to proving path-wise uniqueness of solutions of stochastic differential equations and its application to excursion theory, Bessel processes and pole-seeking Brownian motion, queues, potential theory and Wiener-Hopf theory. Some new results for Brownian motion on how one process determines another are also given. The applications to Mathematical Physics appear first, followed by the papers on local-time, Bessel processes and queues. The papers of the Swansea school are collected together at the end of the volume. We are grateful to SERC for financial support through research grant GR/C52162 and we are especially indebted to James Taylor for invaluable help and advice during the conference. Finally we should like to record our thanks to Mrs E. Williams, Mrs M. Prowse and Mrs M. Brook for making such an excellent job of typing the Swansea contributions.

A. Truman
D. Williams
Swansea
April, 1983

TABLE OF CONTENTS

NEWTONIAN DIFFUSIONS AND PLANETS, WITH A REMARK ON NON-STANDARD DIRICHLET FORMS AND POLYMERS

by

S. Albeverio
Mathematisches Institut
Ruhr-Universität
D-4630 Bochum

Ph. Blanchard
Theoretische Physik
Universität Bielefeld
D-4800 Bielefeld

R. Høegh-Krohn
Université de Provence
Centre de Physique
Théorique, CNRS
F-13288 Marseille
and
Matematisk Institutt
Universitetet i Oslo
Blindern, Oslo

Abstract

We discuss diffusion processes on Riemannian manifolds, for which a Newton law holds (in the stochastic sense). We emphasize the existence of a general mechanism for the formation of impenetrable barriers for these processes, corresponding to the nodes of the density of their distribution. We discuss some applications to natural phenomena like the formation of planetary systems, the morphology of galaxies, the formation of zones of winds in the atmosphere and the formation of spokes in the rings of Saturn. We also relate the recent hyperfinite theory of Dirichlet forms with the theory of local times of Brownian motion, polymer measures and the $(\phi_1^2\ \phi_2^2)_4$-model of quantum field theory.

1. Introduction

In this lecture we shall discuss two topics, which are connected by the theory of diffusion processes. In the first part, consisting of Sections 2, 3 and 4, we shall discuss a class of diffusion processes, which we call "Newtonian diffusion processes", which show the remarkable phenomenon of "barrier formation", leading to a possible explanation of a large class of natural phenomena. In the second part we shall briefly discuss the concept of local time of Brownian motion and a new hyperfinite version of the theory of Dirichlet forms and apply this to the study of polymer measures associated with certain quantum field theoretical models.

In Section 2 we give the definition and basic properties of Newtonian diffusions on manifolds. This theory has been introduced by E. Nelson in connection with stochastic mechanics [3], [4], [29] and developed further particularly by Dankel [1], Dohrn and Guerra [2] and for the case of manifolds recently by Meyer [5] and Morato [19]. We review the basic formalism and discuss the stationary case in particular (this case has been discussed previously by ourselves in [13], [28] and, in connection with Dirichlet forms, in [7], [10] and by Nagasawa [16]).

In Section 3 we discuss the general mechanism in the symmetric case for the barrier formation in Newtonian processes, using previous results obtained in ref. [7] and [10], in the context of the "Dirichlet approach" to quantum mechanics (see e.g. [42], [31], [34], [50], [52], [57]).

In Section 4 we discuss the mentioned applications to natural phenomena like the formation of planetary systems, the morphology of galaxies, zones of winds circulation and the formation of spokes in the rings of Saturn.

In Section 5 we briefly discuss some problems in connection with the so-called polymer measures. This involves the study of "times spent at intersections of Brownian motions", quantities that also have arisen in the very stimulating lecture of Prof. E.B. Dynkin. We mention a couple of central problems in this area and indicate how in dimension 4 we can partially solve these by using a non standard theory of Dirichlet forms.

2. Newtonian Diffusion Processes

In this section we shall briefly describe how an important class of Markov diffusion processes, called "Newtonian processes", shows an interesting phenomenon of formation of barriers on the nodes of the solution of a linear equation of elliptic or Schrödinger type, and how this remarkable property can be used to describe situations in nature, in which regular patterns of "confinement" arise.

Let M be a smooth oriented Riemannian manifold of dimension d. Let X_t, $t \in T \subset \mathbb{R}_+$, be a diffusion process with values in M. The analytic description of X_t is by its infinitesimal generator L_t, which we assume of the form

$$L_t = \frac{1}{2}\Delta + \beta . D \tag{2.1}$$

where $\beta . D \equiv \beta^i D_i$, β^i $i = 1, \ldots d$, is a (non random) C^∞ vector field (the "drift"), which might depend explicitly on the time t. D is the covariant derivative and Δ is the Laplace-Beltrami operator on M. The connection between β and the process X_t is that $\beta(X_t,t)$ is the (mean) forward derivative of X_t at time t in the sense that

$$\beta^i(x,t) = \lim_{\Delta t \downarrow 0} (\Delta t)^{-1} E[Y^i{}_{\Delta t} | X_t = x] \tag{2.2}$$

where $E[\cdot | X_t = x]$ means conditional expectation with respect to $X_t = x$. $Y^i{}_{\Delta t}$ is the vector attached to X_t tangent to geodesics from X_t to $X_{t+\Delta t}$, with length $|Y^i{}_{\Delta t}|$ equals to the geodesics distance of $X_{t+\Delta t}$ and X_t. $\beta(X_t,t)$ is also the forward stochastic derivative in the sense of [5].

Let dx be the Riemannian volume element on M. Due to the assumptions we know that there exists a smooth density $\rho(x,t)$ of the law of X_t with respect to dx, i.e. $dP(X_t \in dx) = \rho(x,t)dx$.

Let $f \in C_0^\infty(M)$, then $E[f \circ X_t] = \int_M f(x)\rho(x,t)dx$ and $\frac{d}{dt} E[f \circ X_t] = \int_M f(x)\frac{\partial \rho}{\partial t}(x,t)dx$. On the other hand, by the definition of L_t, the left hand side is equal to $\int_M (L_t f)(x,t)\rho(x,t)dx$. By partial integrations we arrive at the Kolmogorov forward equation (Fokker-Planck equation)

$$\frac{\partial}{\partial t}\rho = \frac{1}{2}\Delta\rho - \text{div}(\beta\rho) , \tag{2.3}$$

$\frac{1}{2}\Delta$ - div($\beta\cdot$) being the adjoint of L_t and div means divergence operator on vector fields on M .

Let us now denote by $\hat{X}_t$, $t \in -T$, the time reversed process to X_t , i.e. $\hat{X}_{-t}$ has the same law as X_t. It is well known, see e.g. [27], that $\hat{X}_t$ is again a Markov process with infinitesimal generator

$$\hat{L}_t \equiv \frac{1}{2}\Delta - \hat{\beta} . D \tag{2.4}$$

with $\hat{\beta} . D \equiv \hat{\beta}^i D_i$, $\hat{\beta}^i$ being the "backward drift" defined, for $t \in T$, by

$$\hat{\beta}^i(x,t) = \lim_{\Delta t \downarrow 0} (\Delta t)^{-1} E[Y^i_{-\Delta t} | X_t = x] \tag{2.5}$$

with $Y^i_{-\Delta t}$ defined as $Y^i_{\Delta t}$ with $-\Delta t$ replacing Δt . Then $\hat{\beta}$ is the backward stochastic derivative of X_t . By the same procedure as above, one arrives at the Fokker-Planck equation for the reversed process, $t \in T$:

$$-\frac{\partial}{\partial t}\rho = \frac{1}{2}\Delta \rho + \mathrm{div}(\hat{\beta}\rho) \quad . \tag{2.6}$$

Set now $u \equiv \frac{1}{2}(\beta-\hat{\beta})$ and $v \equiv \frac{1}{2}(\beta+\hat{\beta})$. u is called the <u>osmotic velocity</u> and v is called the <u>current velocity</u>. Inserting these expressions in the Fokker-Planck equations (2.3), (2.6) we get the "continuity equation"

$$\frac{\partial}{\partial t}\rho = -\mathrm{div}(\rho v) \tag{2.7}$$

and the "osmotic equation"

$$\frac{1}{2}\Delta \rho = \mathrm{div}(\rho u) \quad . \tag{2.8}$$

As remarked first by Nelson in the case $M = \mathbb{R}^d$ we have also

$$u = \frac{1}{2} \nabla \log \rho \ . \tag{2.9}$$

This follows, see [29], by computing for $f,g \in C_o^\infty(T \times M)$:

$$0 = \int_T \frac{d}{dt} E[f(X_t,t)g(X_t,t)]\,dt$$

$$= \int_T E[D_+f(X_t,t)g(X_t,t)]dt + \int_T E[f(X_t,t)D_-g(X_t,t)]\,dt$$

where $D_+ \equiv \frac{\partial}{\partial t} + \frac{1}{2}\Delta + \beta.\nabla$ (the operator of mean derivative on functions) and $D_- \equiv \frac{\partial}{\partial t} - \frac{1}{2}\Delta + \hat{\beta}.\nabla$. Using partial integrations to bring D_+ to act on $g\rho$ and using Fokker-Planck's equation (2.3), we arrive easily from this to the conclusion that $-D_- = -\frac{\partial}{\partial t} - \beta.\nabla + \nabla(\log\rho)\nabla + \frac{1}{2}\Delta$. Using $\hat{\beta} = D_- X_t$ we then get (2.9). From this equation (2.9) we have, taking the time derivative and using the continuity equation (2.7)

$$\frac{\partial u}{\partial t} = - \text{grad div } v - \text{grad } (v \cdot u) \ . \qquad (2.10)$$

We shall now define the mean acceleration associated with the process X_t . To do this we would like to have the concept of mean forward and backward derivatives of vectors on M . The appropriate definition has been given by Dohrn and Guerra [2], [5]. Let $F = F^i(x,t)$ be a vector field on M , then the mean forward derivative of F is defined by

$$D_+F(x,t) \equiv \lim_{\Delta t \downarrow o} (\Delta t)^{-1} E[T_{X_t, X_{t+\Delta t}} F(X_{t+\Delta t}, t+\Delta t) - F(X_t, t) \,|\, X_t = x] \qquad (2.11)$$

where $T_{y,y+\Delta y}F$ is the vector at $T_{y+\Delta y}M$ obtained from the vector $F \in T_yM$ by Dohrn-Guerra's stochastic parallel transport along the geodesics from y to $y + \Delta y$. We recall briefly the definition of this transport, for more details see [2], [4] and [5]: Let $\gamma(s)$, $s_o \le s \le s_1$, be a segment of geodesics on M . Let F^i be a vector in $T_{\gamma(s_o)}M$. Let $h(t)$, $t \in [0,1]$, be a curve on M such that $\dot{h}(o) = F$. Let us transport in a Levi-Civita parallel way $\dot{\gamma}(s_o) \equiv G(t)$ along $h(t)$, getting a vector field $G(t)$. Let $\gamma_t(s)$ be the geodesic such that $\gamma_t(s_o) = h(t)$, $\dot{\gamma}_t(s_o) = G(t)$, $s < s_1$. The family of geodesics $\{\gamma_t(s)$, $s_o \le s \le s_1$, $t \in [o,1]\}$ is parallel for $s = s_o$. Let $B(s) \equiv \frac{d}{dt} \gamma_t(s)\restriction_{t=o}$; this is a vector field along $\gamma(s)$, with $B(s_o) = F$. By definition $T_{\gamma(s_o),\gamma(s)}F \equiv B(s)$ and $T_{\gamma(s_o),\gamma(s)}F$ differs from the Levi-Civita displacement of F by second order terms in s . This then gives, for $y = \gamma(s_o)$, $y + \Delta y = \gamma(s)$ the transport $T_{y,y+\Delta y}F$ needed in (2.11). One computes easily [2], [19]

$$D_+ = \frac{\partial}{\partial t} + \beta.\nabla + \frac{1}{2}\Delta_{DR}$$
$$\qquad (2.12)$$
$$D_- = \frac{\partial}{\partial t} + \hat{\beta}.\nabla - \frac{1}{2}\Delta_{DR}$$

$\Delta_{DR} \equiv \Delta - R$ being the Laplace-de Rham-Kodaira Laplacian on M , R being the Ricci tensor, acting on vectors.

Let us now define the <u>mean acceleration</u> $a(X_t,t)$ by

$$a(X_t,t) \equiv \frac{1}{2}(D_+D_- + D_-D_+)X_t .$$

Using $D_+X_t = \beta(X_t,t)$ and $D_-X_t = \hat{\beta}(X_t,t)$ we get

$$a(X_t,t) = \frac{1}{2} D_+\hat{\beta}(X_t,t) + \frac{1}{2} D_- \beta(X_t,t)$$

$$= \frac{1}{2}(\partial_t+\beta.\nabla+\frac{1}{2}\Delta_{DR})\frac{v-u}{2} + \frac{1}{2}(\partial_t+\hat{\beta}.\nabla-\frac{1}{2}\Delta_{DR})\frac{v+u}{2}$$

$$= (\partial_t v + v\cdot\nabla v - u\cdot\nabla u - \frac{1}{2}\Delta_{DR}u)(X_t,t) \qquad (2.13)$$

hence

$$\frac{\partial v}{\partial t} = a + u\cdot\nabla u - v\cdot\nabla v + \frac{1}{2}\Delta_{DR} u . \qquad (2.14)$$

Let us also remark that a purely probabilistic description of the process is given by the solution of the stochastic differential equation (in Ito's sense)

$$dX_t = \beta(X_t,t)dt + dW_t ,$$

with W_t the standard Brownian motion on M. Of course this is not an intrinsic description, for such see e.g. [5], [53]. Given β we can, under suitable assumptions, construct X_t and get ρ, hence $\hat{\beta}$ and hence u and v satisfying (2.10), (2.14). Moreover, given X_t and its distribution ρ we can get β and $\hat{\beta}$ as mean forward resp. backward derivatives and then get u, v satisfying (2.10), (2.14), with a defined as mean acceleration of X_t.

For a class of diffusions X_t, which we shall call here "Newtonian diffusions" (they coincide, under regularity assumptions, with Nelson's conservative diffusions [4]), we shall show that one can recover X_t from (2.10), (2.14) and the initial conditions $u(x,o)$, $v(x,o)$.

<u>Definition</u>: A Markov process X_t is called a "<u>conservative Newtonian diffusion</u>" if X_t satisfies "Newton's law in the mean", in the sense that there exists a positive constant m and a real-valued function V on $M \times T$ such that

$$m\, a(X_t,t) = -\nabla V(X_t,t)$$

and such that in addition the corresponding current velocity is a gradient field. We then say that a is the acceleration of the Newtonian diffusion.

We shall see below that there exist conservative Newtonian diffusions. We shall first discuss the forms of their distributions. Let $S(x,t)$ be any function on $M \times T$ such that $v(x,t) = \text{grad}\, S(x,t)$ (clearly S is only defined modulo functions of t alone). Let $\rho(x,t)$ be the density of the distribution of X_t, as above. Under our assumptions ρ is smooth and strictly positive; hence $\log \rho$ is well defined. Setting $\psi \equiv e^{R+iS}$, so that $|\psi|^2 = \rho$, $R \equiv \frac{1}{2} \log \rho$ we have easily from $m\,a = -\nabla V$ and (2.14):

$$\frac{\partial S}{\partial t} - \frac{1}{2}\,\text{div}\, u + \frac{1}{2}\,(|v|^2 - |u|^2) + V = 0 \tag{2.15}$$

(because by taking grad on this and using $\text{div grad} = \Delta$, $\text{grad div} = \Delta_{DR}$, $\text{grad}\,|u|^2 = 2u\cdot\nabla u$, $\text{grad}\,|v|^2 = 2v\cdot\nabla v$, both u and v being gradients, we see that this is just (2.14)). Moreover, from $u = \text{grad}\, R$ we have

$$\frac{\partial}{\partial t} e^{2R} = -\,\text{div}\,(e^{2R}\,\nabla S)$$

hence

$$\dot{R} + \frac{1}{2}\,\text{div grad}\, S + \text{grad}\, R.\text{grad}\, S = 0 \ . \tag{2.16}$$

From (2.15), (2.16) we see that

$$i\,\frac{\partial \psi}{\partial t} = -\,\frac{1}{2}\,\Delta\psi + V\,\psi\,. \tag{2.17}$$

Thus we see that for conservative Newtonian diffusions the probability distribution $\rho(x,t)$ of the process X_t is given, for all $x \in M$ and all $t \in T$, by $\rho = |\psi|^2$, where ψ is a solution of the Schrödinger equation (2.17), with initial condition $\psi(x,0)$ such that $|\psi(x,0)|^2$ gives the initial distribution of the process.

Vice versa if ψ is a solution of (2.17) and if we write $\psi = e^{R+iS}$ and define u and v by $u = \text{grad}\, R$, resp. $v = \text{grad}\, S$, then u and v satisfy (2.10) resp. (2.14). From u and v we can get in particular β and thus the stochastic equation for a proces X_t, the distribution of which is then for all times $\rho(x,t) = |\psi(x,t)|^2$, if at time $t = 0$ it has distribution $|\psi(x,0)|^2$. Moreover, the process statisfies Newton's equation in the mean.

Remark: It has recently been shown by Nelson that the stochastic Newton law in the definition of conservative Newtonian diffusion can be replaced by a variational principle [4] (see also [30], [58]).□

We are particularly interested in the case where there exists a stationary distribution $\rho(x,t) = \rho(x)$ for the process, i.e. $\frac{\partial \rho}{\partial t} = 0$. In this case we have from the above, that $|\psi(x,t)|^2$, and hence R , is independent of t . Assume ψ satisfies (2.17) with initial condition $\psi(x,0) = e^{R(x)+iS(x,0)}$. Then we have, using $\frac{\partial R}{\partial t} = 0$, that $i\frac{\partial \psi}{\partial t} = -\frac{1}{2}\Delta\psi + V\psi$ is equivalent with $-\psi\frac{\partial S}{\partial t} = -\frac{1}{2}\Delta\psi + V\psi$ i.e.

$$-(\frac{\partial S}{\partial t})\, e^{R+iS} = [-\tfrac{1}{2}\Delta R - (\nabla R)^2 + i\Delta S + (\nabla S)^2 + 2i\nabla R.\nabla S + V]\, e^{R+iS}. \tag{2.18}$$

Hence for all x such that $e^{R}(x) \neq 0$:

$$-\frac{\partial S}{\partial t} = -\tfrac{1}{2}\Delta R - (\nabla R)^2 + (\nabla S)^2 + V \tag{2.19}$$

$$\Delta S + 2\,\nabla R.\nabla S = 0 . \tag{2.20}$$

This system of coupled partial differential equations with initial condition $S(x,0) = S_o(x)$ has a solution of the form

$$S(x,t) = -Et + S_o(x) \tag{2.21}$$

iff $-E\psi = -\frac{1}{2}\Delta\psi + V\psi$ has a solution of the form $\psi(x,t) = e^{R(x)}e^{-iEt}e^{iS_o(x)}$. Splitting in real and imaginary parts we get

$$-\frac{1}{2}\,\Delta R - (\nabla R)^2 + (\nabla S_o)^2 + V = E \tag{2.22}$$

$$\Delta S_o + 2\,\nabla R.\nabla S_o = 0 . \tag{2.23}$$

We set $\rho \equiv e^{2R}$ and remark that (2.23) can be written in the form

$$\Delta S_o + \frac{1}{2}\frac{\nabla\rho}{\rho}\cdot\nabla S_o = 0 \tag{2.24}$$

i.e. $\nabla(\rho\nabla S_o) = 0$ for all x such that $\rho \neq 0$. We set $v = \nabla S_o$, then (2.24) is just the continuity equation $\frac{\partial \rho}{\partial t} = \mathrm{div}\,(\rho v)$. Hence if ψ is such that $-\frac{1}{2}\Delta\psi + V\psi = E\psi$ has a solution of the form $\psi = e^{R}e^{-iEt}e^{iS_o}$ then (2.22), (2.23) are satisfied. Setting $u = \frac{1}{2}\nabla\log\rho$ in this case we then have that the system of coupled equations is satisfied. From

u and v we can compute for all times β and hence the process X_t. This process has ρ as invariant distribution. Note that in the stationary case β does not depend explicitly on t.

Remark: The case where $\nabla S = 0$ is the only case where $v = 0$. In this situation (2.23) is clearly satisfied. Setting $u = \frac{1}{2} \nabla \log \rho$ and solving for the process X_t, we find from the continuity equation that $\frac{\partial \rho}{\partial t} = 0$, hence ρ is stationary. $v = 0$ is equivalent with the symmetry in $L^2(M, \rho dx)$ of the Markov semigroup P_t giving the distribution of X_t [12]. In fact, this is equivalent with time reversal invariance [11].

3. Barriers for Newtonian Diffusion Processes

In the case of a symmetric diffusion process a general theory of formation of barriers for the process (non trivial decomposition into time ergodic components) has been given in [13], [16], [7], [10], [28], [51]. Let us take the opportunity to recall here the main facts of this theory.

Let E be a regular Dirichlet form on L^2. Let M be a locally compact space with a countable base for the topology and let m be a Radon measure on M, strictly positive on every non void open subset. It is well known that there exists a one-to-one correspondence between submarkovian semigroups P_t on $L^2(M; m)$ and Dirichlet forms (i.e. positive symmetric closed bilinear forms E on $L^2(M; m)$ with the contraction property

$$E(f^{\#}, f^{\#}) \leq E(f,f) \quad \text{with} \quad f^{\#} \equiv (f \vee 0) \wedge 1$$

$\forall f$ for which $E(f,f) < +\infty$). The correspondence is such that if $-L$ is the infinitesimal generator of P_t then $E(f,f) = (L^{1/2}f, L^{1/2}f)$, where on the right hand side we have the scalar product in $L^2(M; m)$. Moreover, diffusion semigroups are in one-to-one correspondence with local Dirichlet forms, local meaning $E(f,g) = 0$ whenever f,g have disjoint supports. On the other hand, to any regular Dirichlet form (regular meaning that the continuous functions with compact support in the domain of the form i.e. $D(E) \cap C_0(M)$ are dense both in $(D(E), E_1)$ and in $(C_0(M), \|\cdot\|_\infty)$, where $E_1(f,f) \equiv E(f,f) + (f,f)$ and $\|\cdot\|_\infty$ denotes the supremum norm) by a construction of Fukushima and Silverstein there is a Hunt process X_t with m-symmetric transition function precisely P_t, P_t being the symmetric semigroup associated with E. X_t can be taken to be a diffusion (in the sense of having continuous paths, P^x (X_t continuous in $t \in [0,\zeta]$) = 1) iff E is local in addition to being regular. One can show [54] that any regular local Dirichlet form can be written as closed **extension** of the form

$$E(f,f) = \int_{\mathcal{H}} (\nabla f,\, d\nu\, \nabla f) + \int_{\mathcal{H}} |f|^2\, d\nu_0 ,$$

for some Hilbert space $\mathcal{H}$, f being smooth cylinder functions on $\mathcal{H}$, ν, ν_0 being positive Radon measures with values in the cone of positive self-adjoint operators resp. in $\mathbb{R}_+$.

If M is an oriented Riemannian manifold and m is a Radon measure on M then the closure of $E(f,f) = \int_M df(x).df(x)\, m(dx)$ in $L^2(M;m) \times L^2(M;m)$, first defined on $C_0^1(M)$-functions, is a local Dirichlet form. Sufficient conditions for the existence of the closure can be extracted from work by Fukushima [8], Albeverio, Høegh-Krohn and Streit [31] and Röckner and Wielens [9] (the latter reference gives also a survey of this type of results). In particular, if $m(dx)$ has a density ρ with respect to the volume measure dx and if ρ is strictly positive on compacts and locally Lipschitz, then there exists only one closed extension of E [33]. Other results yielding closability (but in general not uniqueness) involve e.g. conditions of the type $\rho > 0$, dx a.e and that the distributional derivatives of $\rho^{1/2}$ with respect to local coordinates are locally in $L^2(dx)$, on any open subset of M whose complement has m-measure zero ([31], see also [9], [46]. For any regular Dirichlet form E on $L^2(M;m)$, (M being again a locally compact second countable Hausdorff space and m a Radon measure strictly positive on non void open sets) one defines the capacity of U by

$$\text{Cap}\, U = \inf\, [\, E(f,f) + (f,f)\,] ,$$

the infimum being taken over the set L_U of functions f in the domain of E, which satisfy $f \geq 1$ m-a-e on U (the infimum is taken to be $+\infty$ if $L_U = \emptyset$). By standard methods one can then extend the definition to any subset $A \subset M$ as an outer capacity yielding a Choquet capacity. L_U being closed and convex for open U there exists a unique element e_U in L_U which minimizes $E(f,f) + (f,f)$, this element being the equilibrium potential of U. One has $0 \leq e_U \leq 1$ and

$$\text{Cap}(U) = E(e_U, e_U) + (e_U, e_U) .$$

e_U is actually a version of the hitting probability P_U^x of U :

$$P_U^x \equiv E^x[e^{-\tau_U}, \tau_U < \infty] ,$$

$\tau_U(\omega) \equiv \inf \{t > 0 \mid X_t(\omega) \in U\}$ being the hitting time of U for the process X_t associated with E . From this it follows that a set $A \subset M$ has zero capacity iff $P^x[(X_t$ or $X_{-t} \in \tilde{A}$ for some $t \geq 0)] = 0$ for any $x \in M\setminus\tilde{A}$, where $\tilde{A}$ is some Borel set $\tilde{A} \supset A$ with $m(\tilde{A}) = 0$. Moreover, a monotone decreasing sequence of open subsets U_n of M has capacity decreasing monotonically to zero iff $P^x[\lim_n \tau_{U_n} < \zeta] = 0$ for quasi-every x (ε: death time).

Let P_t a sub-Markov semigroup on $L^2(M;m)$. A Borel subset B of M is called P_t invariant if $P_t \chi_B = \chi_B P_t$ on all functions of $L^2(M;m)$. B is called invariant for the process X_t if $P^x[\tau_{M\setminus B} < +\infty] = 0$ for any $x \in B$. Following Lejan [32] one calls a Borel set B quasi-open (resp. quasi-closed) if there exists an increasing sequence of closed sets F_n with $\mathrm{Cap}(M\setminus F_n) \to 0$ and $B \cap F_n$ is open (resp. closed). It was proven in [7], [10], [51] that B is P_t invariant iff $\tilde{B}$ is quasi-open and quasi-closed for some Borel set $\tilde{B}$ such that $m(\tilde{B}\setminus B) = 0$. This is the case iff $M = B_1 + B_2 + N$ with $m(N) = \mathrm{Cap}(N) = 0$, $m(B_1\setminus B) = m(B_2\setminus(M\setminus B) = 0)$, B_i X_t-invariant. In this case we have $L^2(M;m) = L^2(B_1;m) \oplus L^2(B_2;m)$ and a corresponding direct decomposition of P_t . In fact, E is irreducible iff there exists non trivial decomposition $L^2(M;m) = L^2(B_1;m) \oplus L^2(B_2;m)$ which reduces P_t . If E is reducible a set N with above properties is said to be a <u>barrier</u> of P_t . One shows [7] and it follows from the above that if O is a non void open set such that there exists a decreasing sequence of open subsets U_n of M such that $\mathrm{Cap}(U_n) \to 0$ as $n \to +\infty$ and $\bar{O}_n \supset \partial U$ then if each U_n separates U and $M\setminus\bar{U}$ in the sense that any continuous path connecting a point in $M\setminus U_n$ and $(M\setminus\bar{U})\setminus U_n$ crosses U_n and $m(\partial U) = 0$, then E is reducible. Viceversa, if E is reducible and $\chi_U g$ for any continuous function with compact support belonging to the domain of E one has that $\chi_U g$ is again in the domain of E , then there exists above sequence of sets U_n .

E is called irreducible if any P_t-invariant set B is trivial in the sense that either $m(B) = 0$ or $m(M\setminus B) = 0$. If P_t is conservative, i.e. $P_t 1 = 1$ (and hence P_t is a Markov semigroup) then this irreducibility is equivalent with the ergodicity of

$P(\cdot) \equiv \int_M P^x(\cdot)\, m(dx)$ and of the process X_t (the latter being understood in the sense that $P[X_t \in A\ ,\ X_t \in B] = 0\ \forall t$ implies $m(A) = m(B) = 0$). In the case where E is not irreducible an ergodic decomposition has been given in [34], see also [6], where it was called T-ergodic decomposition. Let us assume that m is a probability measure. The decomposition relies simply in a direct decomposition of $L^2(M;m)$ with respect to the commutative C^*-algebra A of bounded measurable functions which commute with P_t

$$L^2(M;m) = \int_K L^2(M;\, m(\cdot|\alpha))d\alpha \quad ,$$

with K a compact Gelfand representation space for A (such that A is represented by $C(K)$), $d\alpha$ being the measure induced on K by the integral induced in A by m and $m(\cdot|\alpha)$ is the conditional probability measure conditioned with respect to the σ-subalgebra generated by the functions in A. P_t and its generator L are reduced by the above direct decomposition, hence

$$P_t = \int_K^{\oplus} P_t^{\alpha}\, d\alpha \ , \qquad L = \int_K^{\oplus} L^{\alpha}\, d\alpha \quad ,$$

with L^{α} being the self-adjoint generator of P_t^{α}, associated with a reduced Dirichlet form E^{α} in $L^2(M;\, m(\cdot|\alpha))$. If M is a Riemannian manifold and E is a regular local Dirichlet form on M of the form $E(f,f) = \int_M df(x).df(x)\, m(dx)$, then $E^{\alpha}(f,f) = \int_M df(\alpha).df(\alpha)\, m(dx|\alpha)$.

In the above general case, zero is a simple eigenvalue for L^{α} in $L^2(M;\, m(dx|\alpha))$ and the corresponding eigenfunction is positive $m(dx|\alpha)$ - a·e , the zero eigenspace for L being the closure of A in $L^2(M;m)$. For these results see [6], [34]. The ergodicity of the Markov semigroup P_t is equivalent with 1 being a simple eigenvalue for P_t, and this is also equivalent with the condition that from $f,g \geq 0$, $(f,\, P_t\, g) = 0\ \ \forall t > 0$ it follows $f = 0$ or $g = 0$.

The components P_t^{α} are always ergodic, and $m(\cdot) = \int_K m(\cdot|\alpha)d\alpha$ is the ergodic decomposition of m with respect to the action of the process X_t .

If in M there exists a set N giving rise to the above reducibility of a local regular Dirichlet form E , so that there exists B_1 , B_2

such that $L^2(B_i;m)$ reduce P_t , then we have that the process X_t and the semigroup P_t decompose into ergodic components in correspondence with B_1 , B_2 . The process X_t^x started at $x \in B_1$ never can reach B_2 , i.e. N acts as impenetrable barrier for X_t^x . $m(dx)$ as well as $\chi_{B_i}\, m(dx)$ are in this case invariant measures for X_t . Detailed analytic criteria for deciding whether or not a set acts as a barrier have been discussed in [7], [10], [51], for the case where M is an open subset of $\mathbb{R}^d$. E.g. for $E(f,f) = \frac{1}{2}\int_{\mathbb{R}^2} (\nabla f)^2\, \rho\, dx$, $\rho \in L^1_{loc}(\mathbb{R}^2)$, $\inf\ \rho > 0$ on compacts which do not intersect the plane $C = \{x \in \mathbb{R}^2\, |x^1 = 0\}$ and such that E is closable on $C_0^1(\mathbb{R}^2)$,

E is irreducible if we can find some $\alpha < \beta$, $b > 0$ s.t. $\int_\alpha^\beta \left(\int_{-b}^b \frac{dx^1}{\rho(x^1,x^2)}\right)^{-1} dx^2 > 0$, while it is reducible (in fact the left half plane $x^1 > 0$ and the right half plane $x^1 < 0$ are disconnected) if there exists an open covering J_K of $\mathbb{R}$ such that for each k

and some $b_k > 0$ $\displaystyle\int_0^{b_k} \frac{dx^1}{\int_{J_K} \rho(x_1,x_2)dx^2} = +\infty$ or $\displaystyle\int_{-b_k}^0 \frac{dx^1}{\int_{J_K} \rho(x_1,x_2)dx^2} = +\infty$.

More generally, in $\mathbb{R}^d$ reducibility and irreducibility have to do with "the strength of the zeros" of ρ ; if ρ is smooth each nodal surface acts as a barrier. For a discussion of the nodal surfaces of the eigenfunctions of Schrödinger operators see e.g. [59], [60].

Remark: Part of these results can be extended to the case of non symmetric Newtonian processes. In the case of a 1-dimensional manifold this follows from Feller theory, for results on a connected domain of $\mathbb{R}^d$ (assuming essentially that either V is bounded from below or $|x|^2\, V(x)$ stays bounded as $|x| \to +\infty$) or on a torus the reader is referred to [16] resp. [4] (see also for an extension of latter work ref. in [4]).

4. Some Natural Phenomena Which Can Be Described by the Formation of Barriers for Newtonian Diffusions

Let us consider some phenomena which can be described by a stochastic differential equation of the form

$$dX_t = \beta(X_t,t)dt + D(X_t)dW_t \tag{4.1}$$

with W_t the standard Wiener process in $\mathbb{R}^d$, β and D being the drift resp. diffusion coefficients. By introducing the metric $ds^2 = g_{ij}\, dx^i\, dx^j$ with $g_{ij} \equiv (D^t D)^{-1}_{ij}$ in $\mathbb{R}^d$, we can rewrite the above equation as a stochastic differential equation for a process $\hat{X}_t$ with values in the Riemannian manifold $M = (\mathbb{R}^d, g)$:

$$d\hat{X}_t = \hat{\beta}(\hat{X}_t, t)dt + d\hat{W}_t \quad , \tag{4.2}$$

$\hat{W}_t$ being the standard Wiener process on M and $\hat{\beta}^i = \beta^i - \frac{1}{2} g^{k\ell} \Gamma^i_{k\ell}$, Γ being the Christoffel symbol associated with the metric g. This is the class of processes we discussed in Section 2, where we showed that if $\hat{X}_t$ is a conservative Newtonian diffusion with mean Newton law $m\,a = -\nabla V$ for some V on $\mathbb{R} \times M$ and current velocity given by $v = \nabla S$, for some real function S on $\mathbb{R} \times M$, then $\psi \equiv \rho^{1/2} e^{iS}$, ρ being the density associated with $\hat{X}_t$, solves the Schrödinger equation (2.17). Having ψ we obtain the distribution of the original process X_t by $\rho \equiv |\mathrm{Det}\,(D^t D)^{1/2}|\ |\psi|^2$. The formation of barriers for $\hat{X}_t$, due to nodal sets of ψ, generates the formation of barriers for processes solving (4.1).

If in particular $|\psi|^2$ is time-independent, then we have the stationary situation discussed at the end of Section 2. If in addition $v = 0$ we have a symmetric situation. In this case we could have started more generally, instead of with (4.1) resp. (4.2), with a symmetric sub-Markov semigroup P_t on $L^2(\mathbb{R}^d, \rho\, dx)$ and could have used the stochastic calculus for Dirichlet forms of Albeverio, Høegh-Krohn, Streit [31] and Fukushima [8] to define a diffusion process satisfying a stochastic equation (in an extended sense, with singular coefficients, cf. Sect. 3). The general theory of barrier formation discussed in Section 3 leads then directly to this mechanism for the diffusion process associated with P_t.

We shall now discuss some examples in which the formation of barriers in Newtonian diffusion processes leads to an explanation of natural phenomena.

Example 1: It is an old hypothesis that the solar system was formed from a protosolar nebula consisting essentially of a gas of small particles (dust) of stellar or interstellar origin. In one form or another this hypothesis was discussed originally by Descartes (1644), Kant (1755) and Laplace (1796) and has been steadily accompanying all later developments in the discussion of the origin of our planetary system (see e.g. [25]). There have been many earlier attempts to explain the origin of the regularity in the distances r_n of the planets from the sun, a regularity which was described classically by the Titius-Bode law (1766) givin r_n in the form $r_n = a + b\,c^n$, for suitable constants a, b, c. We refer to the books of Nieto [35] and Jaki [61] for the history of attempts to explain Titius-Bode's law. It seems fair to say that several of these attempts were of merit but none of them has really prevailed. Many models link the origin of the law with the evolution of the protosolar nebula. One principal idea which has been intensively discussed recently is a sort of modern version of the Laplace ring formation: namely that, before the aggregation into planets, concentric roughly planar rings of dust where formed (or rather regions confined between concentric spheres centered at the origin and radii R_n and two cones of the form $\{(r,\Theta,\phi)\,|\,-\frac{\pi}{2}+\Theta_o \le \Theta \le \Theta_o + \frac{\pi}{2}\}$ in spherical coordinates (r,Θ,ϕ), with small Θ_o, Θ being the latitude on the unit sphere centered at the center of the sun). The rings consist of gas, ice particles and dust, circulating inside the rings but with no communication with neighbouring rings. In Alfven's theory these rings, the so-called "jet-streams" are essentially of electromagnetic origin. The formation of the planets should then have happened in a later state by aggregation (condensation) from the jet-streams, under the influence of some new effects like e.g. nuclear processes in the sun. For the process of aggregation for jet-streams see e.g. [25], [36]. We shall now explain (see also [14], [15], [28]) the formation of these jet-streams from Newtonian diffusion, resulting in distances in accordance with the actual distances of the planets (and hence, to a good extent, with Titius-Bode law).

The basic idea consists in thinking of a typical "particle" in the protosolar nebula as performing, under the steady influence of the attraction of a central body (the sun) and the innumerous chaotic collisions with other particles, a diffusion process given by stochastic differential equation of the type (4.1). There actually is hope that such an equation can indeed be obtained from classical mechanics in a suitable limit (see e.g. for related problems [37], [38] and reference therein). It is reasonable to assume that the diffusion is Newtonian, with a potential V given approximately by the gravitational attraction, since it arises from classical motion, and that there exists an invariant distribution, as the potential is attractive and the time scale involved is large (of course the invariant distribution is thought to hold as long as the diffusion approximation is valid). From the results of Section 2 and 3 we then know that the invariant distribution is given by the solution of a Schrödinger equation $H\psi = E\psi$, with $H = -\frac{\sigma}{2}\Delta + V$. V being attractive the negative part of the spectrum of H in L^2 consists of discrete eigenvalues with eigenfunctions which all have, except for the one belonging to the lowest eigenvalue, nodes. From Section 3 we know that these nodes act as barriers for the diffusion process. The node structure is particularly easy to discuss for $\sigma = 1$ and V the Newton gravitational potential. There are solutions whose nodes coincide very well with the borders of the jet-streams of the above form, hence yielding, by virtue of the barrier effect, an explanation of the non communicating rings in the proto-solar nebula. Using these solutions we obtain for the distances $r_n = 0.9\ (1.75)^n$, in very good agreement with the actual distribution of the planets. The same kind of ideas can be applied to the formation of jet-streams around the Jovian planets, leading to results in good agreement with the Titius-Bode law for the satellites of the Jovian planets. For more details see [14], [17].

Example 2: The exploitation of the mechanism of barrier formation can be extended to several other natural phenomena. An application to the discussion of the morphology and rigid rotation of galaxies is given in [18]. The idea is to approximate the motion of stars in a galaxy by a diffusion process and to look at suitable non stationary solutions ψ of the equation (2.17).

Example 3: Recently these ideas have been applied to the confinement of winds in zones on the surface of planets, the idea being that the motion of a typical particle (molecule, ice particle, dust) is well approximated by a Newtonian diffusion process (the attraction from the planet acting as potential V). Among the possible stationary solutions and confinement regions are those given in the angular variables by the associated Legendre polynomials $P_6^1(\cos\Theta)$, the nodes being given in this case by 5 different latitudes Θ . This gives a very good description of the observed cells of confinement for the Earth (2 Hadley cells at the equator, 2 Ferrel cells and 2 polar cells), see [24].

Example 4: It has also been observed [28], [47] that the formation of spokes in uniform rotation in the rings of Saturn can be explained by the confinement regions given by the possible stationary solutions, depending on the longitude angle, of the stationary equation $-\frac{1}{2}\Delta\psi + V\psi = E\psi$, with V the potential created by the gravitational attraction of Saturn, on a typical particle in its rings. For other discussions see e.g. [62].

Finally, let us remark that similar ideas have been used by Nagasawa [16], [48], [49], to give an interesting description of segregation phenomena in biological and sociological systems, as well as the formation of hadron spectra in elementary particle physics. In these cases the potential V is not thought of as the driving force for a "mean Newton law", but the equation for the density distribution is derived from an "equilibrium principle".

5. Non Standard Dirichlet Forms and Polymers

In this section we make a brief remark on the application of a recent theory of non standard Dirichlet forms to the construction of polymer measures. This remark has a connection with Prof. Dynkin's talk at this conference. Let us consider the multiple regularized local time of Brownian motion $b_t(\omega)$ started at time zero at the origin on $\mathbb{R}^d$ defined by

$$L_\varepsilon(b) = \int dz \int_I \int_I \delta_{\varepsilon,z}(b_{t_1})\delta_{\varepsilon,z}(b_{t_2})\, dt_1\, dt_2,$$

$$:L_\varepsilon(b): = L_\varepsilon(b) - E(L_\varepsilon(b))$$

$\delta_{\varepsilon,z}$ being a smooth approximation for the Dirac measure at z, $\int \delta_{\varepsilon,z} = 1$

$\delta_{\varepsilon,z}(x) = c\,\varepsilon^{-d}\,\chi_1\left(\frac{|x-z|}{\varepsilon}\right)$, χ_1 being the characteristic function of the unit sphere centered at the origin, with $I = [0,t]$. The weak limit ν_λ as $\varepsilon \downarrow 0$ of $e^{-\lambda :L_\varepsilon:}\, dP/E[e^{-\lambda :L_\varepsilon:}]$, with P the Wiener measure and E the expectation value with respect to it, has been shown to exist for all $\lambda \geq 0$ and for $d = 2,3$ (by Varadhan [44], for $d = 2$, in which case one has even $:L_\varepsilon: \to L$ in $L^2(P)$, and by Westwater [41], [55], for $d = 3$, who also showed $\nu_{\lambda_1} \perp \nu_{\lambda_2}$ for $\lambda_1 \neq \lambda_2$).

Recently Kusuoka [43] has shown that the process X_t with distribution ν_λ is for $d = 3$ the sum of Brownian motion and a process Z_t, adapted to X_t and of "zero energy" variation (a Dirichlet process [72]).

The measure ν_λ gives a model for the formation of long chains of flexible molecules (polymers), the chains being simulated by a sample path of X_t (this is the so-called "Edwards model", see references in [55]). The problem of the asymptotics as $t \to +\infty$ of $E[X_t^2]$ has only been solved for $d = 1$ [41], [63] (it is conjectured that $E[X_t^2] \sim t^{\frac{6}{2+d}}$ $d \leq 3$, $\sim t$ or $t(\log t)$ for $d = 4$, $\sim t$ for $d \geq 5$). No answer is known up to now regarding the existence of ν_λ for $d \geq 4$ (although it is conjectured that $\nu_\lambda = P$ for $d \geq 5$, in connection with results of [64], [65], [66], [56], [71]. For $d = 4$ the problem of the construction of ν_λ is very interesting because of its connection with the one of the construction of ϕ_4^4 models of quantum fields, see [44], [41], [56], [67], [64]. Related problems are the ones posed by

$$\tilde{L}_\varepsilon(b,\tilde{b}) \equiv \int dz \int_I \int_I \delta_{\varepsilon,z}(b_{t_1})\, \delta_{\varepsilon,z}(\tilde{b}_{t_2})\, dt_1\, dt_2 ,$$

b, $\tilde{b}$ being independent Brownian motions in $\mathbb{R}^d$ issued from the origin at time zero.

From results of Wolpert [39], Dynkin [40] and Westwater [41] one has that for $d \leq 3$, $\tilde{L}(b,\tilde{b}) \equiv \lim_{\varepsilon \downarrow 0} \tilde{L}_\varepsilon(b,\tilde{b})$ exists in $L^2(P \times \tilde{P})$, P, $\tilde{P}$ being the Wiener measures to b resp. $\tilde{b}$, and one has $\tilde{L} \in L^p$ for all $1 < p < +\infty$ (in fact $c_1\, p^{3/2} \leq \|\tilde{L}\|_p \leq c_2\, p^{3/2}$). What about $d \geq 4$?

By a reasoning based on the Feynman-Kac formula the problem of constructing $d\tilde{\nu}_\lambda \equiv e^{-\lambda\tilde{L}(b,\tilde{b})} dP d\tilde{P}.(E[e^{-\lambda\tilde{L}}])^{-1}$ for all d is solved once we are able to give a sense to the stochastic Hamiltonian

$$H_\lambda^\omega \equiv -\Delta + \int_0^t \lambda\delta_z\,(b_s)\, ds$$

in $L^2(\mathbb{R}^d, dx)$. The following result is proven by non standard analytic means in [56].

<u>Theorem</u>: For $d = 4,5$ there exists a choice $\lambda_\varepsilon(\omega)$, depending on the path ω of $b_s(\omega)$, $0 \leq s \leq t$, of the form

$$\lambda_\varepsilon(\omega) = -c_1(\omega)\varepsilon + O(\varepsilon^2),$$

with ε infinitesimal, s.t. for $\varepsilon > 0$ infinitesimal

$$H_{\varepsilon,\lambda}^\omega = -\Delta + \int_0^t \lambda_\varepsilon \delta_{\varepsilon,z}(b_s)ds$$

is near standard and defines, by standard parts, a self-adjoint lower bounded operator $H_\lambda^\omega \neq -\Delta$ in $L^2(\mathbb{R}^d, dx)$. For $d \leq 3$ one can choose $\lambda_\varepsilon(\omega) = \lambda'$ a real constant independent of ε and ω.

The proof uses a realization of the stochastic Hamiltonian $H^{\omega}_{\varepsilon,\lambda}$ as a non-standard self-adjoint operator in $L^2(\mathbb{R}^d, dx)$, with b_s Anderson's Brownian motion [56], the integral $\int_0^t \delta_{\varepsilon,z}(b_s)ds$ being then replaced by $\sum_{s=\Delta t}^{t} \delta_{\varepsilon,z}(b_s)$, with $I = [0,t]$ replaced by $T \equiv [0, \Delta t, 2\Delta t, \ldots, t]$, $\Delta t = \frac{t}{n}$, where n is a positive infinite integer, Correspondingly the underlying probability space is replaced by the hyperfinite Anderson sample path $\{-1, +1\}^T$ with the algebra of internal subsets. A computation of the resolvent kernel of $H^{\omega}_{\varepsilon,\lambda}$ shows that it will be near standard and $\neq -\Delta$ if λ is suitably choosen. The critical point of the discussion, which also involves restricting $d \leq 5$, is the one of showing that $\frac{1}{\lambda}\delta_{b_s b_{s'}} + G_k(b_s(\omega) - b_{s'}(\omega))$, $G_k(x-y) = (-\Delta+k^2)^{-1}(x,y)$, $\mathrm{Im}\, k^2 \neq 0$, exists as a self-adjoint operator in $^*\ell^2(T)$, for P-a.e ω. This involves in particular giving sense to the Dirichlet form

$$E(f,f) = \Sigma\,\Sigma\, G_k\ (b_s(\omega) - b_{s'}(\omega))f(s)f(s')\ .$$

The theory of such hyperfinite Dirichlet forms is developed in [56], which also gives a hyperfinite version of the Fukushima-Silverstein theory of Dirichlet forms we mentioned briefly in Section 3. The reason for the possibility of the above choice for λ is a combination of the reason why a non trivial perturbation of $-\Delta$ by a potential with support in a point is possible for $d \leq 3$ (see e.g. [45] and references therein) and the fact that the sample paths of Brownian motion are of Hölder index $< \frac{1}{2}$.

In view of the fact that the measure $\tilde{\nu}_\lambda$ has the same relation with the $(\phi_1^2\ \phi_2^2)_4$-model, ϕ_1 ϕ_2 being two independent scalar fields, as the ν_λ measure has with the ϕ_4^4 - models of quantum fields, it would be nice to know whether it is possible for $d = 4,5$ to choose λ indepen- of ω in the above Theorem (for $d = 3$ this is possible). A result interesting in itself and going rather in the opposite direction is the following

Theorem (S. Kusuoka): Let f be a symmetric function in $C([0,t] \times [0,t])$. For any $\varepsilon > 0$ let

$$X^f(\cdot) \equiv \int_0^t \int_0^t \frac{f(t,s)\ ds\ ds'}{|b(s)-b(s')|^2+\varepsilon}$$

$b(s)$ being the standard Brownian motion in $\mathbb{R}^4$.

Then $\lim_{\varepsilon \downarrow 0} X_\varepsilon^f - E[X_\varepsilon^f]$ exists in L^p, $1 < p < 2$, but

$\sup X^f = - -\inf X^f = +\infty$ P a.s , the supremum and the infinumum being taken over all f of the form $f = g \otimes g$, with $g \in C_0^\infty[o,t]$ such that $\int_0^t |g|^2 \, ds = 1$.

More details on these results will be given elsewhere.

Acknowledgments

One of us (S.A.) would like to take the opportunity to thank Professors Aubrey Truman and David Williams for a very kind invitation to participate in the workshop. The hospitality and support of the following institutions are also gratefully acknowledged: The Mathematics Institutes of the Ruhr-University Bochum and of the Oslo University, the Institute of Theoretical Physics and ZiF, University of Bielefeld, the Centre de Physique Théorique, CNRS, Marseille and the Université d'Aix-Marseille II.

References

[1] Dankel, Th.G.: Mechanics on manifolds and the incorporation of spin into Nelson's stochastic mechanics. Arch.Rat.Mech.Anal. 37, 1971, 192-221

[2] Dohrn, D., Guerra, F.: Nelson's stochastic mechanics on Riemannian manifolds. Lett. al Nuovo Cimento 22, 1978, 121-127

[3] Nelson, E.: Dynamical Theories of Brownian Motion. Princeton University Press 1967

[4] Nelson, E.: a) Quantum Fluctuations, Cours de Troisième Cycle Ecole Polytechnique de Lausanne and book in preparation, b) "Quantum Fluctuations - An introduction", Proceedings Boulder IAMP Conference Boulder Plenum Press 1983

[5] Meyer P.A.: Géometrie différentielle stochastique (bis) Séminaire de Probabilité XVI, 1980/81, Supplément: Géometrie Différentielle Stochastique. Lecture Notes in Mathematics 921, Springer 1982

[6] Albeverio, S., Høegh-Krohn, R.: Dirichlet Forms and Diffusion Processes on Rigged Hilbert Spaces. Z. Wahrscheinlichkeitstheorie verw. Gebiete 40, 1-57 (1977).

[7] Albeverio, S., Fukushima, M., Karwowski, W., Streit, L.: Capacity and Quantum Mechanical Tunneling Commun. Math. Phys. 81, 501-513 (1981)

[8] Fukushima, M.: Dirichlet Forms and Markov processes. North Holland Kodansha 1980

[9] Röckner, M., Wielens, N.: Dirichlet Forms - Closability and Change of Speed Measure. Preprint Bielefeld 1983.

[10] Fukushima, M.: A note on Irreducibility and Ergodicity of Symmetric Markov Processes in "Stochastic Processes in Quantum Theory and Statistical Physics". Lecture Notes in Physics 173, Eds. S. Albeverio, Ph. Combe, and M. Sirugue-Collin, Springer 1982

[11] Kent, J.: Time Reversible Diffusions. Adv.Appl.Prob. 10, 819-835, 1978; 11,888, 1978

[12] Kolmogoroff, A.: Zur Umkehrbarkeit der statistischen Naturgesetze, Math. Ann. 112, 155-160, 1936

[13] Albeverio, S., Høegh-Krohn, R.: A remark on the connection between stochastic mechanics and the heat equation. Journal of Math. Phys. 15, 1745-1748, 1974

[14] Albeverio, S., Blanchard, Ph., Høegh-Krohn, R.: A stochastic model for the orbits of planets and satellites: an interpretation of the Titius-Bode law. Expositiones Mathematicae 1, 365-373 (1983)

[15] Albeverio, S., Blanchard, Ph., Høegh-Krohn, R.: Processus de diffusion, confinement et formation de "jet-streams" dans la nébuleuse protosolaire. CERN Preprint TH 3536, Février 1983

[16] Nagasawa, M.: Segregation of a population in an environment. J. Math. Biology 9, 213-235, 1980

[17] Albeverio, S., Blanchard, Ph., Høegh-Krohn, R., Schneider, W.: in preparation

[18] Albeverio, S., Blanchard, Ph., Høegh-Krohn, R., Ferreira, L., Streit, L.: in preparation

[19] Morato, L.M.: On the dynamics of diffusions and the related general electromagnetic potentials. J. Math. Phys. 23, 1020-1024, 1966

[20] Nelson, E.: Derivations of the Schrödinger Equation from Newtonian Mechanics. Phys.Rev. 150, 1079-1085, 1966

[21] Nelson, E.: The adjoint Markoff process. Duke Math. J. 25, 671-690, 1968

[22] Chung, K.L., Walsh, J.B.: To reverse a Markov process. Acta Mathematica 123, 225-251 (1969)

[23] Meyer, P.A.: Le retournement du temps d'après Chung et Walsh. Séminaire de Probabilité, Université de Strasbourg, 213-236, 1969-1970

[24] Albeverio, S., Blanchard, Ph., Høegh-Krohn, R., Combe, Ph., Rodriguez, R., Sirugue, M. Sirugue-Collin, M., in preparation

[25] Alfven, H., Arrhenius, G.: Structure and Evolutionary History of the Solar System. D. Reidel 1975

[26] Nagasawa, M.: Time reversions of Markov processes. Nagoya Math. Journal 24, 117-204, 1964

[27] Doob, J.L.: Stochastic processes. John Wiley & Sons, New York 1953 (§ 6, p. 83)

[28] Albeverio, S., Blanchard, Ph., Høegh-Krohn, R.: Diffusions sur une variété riemanienne: barrières infranchissables et applications, Colloque en l'honneur de Laurent Schwartz, Astérique Société Mathématique de France, 1984 and ZiF preprint 1983

[29] Nelson, E.: Connection between Brownian Motion and Quantum Mechanics. In: Einstein Symposion, Berlin. Lecture Notes in Physics, Springer 1979

[30] Yasue, K.: Stochastic calculus of variation. Journal of Functional Analysis 41, 1981, 327-340

[31] Albeverio, S., Høegh-Krohn, R., Streit, L.: Energy Forms, Hamiltonians and Distorted Brownian Paths. J. Math. Phys. 18, 907, 1977

[32] Lejan, Y., Quasi-continuous functions and Hunt processes. Paris VI preprint 1981

[33] Wielens, N.: Eindeutigkeit von Dirichletformen und wesentliche Selbstadjungiertheit von Schrödingeroperatoren mit stark singulären Potentialen. Diplomarbeit, Univer. Bielefeld, 1982

[34] Albeverio, S., Høegh-Krohn, R.: Quasi invariant measures, symmetric diffusions processes and Quantum Fields in "Les méthodes mathématiques de la théorie quantique des champs", Marseille, Juin 1975, Editions du CNRS 1976

[35] Nieto, M.M.: The Titius-Bode Law of Planetary Distances, its History and Theory. Pergamon Press, Oxford 1972

[36] McCrea, W.H., Williams, I.P.: Segregation of materials in cosmonogy. Proc. R. Soc. A287, 143-164, 1975

[37] Dürr, D.: All that Brownian Motion in Stochastic Processes in Quantum Theory and Statistical Physics. Marseille 1981, Lecture Notes in Physics 173, 1983

[38] Dürr, D., Goldstein, S., Lebowitz, J.L.: A mechanical model of Brownian Motion. Comm. Math. Phys. 78, 507, 1981

[39] Wolpert, R.L.: Local time and particle picture for Euclidean Field Theory, J. Functional Anal. 30, 341-357, 1978

[40] Dynkin, E.: a) Markov processes as a tool in field theory, J.funct.Anal. 50, 167-187, 1983; b) Gaussian and non-Gaussian random fields associated with Markov processes. Cornell Preprint 1983

[41] Westwater, J.: On Edwards model for polymer chains. To appear in Bielefeld Encounters in Mathematics and Physics IV, Eds. S. Albeverio, Ph. Blanchard, World Scientific Publishing (1984).

[42] Streit, L.: Energy Forms: Schrödinger Theory, Processes, Physics Reports 77, 363-375, 1981

[43] Kusuoka, S.: On the path property of Edward's model for long polymer chains in three dimensions. Tokyo Preprint 1983 (to appear in 63)

[44] Symanzik, K., Varadhan, S.R.S.: Euclidean Quantum Field Theory in Teoria Quantistica Locale Varenna 1968, Ed. R. Jost, Academic Press 1969

[45] Albeverio, S., Høegh-Krohn, R.: Schrödinger Operators with point interactions and short range expansions. To appear in Proceedings of the VII Intern. Congress of Mathematical Physics, Boulder, Colorado, August 1983

[46] Silverstein, M.L.: On the closability of Dirichlet forms. Z. Wahrsch. verw. Gebiete 51, 185-200, 1980.

[47] Albeverio, S. Blanchard, Ph., Høegh-Krohn, R., work in preparation.

[48] Nagasawa, M., An application of the segregation-model for septation of Esterichia coli. J. Theor. Biol. 90, 445-455 (1981).

[49] Nagasawa, M. Yasue, K., A statistical model for systems of interacting particles and synthesis of the family of mesons. Zürich Preprint, 1983.

[50] Albeverio, S., Høegh-Krohn, R., Some Markov processes and Markov fields in quantum theory, group theory, hydrodynamics, and C*-algebras, pp. 492-540 in Stochastic Integrals, Proceed. London Mathematical Society Symposium, 1980, Ed. D. Williams, Lect. Notes in Maths. 851, Springer, Berlin (1981).

[51] Fukushima, M., Markov processes and functional analysis. Proc. Int. Math. Conf., Singapore, 1981, North Holland, Amsterdam

[52] Albeverio, S., Høegh-Krohn, R., Diffusion fields, quantum fields, fields with values in Lie groups, to appear in Adv. in Probability Ed. M. Pinsky, M. Dekker, New York, 1983.

[53] Ikeda, N., Watanabe, S., Stochastic Differential Equations and Diffusion Processes, North Holland/Kodarsha, Amsterdam, 1981.

[54] Albeverio, S., Høegh-Krohn, R., The structure of diffusion processes, Bielefeld Preprint (to be published).

[55] Westwater, J., On Edward's model for long polymer chains. Comm. Math. Phys. 72, 131-174 (1980).

[56] Albeverio, S., Fenstad, I.E., Høegh-Krohn, R., Lindstrøm, T., Non standard methods in probability theory and mathematical physics, book in preparation.

[57] Albeverio, S., Høegh-Krohn, R., Some remarks on Dirichlet forms and their applications to quantum mechanics and statistical mechanics, pp. 120-133 in "Functional Analysis in Markov Processes", Proc. Katata and Kyoto 1981, Ed. M. Fukushima, Lect. Notes Maths. 923, Springer-Verlag, Berlin (1982).

[58] Guerra, F., Morato, L.M., Quantization of dynamical systems and stochastic control theory. Phys. Rev. D27, 1774-1786, 1983

[59] Cheng, S.Y., Eigenfunctions and nodal sets, Comment. Math. Hev. 51, 43-55 (1976).

[60] Uhlenbeck, K., Generic properties of eigenfunctions, Amer. J. of Math. 98, 1059-1078 (1976).

[61] Jaki, S.L., Planets and Planetarians, Scottish Acad. Press (1978).

[62] Morfill, G.E., Grün, E., Goertz, C.K., Johnson, T.V., On the evolution of Saturn's "spokes" theory, Icarus 53, 230-235 (1983).

[63] Kusuoka, S., Asymptotics of polymer measures in one dimension, Tokyo Preprint 1983, to appear. in Proc. Bielefeld Conf. Infinite dimensional analysis and stochastic processes, Ed. S. Albeverio

[64] a) Aizenman, M., Proof of the triviality of ϕ^4_d field theory and some meanfield features of Ising models for d>4 ,P.R.L. 47, 1-4, 1981
b) Aizenman, M., Geometric analysis of ϕ^4_d fields and Ising models. Comm. Math. Phys. 86, 1-48 (1982)

[65] Fröhlich, J., On the triviality of $\lambda\phi^4_d$ theories and the aproach to the critical point in $d \geq 4$ dimensions, Nucl.Phys. B200, 281-296, 1982

[66] a) Lawler, G.F., A self-avoiding random walk, Duke Math. J. 47, 655-693 (1980)
b) Lawler, G.F., The probability of intersection of independent random walks in four dimensions, Comm. Math. Phys. 86. 539-554 (1982).

[67] Albeverio, S., Blanchard, Ph., Høegh-Krohn, R., Some applications of functional integration. Proc. Int. AMP Conf., Berlin 1981, Ed. R. Schrader, R. Seiler, D.A. Uhlenbrock, Lect. Notes Phys. 153, Springer, Berlin (1982).

[68] Gallavotti, G. Rivasseau, V., A comment on ϕ^4_4 Euclidean field theory, Phys. Lett. 122B, 268-270 (1983).

[69] Nelson, E., A remark on the polymer problem in four dimensions to appear in vol. dedicated to I. Segal 1983.

[70] Albeverio, S., Gallavotti, G., Høegh-Krohn, R., Some results for the exponential interaction in two or more dimensions, Comm. Math. Phys. 70, 187-192 (1979).

[71] Bover, A., Felder, G., Fröhlich, J., On the Critical Properties of the Edwards and the Self-Avoiding Walk Model of Polymer Chains. ETH Zürich Preprint 1983

[72] Föllmer, H., Dirichlet Processes, pp. 476-478 in 50.

THE EQUIVALENCE OF ENSEMBLES IN STATISTICAL MECHANICS

J.T. Lewis and J.V. Pule
Dublin Institute for Advanced Studies
10 Burlington Road
Dublin 4, Ireland

§1. Introduction

In this lecture we describe some probabilistic aspects of a device which Gibbs introduced in statistical mechanics around 1900: the use of the grand canonical ensemble to investigate the bulk limit of certain thermodynamic functions associated with the canonical ensemble. Since this is addressed to an audience of probabilists, the next section (§2) is devoted to an informal description of Gibbs strategy; it may seem familiar to many because we suspect that it was re-invented by probabilists; something very like it, the technique of associated distributions, is used at least four times in Feller's treatise [1] (see especially p. 549 of volume two). Gibbs stratagem relies on the weak law of large numbers holding; then the grand canonical ensemble and the canonical ensemble are said to be equivalent. Situations in which the law of large numbers fails to hold are much more interesting. In the context of models of lattice gases, Berezin and Sinai [2] showed that a violation of the law of large numbers is a sufficient condition for the existence of a first-order phase transition, and Dobrushin [3] sharpened this to give a necessary and sufficient condition for a first-order phase-transition in terms of the rate of convergence in the weak law of large numbers. We discuss all this in §3. In a manuscript circulated privately in 1971 Kac sketched the first rigorous proof of Bose-Einstein condensation; among the difficulties he encountered were: (1) the fact that the grand canonical partition function exists only for negative values of the chemical potential in the case of the free boson gas. (2) the non-equivalence of the grand canonical and canonical ensembles when the mean particle number density exceeds a critical value. We describe this in §4. In §5 we present a lemma similar to Dobrushin's but strong enough to cover the circumstances which arise in the free boson gas, and discuss some of its consequences. It determines the rate of convergence in the weak law of large numbers under the hypothesis that the free energy density converges in the bulk limit; it is of interest to know what can be said under the weaker hypothesis that the grand canonical pressure converges in the bulk limit; this is settled by a lemma presented in §6.

§2. Gibbs' Stratagem

In this section we give an informal description of Gibbs stratagem, introducing the terminology of statistical mechanics. For continuous classical systems it may be described as follows: consider a sequence $\{\Lambda_\ell : \ell = 1,2,\dots\}$ in $\mathbb{R}^d$; associate with each region Λ_ℓ and each positive integer n a Hamiltonian function $H_\ell^{(n)}$ defined on the corresponding phase-space $\Gamma_\ell^{(n)}$, being the total energy of n interacting particles confined to the region Λ_ℓ; the canonical partition function $Z_\ell(n,\beta)$ at inverse temperature $\beta = 1/kT$ is defined as

$$Z_\ell(n,\beta) = \int_{\Gamma_\ell^{(n)}} e^{-\beta H_\ell^{(n)}(\gamma)} d\gamma , \tag{2.1}$$

and we put $Z_\ell(0,\beta) = 1$ for convenience. In Gibbs scheme, the thermodynamic functions can be obtained from the canonical partition function. However, the thermodynamic functions describe the properties of matter in bulk, and to obtain these it is necessary to eliminate surface effects from the model; this is done by taking the *bulk limit* (also called the *thermodynamic limit*) in which the volume $|\Lambda_\ell|$ of the region Λ_ℓ tends to infinity with increasing ℓ and the number n_ℓ of particles in the region is made to depend on ℓ in such a way that the particle number density

$x = n_\ell/|\Lambda_\ell|$ remains fixed; the function $f(x,\beta)$ defined by

$$f(x,\beta) = \lim_{\ell\uparrow\infty} \{ -(\beta|\Lambda_\ell|)^{-1} \operatorname{Log} Z_\ell(n_\ell,\beta) \} \tag{2.2}$$

is then identified with the thermodynamic *free energy density* of the system. Now the bulk limit is not easy to compute directly; Gibbs introduced the following stratagem: regard the number of particles as a random variable N and, for each ℓ, introduce a one-parameter family $\mathbb{P}_\ell^\mu$ of probability distributions for N given by

$$\mathbb{P}_\ell^\mu[N=n] = e^{\beta\mu n} Z_\ell(n,\beta) \{\sum_{n\geq 0} e^{\beta\mu n} Z_\ell(n,\beta)\}^{-1} ; \tag{2.3}$$

the parameter μ (interpreted as the *chemical potential*) can be chosen to give the expectation value $\mathbb{E}_\ell^\mu[X_\ell]$ of the *particle number density* $X_\ell = N/|\Lambda_\ell|$ a prescribed value; this can be seen by introducing the *grand canonical pressure* $p_\ell(\mu,\beta)$ defined by

$$e^{\beta|\Lambda_\ell| p_\ell(\mu,\beta)} = \sum_{n\geq 0} e^{\beta\mu n} Z_\ell(n,\beta) ; \tag{2.4}$$

the function $\mu \mapsto p_\ell(\mu,\beta)$ is closely related to the cumulant generating function of the particle number density X_ℓ and we have

$$\mathbb{E}_\ell^\mu[X_\ell] = \frac{\partial}{\partial\mu} p_\ell(\mu,\beta) , \tag{2.5}$$

while the variance is given by

$$\mathbb{D}_\ell^\mu[X_\ell] = (\beta|\Lambda_\ell|)^{-1} \frac{\partial^2}{\partial\mu^2} p_\ell(\mu,\beta) ; \tag{2.6}$$

since the variance is strictly positive, the function $\mu \mapsto \frac{\partial}{\partial\mu} p_\ell(\mu,\beta)$ is strictly increasing and the function $\mu \mapsto \mathbb{E}_\ell^\mu[X_\ell]$ may be inverted; provided the sequence $\{\frac{\partial^2}{\partial\mu^2} p_\ell(\mu,\beta) : \ell = 1,2,\dots\}$ converges to a finite limit the variance converges to zero by (2.6); it follows by Tchebechev's inequality that the weak law of large numbers holds for the particle number density X_ℓ so that for large ℓ its probability distribution is concentrated around its mean value; a second glance at (2.4) which can be re-written as

$$e^{\beta|\Lambda_\ell| p_\ell(\mu,\beta)} = \sum_{n\geq 0} e^{\beta|\Lambda_\ell| \{ \mu \frac{n}{|\Lambda_\ell|} - f_\ell(\frac{n}{|\Lambda_\ell|},\beta)\}} \tag{2.7}$$

suggests that

$$p(\mu,\beta) = \lim_{\ell\uparrow\infty} p_\ell(\mu,\beta) = \sup_x \{ \mu x - f(x,\beta)\} \tag{2.8}$$

so that the grand canonical pressure is the Legendre transform of the free energy density; inverting we get

$$f(x,\beta) = \sup_\mu \{ \mu x - p(\mu,\beta)\} . \tag{2.9}$$

This argument was made rigorous in the 1960s by Ruelle and Fisher for continuous systems, by Griffiths and Dobrushin for lattice systems; details and references may be found in Ruelle's book [4].

§3. Phase-Transitions in the Lattice-Gas

It wasn't until the late 1930s that it was understood that the bulk limit is necessary for the sharp mathematical manifestation of a phase-transition. It became clear that although the functions $\mu \mapsto p_\ell(\mu,\beta)$ are infinitely differentiable for each ℓ, the limit function need not be; on the other hand since each function $\mu \mapsto p_\ell(\mu,\beta)$ is convex, the limit function $\mu \mapsto p(\mu,\beta)$ is convex and so must be differentiable except at most at a countable number of points; a point μ at which the left-hand derivative $p'_-(\mu)$ is not equal to the right-hand derivative $p'_+(\mu)$ is identified with a *first-order phase-transition*; from (2.9) it is clear that, in terms of the free energy density $f(x,\beta)$, a first-order phase-transition corresponds to an interval $[x_1,x_2]$ on which $x \mapsto f(x,\beta)$ is linear.

The best understood model which displays a phase-transition is that of the lattice-gas. In the interests of clarity we describe the model on a two-dimensional lattice, but everything goes through in higher dimensions. Now Λ_ℓ is the set of points $\{X=(x_1,x_2): x_i = 1,2,\ldots,\ell;\ i=1,2\}$; call the subset $\omega=\{X_1,\ldots,X_n\}$ the *configuration* of n particles in the square Λ_ℓ and denote the set of all such configurations by $\Omega_\ell^{(n)}$. It is often useful to think of Λ_ℓ as a square grid with ℓ^2 unit square cells; we can do this by identifying a point X with the unit cell having X as its centre; in this way we can think of a configuration ω as a declaration that n specified cells are *occupied* and the remaining ℓ^2-n are *empty*. The potential U is a function defined for all Y in $\mathbb{Z}^2$, with the convention that $U(0)=0$, and depending only on the length $|Y|$ of the vector Y. The canonical partition function in this case is given by

$$Z_\ell(n,\beta) = \sum_{\omega\in\Omega_\ell^{(n)}} \exp\{-\beta\sum_{i<j} U(X_i-X_j)\}. \tag{3.1}$$

It is known that if for some $C<\infty$ and some $\varepsilon>0$ the condition

$$|U(Y)| \leq C|Y|^{-(2+\varepsilon)} \tag{3.2}$$

holds, then the limit function $f(x,\beta)$ exists and $x \mapsto f(x,\beta)$ is a closed convex function on $[0,1]$. Dobrushin [3] has shown that under conditions on U which, crudely speaking, ensure that the negative part of the potential outweighs the positive part (see [3] for a precise statement) there exists a critical value $\beta_c<\infty$ such that for $\beta>\beta_c$ there is a non-empty phase-transition segment $[x_1,x_2]$. From our present point of view the interest lies in his method of proof. He used the following result of Berezin and Sinai [2]:

Berezin-Sinai Lemma: *In order that a non empty phase-transition segment with chemical potential* $\mu = 1/2\sum U(Y)$ *exist for some* β, *it is sufficient that for some* $\delta>0$ *and* $\gamma>0$ *and all sufficiently large* ℓ,

$$\mathbb{P}_\ell^\mu[\,|X_\ell - \tfrac{1}{2}| \geq \delta\,] > \gamma. \tag{3.3}$$

In this case, the interval $[\frac{1}{2}-\delta, \frac{1}{2}+\delta]$ *is a phase-transition segment.*

In other words, a first-order phase-transition can be detected as a violation of the law of large numbers in the grand canonical ensemble. (Griffiths [5] showed that for sufficiently large β and ℓ the mean-value of X_ℓ is less than $\frac{1}{2}-\delta$ from which

(3.3) certainly follows.) Dobrushin gave a proof of the Berezin-Sinai Lemma which is simpler than the one given in [2]. He deduced it from the following

Dobrushin Lemma: *For* $\delta>0$ *and* $\mu=\frac{1}{2}\Sigma U(Y)$ *we have*

$$\lim_{l\uparrow\infty}\frac{1}{|\Lambda_l|}\log \mathbb{P}_l^{\mu}[|X_l-\tfrac{1}{2}|\geq\delta] = \max_{|x-\frac{1}{2}|\geq\delta} g(x,\beta) - \max_{0\leq x\leq 1} g(x,\beta), \tag{3.4}$$

where $g(x,\beta) = \beta(\mu x - f(x,\beta))$.

The function $x\mapsto g(x,\beta)$ is convex; in the particular case of the lattice-gas model it satisfies the symmetry condition

$$g(\tfrac{1}{2}-\delta,\beta) = g(\tfrac{1}{2}+\delta,\beta)\ ; \tag{3.5}$$

it follows that the right-hand side of (3.4) is equal to $g(\frac{1}{2}-\delta)-g(\frac{1}{2})$. Then (3.3) implies that $[\frac{1}{2}-\delta,\frac{1}{2}+\delta]$ is a phase-transition segment.

§4. The Bose-Einstein Phase-Transition

The traditional description of Bose-Einstein condensation is this: in a system of non-interacting bosons in thermal equilibrium the excited states saturate at a critical value ρ_c of the density; when the density ρ is increased beyond this value the excess $\rho-\rho_c$ goes into the zero-energy state. The phenomenon is sometimes described as 'condensation in momentum space'. The condensate has zero entropy as well as zero energy, and so makes no contribution to the pressure. Consequently, the pressure-density isotherm has a flat part: the pressure increases with increasing density for densities below ρ_c and thereafter remains constant. There is a basic difficulty which we have to face if we attempt a rigorous proof of these statements: a phase-transition manifests itself sharply in the mathematical behaviour of thermodynamic functions only in the bulk limit, but in this limit there is no unique precise formulation of the zero-energy state. For non-interacting particles in a box of finite volume, the single-particle energy-levels are well-defined and there is a unique ground state; as the volume increases, every energy-level tends to zero; for the infinite system, the single-particle energy-spectrum is a continuum filling the half-line but there are no eigenstates. There are two good candidates for the concept of macroscopic occupation of the zero-energy state: *macroscopic occupation of the ground state* is said to occur when the number of particles in the ground state becomes proportional to the volume; *generalized condensation* is said to occur when the number of particles whose energy levels lie in an arbitrarily small band above zero becomes proportional to the volume. Obviously, the first implies the second. However, the second can occur without the first; this is called *non-extensive condensation*. These matters are discussed in [6] where it is proved that there are, in general, two critical densities: there is ρ_c which is the density at which singularities in the thermodynamic functions occur; there is ρ_m which is the minimum density for macroscopic occupation of the ground state. Generalized condensation occurs whenever ρ is greater than ρ_c; macroscopic occupation of the ground state occurs if and only if the weak law of large numbers for the particle number density is violated.

As far as we know, the first rigorous proof of the macroscopic occupation of the ground state of the Laplacian when the bulk-limit is taken by dilating an arbitrary star-shaped region was sketched by Kac in 1971; his manuscript remained unpublished until 1977 when it was incorporated in the review by Ziff, Uhlenbeck and Kac [7]. The mathematical details were supplied in the thesis of PULÈ [8] and in the papers of Cannon [9] and LEWIS and PULÈ [10]; the connection with the work of Araki and Woods [11] was discussed by LEWIS [12]. Kac obtained the limiting distribution $K(x;\rho)$ (now known as the Kac distribution) of the particle number density $X_\ell = N/|\Lambda_\ell|$ at fixed mean density ρ by computing its Laplace transform:

$$\int_{[0,\infty)} e^{-sx} dK(x;\rho) = \lim_{\ell\uparrow\infty} E^\rho[e^{-sX_\ell}] . \tag{4.1}$$

He found that, when ρ exceeds ρ_c, the distribution is exponential; details may be found in [6] where it is shown that, in general, the distribution is infinitely divisible.

In the mean-field model of a system of interacting bosons, the interaction energy is represented by a term $aN^2/2|\Lambda_\ell|$ which is added to the hamiltonian of the free boson gas, where a is a strictly positive constant representing the strength of the interaction. This crude model of a system of interacting bosons is commonly called the *imperfect boson gas*; it is of interest because the pathological aspects of the free boson gas are removed by the mean-field interaction: the grand canonical partition function converges for all real values of the chemical potential [4]; the weak law of large numbers holds for the particle number density for all values of the chemical potential [13] (see also [14] and [15]). However, it is proved in [16] that generalized condensation persists in the imperfect boson gas: generalized condensation is stable with respect to a mean-field perturbation of the free-particle hamiltonian.

§5. An Extension of Laplace's Method for Integrals

In this section, we present a version of Dobrushin's Lemma which holds under conditions which are satisfied by a wide class of continuous systems in statistical mechanics, both classical and quantum. We do this by means of a version of Laplace's method for integrals which, unlike the standard treatments (see Copson [17], for example), makes no hypothesis of differentiability concerning the integrand.

Lemma 1 (Laplace's method)

Let $\{f_n : n=1,2,\dots\}$ *be a sequence of lower semi-continuous functions,* $f_n : [0,\infty)\to\mathbb{R}$; suppose *that on each compact the sequence* $\{f_n\}$ *is bounded below and converges uniformly to* f, *and that* $f(0)=0$. *Let* $\{t_n : n=1,2,\dots\}$ *be a sequence of positive real numbers diverging to* $+\infty$ *and let*

$$\mu_\infty = \lim_{n\uparrow\infty} \left(\liminf_{x\uparrow\infty} \left(\frac{1}{x} \inf_{k\geq n} f_k(x) \right) \right) . \tag{5.1}$$

Then for each $\mu < \mu_\infty$ *we have*

$$\lim_{n\uparrow\infty} \frac{1}{t_n} \log \int_I e^{t_n(\mu x - f_n(x))} dm(t_n x) = f_I^*(\mu) < \infty , \tag{5.2}$$

where m *is either Lebesgue measure or counting measure,* I *is* $\mathbb{R}^+$ *or a compact subset of it, and*

$$f_I^*(\mu) = \sup_{s\in I} \{ \mu s - f(s) \} . \tag{5.3}$$

Proof: Put $g_n(x) = \mu x - f_n(x)$ and $g(x) = \mu x - f(x)$. Choose A such that $\mu < A < \mu_\infty$; choose m such that

$$\liminf_{x \uparrow \infty} \left(\frac{1}{x} \inf_{k \geq m} f_k(x) \right) > A ;$$

then there exists x_1 such that $f_n(x) > Ax$ for $x > x_1$ and $n > m$; hence

$$g_n(x) \leq -(A-\mu)x \quad \text{for all } x > x_1 \text{ and } n > m, \tag{5.4}$$

and $g(x) \leq -(A-\mu)x$ for all $x > x_1$. But $g(0) = 0$ so that

$$\sup_{x \in \mathbb{R}^+} g(x) = \sup_{x \in [0, x_1]} g(x) . \tag{5.5}$$

Now g is upper semi-continuous and bounded above on compacts so that the supremum on the right hand side of (5.4) is attained at a point x_0 in $[0, x_1]$; hence

$$f^*(\mu) = \sup_{x \in \mathbb{R}^+} g(x) = g(x_0) < \infty . \tag{5.6}$$

Furthermore, given $\epsilon > 0$, there exists $\delta > 0$ such that $g(x_0) - g(x) < \frac{\epsilon}{2}$ in $[x_0 - \delta, x_0 + \delta]$; by the uniformity of convergence on compacts of $\{f_n\}$ we have $g(x_0) - g_n(x) < \epsilon$ for all x in $[x_0 - \delta, x_0 + \delta]$ and all n greater than some m'. Thus we have, for

$$\int_{\mathbb{R}^+} e^{t_n g_n(x)} dm(t_n x) > \int_{x_0 - \delta}^{x_0 + \delta} e^{t_n g_n(x)} dm(t_n x) > (2 t_n \delta + 1) e^{t_n \{g(x_0) - \epsilon\}}$$

for n sufficiently large, so that

$$\liminf_{n \uparrow \infty} \frac{1}{t_n} \log \int_{\mathbb{R}^+} e^{t_n g_n(x)} dm(t_n x) \geq g(x_0). \tag{5.7}$$

On the other hand, by (5.4), we have

$$\int_{\mathbb{R}^+} e^{t_n g_n(x)} dm(t_n x) < (t_n x_1 + 1) e^{t_n g(x_0)} + \int_{x_1}^{\infty} e^{-t_n (A-\mu) x} dm(t_n x)$$

$$< \left\{ (t_n x_1 + 1) + \frac{t_n e^{-t_n g(x_0)}}{A - \mu} \right\} e^{t_n g(x_0)}$$

for $n > m$ hence

$$\limsup_{n \uparrow \infty} \frac{1}{t_n} \log \int_{\mathbb{R}^+} e^{t_n g_n(x)} dm(t_n x) \leq g(x_0). \tag{5.8}$$

Putting together (5.6), (5.7) and (5.8), the claim is proved in the case where $I = \mathbb{R}^+$; the case where I is a compact subset of $\mathbb{R}^+$ is straightforward and is left to the reader.

To apply this to statistical mechanics, we suppose that we are given a sequence $\{\Lambda_\ell : \ell = 1, 2, \ldots\}$ of regions in $\mathbb{R}^d$ such that the sequence $\{|\Lambda_\ell| : \ell = 1, 2, \ldots\}$ diverges to infinity where $|\Lambda_\ell|$ is the volume of the region Λ_ℓ; that for each region Λ_ℓ we are given a sequence $\{Z_\ell(n) : n = 0, 1, 2, \ldots\}$ of positive real numbers with $Z_\ell(0) = 1$,

$Z_\ell(n)$ being the n-particle canonical partition function (in this section we suppress reference to the inverse temperature β in order to simplify the notation). For each ℓ, we define a function f_ℓ, the free energy density, by

$$f_\ell(x) = \begin{cases} 0 , & x = 0, \\ -(\beta|\Lambda_\ell|)^{-1} \log Z_\ell(n) , & \frac{n}{|\Lambda_\ell|} < x \leq \frac{n+1}{|\Lambda_\ell|} ; \end{cases} \tag{5.9}$$

the function f_ℓ is lower semi-continuous. In Ruelle's book one may find details of conditions on the potentials used to define the canonical partition functions which ensure that on each compact the sequence $\{f_\ell\}$ is bounded and converges uniformly to a function f satisfying $f(0)=0$. The grand canonical pressure p_ℓ is defined by

$$p_\ell(\mu) = (\beta|\Lambda_\ell|)^{-1} \log \{ \sum_{n\geq 0} e^{\beta\mu n} Z_\ell(n) \} \tag{5.10}$$

for all values of μ for which the infinite series on the right hand side of (5.10) converges. Putting $t_\ell = \beta|\Lambda_\ell|$ and letting m be counting measure (the measure which is concentrated on the integers and which assigns unit weight to each), we may re-write (5.10) as

$$p_\ell(\mu) = \frac{1}{t_\ell} \log \int_{\mathbb{R}^+} e^{t_\ell\{\mu x - f_\ell(x)\}} dm(t_\ell x) ; \tag{5.11}$$

Lemma 1 now ensures that, for all $\mu < \mu_\infty$,

$$p(\mu) = \lim_{\ell\uparrow\infty} p_\ell(\mu) = f^*(\mu) . \tag{5.12}$$

Remark 1: $[p'_-(\mu), p'_+(\mu)] = \{x : \mu x - f(x) = f^*(\mu)\}$.

Remark 2: Put $\mu^* = \sup\{\mu : f^*(\mu) < \infty\}$; it is clear from the proof of lemma 1 that $\mu_\infty \leq \mu^*$; for systems, both classical and quantum, with superstable interactions we have $\mu_\infty = \infty$; for classical systems with stable interactions, we have $\mu_\infty = \infty$; for the free boson gas we have $\mu_\infty = \mu^* = 0$.

In the grand canonical ensemble at chemical potential μ, the probability distribution of the particle number N is defined to be

$$\mathbb{P}^\mu_\ell[N=n] = e^{\beta\mu n} Z_\ell(n) \Big(\sum_{n\geq 0} e^{\beta\mu n} Z_\ell(n) \Big)^{-1} . \tag{5.13}$$

To discuss the probability distribution of the particle number density $X_\ell = N/|\Lambda_\ell|$, it is convenient to introduce the conditional pressure $p_\ell(\mu|I)$ the pressure conditioned on X_ℓ taking values in I, given by

$$p_\ell(\mu|I) = (\beta|\Lambda_\ell|)^{-1} \log \{ \sum_{\{n:\, n/|\Lambda_\ell| \in I\}} e^{\beta\mu n} Z_\ell(n) \} ; \tag{5.14}$$

in the notation introduced above, this can be re-written as

$$p_\ell(\mu|I) = \frac{1}{t_\ell} \log \int_I e^{t_\ell\{\mu x - f_\ell(x)\}} dm(t_\ell x) . \tag{5.15}$$

Hence

$$\mathbb{P}_\ell^\mu[X_\ell \in I] = e^{\beta|\Lambda_\ell|\{p_\ell(\mu|I) - p_\ell(\mu)\}} \tag{5.16}$$

The following generalization of Dobrushin's Lemma now follows from Lemma 1:

Lemma 2:

Suppose that on each compact the sequence $\{f_\ell : \ell = 1,2,\dots\}$ *of free energy densities is bounded below and converges uniformly to* f; *then for each* $\mu < \mu_\infty$ *we have*

$$\lim_{\ell\uparrow\infty} \frac{1}{|\Lambda_\ell|} \log \mathbb{P}_\ell^\mu[X_\ell \in I] = \beta\{f_I^*(\mu) - f^*(\mu)\} .$$

Remark 3: It follows that a necessary and sufficient condition for the interval I to be a phase-transition segment with chemical potential μ is that $\mathbb{P}_\ell^\mu[X_\ell \in I]$ should not approach zero exponentially rapidly as $\ell \to \infty$.

Remark 4: In the case of a first-order phase-transition, there is no guarantee that the sequence of distribution functions of X_ℓ converges; by the Helly selection theorem there is at least one convergent subsequence; each convergent subsequence converges to a distribution function concentrated on $[p_-'(\mu), p_+'(\mu)]$; distinct limits correspond to distinct ways of taking the bulk limit; see [6] for examples;

Remark 5: Lemma 2 provides a simple explanation of the result of Davies [13] which was discussed in §4: the mean-field perturbation adds a term $ax^2/2$ to the free energy density of the free boson gas; this destroys the phase-transition segment and so the distribution of X_ℓ becomes concentrated at its mean-value.

§6. Large Deviations and the Grand Canonical Pressure

Often it is convenient to prove the existence of the grand canonical pressure by first establishing the existence of the free energy density. In some cases, however, it is possible to prove directly the existence of the pressure (see [6] and [18] for examples). It is therefore natural to ask if there is a result along the lines of Lemma 2 but which requires only the existence of the pressure; in this section we present such a result.

The crucial fact is that the grand canonical pressure $p_\ell(\mu)$ is a strictly convex function of μ; it follows that

$$\mu \mapsto p(\mu) = \lim_{\ell\uparrow\infty} p_\ell(\mu)$$

is convex for all values of μ for which the limit exists; hence the one-sided derivatives $p_-'(\mu)$ and $p_+'(\mu)$ exist for all values of μ in the domain of p, and they are equal except on a set which is at most countable. The following generalization of Griffiths' Lemma [19] is frequently useful in statistical mechanics; the proof is straightforward and so we omit it.

Lemma 3 (Generalization of Griffiths' Lemma)

Let f *be a closed proper convex function on* $\mathbb{R}$; *let* $\{f_\ell : \ell = 1,2,\dots\}$ *be a sequence of convex functions such that*

(1) $\operatorname{dom} f \subset \bigcap_\ell \operatorname{dom} f_\ell$,

(2) $f(x) = \lim_{\ell\uparrow\infty} f_\ell(x)$ *for all* x *in* $\operatorname{dom} f$.

Let $\{x_\ell \in \operatorname{dom} f_\ell : \ell = 1,2,\dots\}$ *be a sequence converging to a point* x *of* $\operatorname{dom} f$
Then

$$f'_-(x) \leq \liminf_{\ell \uparrow \infty} (f_\ell)'_-(x_\ell) \leq \limsup_{\ell \uparrow \infty} (f_\ell)'_+(x_\ell) \leq f'_+(x) .$$

In what follows we shall assume that

$$p(\mu) = \lim_{\ell \uparrow \infty} p_\ell(\mu)$$

exists and is a strictly convex function of μ ; for simplicity of exposition, we shall also assume that there is at most one point at which p is not differentiable. Write

$$K^\mu_\ell(x) = \mathbb{P}^\mu_\ell[X_\ell \leq x] , \tag{6.1}$$

$$C^\mu_\ell(t) = \log \int_{[0,\infty)} e^{tx}\, dK^\mu_\ell(x) ; \tag{6.2}$$

the cumulant generating function $C^\mu_\ell(t)$ can be expressed in terms of $p_\ell(\mu)$ as

$$C^\mu_\ell(t) = \beta|\Lambda_\ell| \{ p_\ell(\mu + t/_{\beta|\Lambda_\ell|}) - p_\ell(\mu) \} . \tag{6.3}$$

<u>Lemma 4</u>:

If p is differentiable at μ then $\{K^\mu_\ell : \ell = 1,2,\ldots\}$ converges to a distribution concentrated at the mean value ρ , $\rho = \frac{d}{d\mu} p(\mu)$.

<u>Proof</u>: Since p is differentiable at μ , it follows from Lemma 3 that

$$\lim_{\ell \uparrow \infty} C^\mu_\ell(t) = t \frac{d}{d\mu} p(\mu). \tag{6.4}$$

The claim then follows using the continuity and uniqueness theorems for the Laplace transform.

If p is not differentiable at μ then there is no guarantee that the sequence $\{K^\mu_\ell : \ell = 1,2,\ldots\}$ converges; nevertheless, by the Helly Selection Theorem, there exists at least one convergent subsequence; we prove that if K is the limit of a convergent subsequence of $\{K^\mu_\ell : \ell = 1,2,\ldots\}$ then K is concentrated on the interval $[p'_-(\mu), p'_+(\mu)]$ and outside that interval the distribution converges to zero exponentially rapidly.

<u>Lemma 5</u>:

Fix $\rho_1 < p'_-(\mu)$; let α_1 be the unique solution of $p'(\mu - \alpha_1) = \rho_1$ and put

$$\Delta_1 = \beta \{ p(\mu) - p(\mu - \alpha_1) - \alpha_1 p'(\mu - \alpha_1) \} . \tag{6.5}$$

Then

$$\limsup_{\ell \uparrow \infty} \frac{1}{|\Lambda_\ell|} \log \mathbb{P}^\mu_\ell [X_\ell \leq \rho_1] \leq -\Delta_1 . \tag{6.6}$$

Fix $\rho_2 > p'_+(\mu)$; let α_2 be the unique solution of $p'(\mu + \alpha_2) = \rho_2$ and put

$$\Delta_2 = \beta \{ p(\mu) - p(\mu + \alpha_2) + \alpha_2 p'(\mu + \alpha_2) \} . \tag{6.7}$$

Then

$$\limsup_{l\uparrow\infty} \frac{1}{|\Lambda_l|} \log \mathbb{P}_l^\mu [X_l \geq \rho_2] \leq -\Delta_2 . \tag{6.8}$$

Proof: By the Markov inequality, we have, for $t > 0$,

$$\mathbb{P}_l^\mu [X_l \geq \rho_2] \leq e^{-\rho_2 t}\, \mathbb{E}_l^\mu [e^{tX_l}] = e^{-(\rho_2 - C_l^\mu(t))} ,$$

But α_2 is strictly positive since p is strictly convex so that, putting $t = \alpha_2 \beta |\Lambda_l|$ we get

$$\limsup_{l\uparrow\infty} \frac{1}{|\Lambda_l|} \log \mathbb{P}_l^\mu [X_l \geq \rho_2] \leq \limsup_{l\uparrow\infty} \{-\rho_2 \alpha_2 \beta + \frac{1}{|\Lambda_l|} C_l^\mu(\alpha_2 \beta |\Lambda_l|)$$

$$= -\Delta_2 ,$$

which is (6.8); we get (6.6) in an analagous fashion.

Lemma 6:

In the notation of Lemma 5 we have

$$\liminf_{l\uparrow\infty} \frac{1}{|\Lambda_l|} \log \mathbb{P}_l^\mu [X_l \leq \rho_1] \geq -\Delta_1 , \tag{6.9}$$

$$\liminf_{l\uparrow\infty} \frac{1}{|\Lambda_l|} \log \mathbb{P}_l^\mu [X_l \geq \rho_2] \geq -\Delta_2 . \tag{6.10}$$

Proof: To demonstrate (6.10) we introduce the associated distribution function $\tilde{K}_l^\mu$ defined by

$$d\tilde{K}_l^\mu(x) = e^{-C_l^\mu(t_l)} e^{t_l x} dK_l^\mu(x) ,$$

where $t_l = \alpha_2 \beta |\Lambda_l|$, whose moment generating function is given by

$$\int_{[0,\infty)} e^{tx} d\tilde{K}_l^\mu(x) = e^{C_l^\mu(t + t_l) - C_l^\mu(t)} ;$$

as $l \uparrow \infty$ this converges to $e^{t p'(\mu + \alpha_2)}$ since p is differentiable at $\mu + \alpha_2$ for $\alpha_2 > 0$; hence $\tilde{K}_l^\mu$ converges to a distribution concentrated at ρ_2 . For each $\epsilon > 0$ we have

$$\mathbb{P}_l^\mu [X_l \geq \rho_2] \geq \int_{[\rho_2,\rho_2+\epsilon]} dK_l^\mu(x) = e^{C_l^\mu(t_l)} \int_{[\rho_2,\rho_2+\epsilon]} e^{-t_l x} d\tilde{K}_l^\mu(x)$$

$$\geq e^{C_l^\mu(t_l) - t_l(\rho_2+\epsilon)} \int_{[\rho_2,\rho_2+\epsilon]} d\tilde{K}_l^\mu(x) ;$$

now $\int_{[\rho_2,\rho_2+\epsilon]} d\tilde{K}_l^\mu(x) > \frac{1}{2}$ for l sufficiently large, so that

$$\liminf_{l\uparrow\infty} \mathbb{P}_l^\mu [X_l \geq \rho_2] \geq -\Delta_2 ,$$

which is (6.10); the inequality (6.9) follows in an analagous fashion.

Putting together Lemma 5 and Lemma 6, we have

Lemma 7:

Suppose that $\mu \mapsto p(\mu) = \lim_{l\uparrow\infty} p_l(\mu)$ exists, is strictly convex and differentiable except possibly at one point μ ; then there exist strictly positive constants Δ_1 , Δ_2 (given by (6.5) and (6.7)) such that

$$\lim_{l\uparrow\infty} \frac{1}{|\Lambda_l|} \log \mathbb{P}^{\mu}_{l}[X_l \leq \rho_1] = -\Delta_1 \quad \text{for } \rho_1 < p'_-(\mu),$$

$$\lim_{l\uparrow\infty} \frac{1}{|\Lambda_l|} \log \mathbb{P}^{\mu}_{l}[X_l \geq \rho_2] = -\Delta_2 \quad \text{for } \rho_2 > p'_+(\mu).$$

References

1. W. Feller, An Introduction to Probability Theory and its Applications (two volumes)(Wiley and Sons, New York, 1966).
2. F.A. Berezin and Ya. G. Sinai, Trudy Mosk. Mat. Obshch. 17, 197-212 (1967).
3. R.L. Dobrushin, Proc. Fifth Berkeley Symposium, III, 73-87 (1967).
4. D. Ruelle, Statistical Mechanics: Rigorous Results (New York, Amsterdam, Benjamin, 1969).
5. R.B. Griffiths, Phys. Rev. 136, 437-439 (1964).
6. M. van den Berg, J.T. Lewis and J.V. Pulè, J. Math. Anal. and Appl. (in press).
7. R.M. Ziff, G.E. Uhlenbeck and M. Kac, Physics Reports 32C, 169-248 (1977).
8. J.V. Pulè, D. Phil. Thesis, (Oxford, 1972).
9. J.T. Cannon, Commun. Math. Phys. 29, 89-104 (1973).
10. J.T. Lewis and J.V. Pulè, Commun. Math. Phys. 36, 1-18 (1974).
11. H. Araki and E.J. Woods, J. Math. Phys. 4, 637 (1963).
12. J.T. Lewis, The Free Boson Gas, in Mathematics of Contemporary Physics, Ed. R.F. Streater, (Academic Press, London, 1972).
13. E.B. Davies, Commun. Math. Phys. 28, 69 (1972).
14. M. Fannes and A. Verbeure, Phys. Lett. 76A, 31 (1980).
15. E. Buffet and J.V. Pulè, J. Maths. Phys. 24, 1608 (1983).
16. M. van den Berg, J.T. Lewis and P. de Smedt, J. Stat. Phys. (to appear in Dec., 1984).
17. E.T. Copson, Asymptotic Expansions, (C.U.P., Cambridge, 1965).
18. D. Ruelle, Lectures in Theoretical Physics, Ed. W.E. Brittin and W.R. Chappell, VI, 73, (Univ. of Colorado Press, Boulder, 1964).
19. R.B. Griffiths, J. Math. Phys. 5, 1215 (1964).

THE UNIQUENESS OF REGULAR DLR MEASURES FOR CERTAIN ONE-DIMENSIONAL SPIN SYSTEMS

F. Papangelou

Random fields are not always uniquely determined by their specifications, i.e. their systems of conditional distributions. A general result is presented here, giving sufficient conditions under which a one-dimensional specification admits at most one random field (up to equivalence in distribution), within a specified class of such fields. In a specific application this result implies that certain one-dimensional spin systems with long range interaction admit unique regular DLR measures, regardless of "temperature".

The present article has been written in the informal style of the talk given at the conference and begins with a brief introduction to some known results from the literature, selected for their immediate relevance to our subject. A sketchy outline of the proofs of new results is given in the last section; for full details the reader is referred to [9] and [10].

§1. The background.

Although the systems to be considered here are ordinary stochastic processes parametrized by the integers, it is useful to think of them as one-dimensional random fields. By a d-dimensional *random field* we mean a collection of random variables X_i defined on the same probability space $(\Omega,\mathfrak{X},P)$ and parametrized by the elements i of the d-dimensional lattice $\mathbb{Z}^d$. If one thinks of the elements of $\mathbb{Z}^d$ as "sites", then X_i is some variable (say a spin) associated with site i. The *specification* of a random field is the system of conditional distributions

$$Q^\Lambda(B|\zeta_j,j\notin\Lambda) = P((X_i)_{i\in\Lambda} \in B|X_j = \zeta_j,j\notin\Lambda)$$

$(B \in \mathcal{B}(\mathbb{R}^\Lambda))$, where Λ ranges over the finite subsets of $\mathbb{Z}^d$. Whenever we refer to a specification, we will assume that the versions of all conditional distributions in it are regular probability kernels satisfying in a strict sense the obvious consistency conditions implied by the definition.

In statistical mechanics and other areas, where random fields have been widely employed, it often happens that the nature of the inter-relationship between the X_i's is best described in terms of the specification. However, a specification does not always uniquely determine an (unconditional) distribution for a random field and, given a particular specification, two natural questions immediately arise: (i) is there a random field admitting the specification? (ii) if there is such a random field, is it unique up to equivalence in distribution? If we identify a random field with its distribution, these questions reduce to the existence and uniqueness of a probability measure Π, on the product σ-field $\mathcal{B}(\mathbb{R}^{\mathbb{Z}^d})$ of $\mathbb{R}^{\mathbb{Z}^d}$, admitting the specification. If there are two or more probability measures admitting the same specification, we say that *phase transition* occurs.

This is in summary the approach adopted by Dobrushin and by Lanford and Ruelle in the late sixties, in their treatment of equilibrium states for infinite Gibbs systems.

Before describing the particular class of statistical mechanical models we will be concerned with, we state three early theorems due to Dobrushin, which give sufficient conditions for uniqueness of the measure admitted. The theorems are not stated in full generality here but only in a form which is relevant to our subsequent discussion. In particular, we only consider translation invariant specifications.

Suppose that a translation invariant specification is given. Take $\Lambda = \{0\}$, where 0 is the null element of $\mathbb{Z}^d$, and for $k \in \mathbb{Z}^d \setminus \{0\}$ define

$$\rho_k = \tfrac{1}{2} \sup \|Q^{\{0\}}(\cdot|\zeta_j, j\neq 0) - Q^{\{0\}}(\cdot|\bar{\zeta}_j, j\neq 0)\|$$

where $\|\cdot\|$ denotes total variation and the supremum is taken over all pairs $(\zeta_j)_{j\neq 0}$, $(\bar{\zeta}_j)_{j\neq 0}$ that differ only at site k.

1.1 Theorem ([4]). If $\sum_{k\neq 0} \rho_k < 1$, then the specification admits at most one random field.

Dobrushin later extended this theorem, replacing the total variation by the Vasserstein distance between probability measures. If we take the usual metric $|\zeta - \eta|$ on the real line, the Vasserstein distance of two probability measures π_1, π_2 on $\mathcal{B}(\mathbb{R})$ can be defined as

$$d(\pi_1,\pi_2) = \inf \iint |\zeta-\eta| \pi(d\zeta d\eta)$$

where the infimum is taken over those two-dimensional distributions π (i.e. probability measures on $\mathcal{B}(\mathbb{R}^2)$) having one-dimensional marginals π_1 and π_2. Given a specification as above, define

$$\rho_k^* = \sup \frac{d(Q^{\{0\}}(\cdot|\zeta_j, j\neq 0), Q^{\{0\}}(\cdot|\bar{\zeta}_j, j\neq 0))}{|\zeta_k - \bar{\zeta}_k|}$$

$(k\neq 0)$, where the supremum is again taken over all pairs $(\zeta_j)_{j\neq 0}$, $(\bar{\zeta}_j)_{j\neq 0}$ that differ only at site k.

1.2 Theorem ([5]). If $\sum_{k\neq 0} \rho_k^* < 1$, then the specification admits at most one random field such that $\sup_i E|X_i| < \infty$.

Of particular interest to us here is the case of one-dimensional random fields

X_i, $i \in \mathbb{Z}$. For this case Dobrushin proved the following theorem. Write $\Lambda = [s,t]$ if $\Lambda = \{s,s+1,\dots,t\}$.

1.3 Theorem ([4]). Suppose there is a sequence of positive numbers β_k, $k = 0,1,2,\dots$ such that

(i) $\beta_k < 2$ for all $k \geq 0$,

(ii) $\lim_{k\to\infty} \beta_k = 0$ and

(iii) for arbitrary $\Lambda = [s,t]$, $(\zeta_j)_{j\notin\Lambda}$ and $(\bar{\zeta}_j)_{j\notin\Lambda}$, if either $\zeta_j = \bar{\zeta}_j$ for $j = s-k,s-k+1,\dots,s-1$ and all $j > t$, or $\zeta_j = \bar{\zeta}_j$ for $j = t+1,t+2,\dots,t+k$ and all $j < s$, then

$$\|Q^\Lambda(\cdot|\zeta_j,j\notin\Lambda) - Q^\Lambda(\cdot|\bar{\zeta}_j,j\notin\Lambda)\| \leq \beta_k.$$

Then the specification admits a unique random field.

If the true state space of a random field arising from the given specification is a compact subset of $\mathbb{R}$ and the conditional probabilities have continuous densities, the conditions of Theorem 1.3 are not unreasonable. In the unbounded case however they are too severe, even for Markovian specifications. For instance $\beta_0 < 2$ (which is an essential hypothesis) means

$$\sup \|Q^\Lambda(\cdot|\zeta_j,j\notin\Lambda) - Q^\Lambda(\cdot|\bar{\zeta}_j,j\notin\Lambda)\| < 2$$

where the supremum is taken over arbitrary $\Lambda = [s,t]$, $(\zeta_j)_{j\notin\Lambda}$ and $(\bar{\zeta}_j)_{j\notin\Lambda}$.

Note that a one-dimensional specification is said to be Markovian if $Q^{[s,t]}(\cdot|\zeta_j,j\notin[s,t])$ depends on $\zeta_j,j\notin[s,t]$ only through ζ_{s-1} and ζ_{t+1}.

§2. Spin systems.

The spin systems discussed in the present article are systems of random variables X_i, $i \in \mathbb{Z}$ taking arbitrary real values. They are special cases of a more general class of (multi-dimensional) random fields considered by Ruelle ([12]) and by Lebowitz and Presutti ([7]), and their specifications are of the following form

$$P(X_i \in d\zeta_i,\ i\in\Lambda | X_j = \zeta_j, j\notin\Lambda) = \text{const.} \exp\{-\beta(\sum_{i\in\Lambda} F(\zeta_i) - \tfrac{1}{2}\sum_{\substack{i,j\in\Lambda \\ i\neq j}} J(|i-j|)\zeta_i\zeta_j - \sum_{i\in\Lambda, j\notin\Lambda} J(|i-j|)\zeta_i\zeta_j)\} \prod_{i\in\Lambda} d\zeta_i \tag{1}$$

Here $\frac{1}{\beta} > 0$ is the "temperature" (the term arising from the statistical mechanical associations), the pair potential $-J(|i-j|)\zeta_i\zeta_j$ represents the interaction between spin ζ_i at site i and spin ζ_j at site j, while F is a continuous function representing the self-interaction. We will make the following assumptions.

(a) $|J(1)| \geq |J(2)| \geq |J(3)| \geq \dots$

(b) $\sum_{i=1}^{\infty} |J(i)| < \infty$.

(c) There are constants $a > 0$ and c such that, for arbitrary $\Lambda = [s,t]$ and $(\zeta_i)_{i\in\Lambda}$,

$$\sum_{i\in\Lambda} F(\zeta_i) - \tfrac{1}{2} \sum_{\substack{i,j\in\Lambda \\ i\neq j}} J(|i-j|)\zeta_i\zeta_j \geq \sum_{i\in\Lambda} (a\zeta_i^2 - c).$$

Assumption (c) is Ruelle's superstability condition. A DLR measure (equilibrium measure) is a probability measure Π on $\mathcal{B}(\mathbb{R}^{\mathbb{Z}})$ admitting the specification (1). Note that the conditional densities in (1) may not be defined for all configurations $(\zeta_j)_{j\notin\Lambda}$, but it is assumed that a DLR measure assigns probability zero to the set of exceptional configurations. The existence of such DLR measures was established in a more general (and multi-dimensional) context by Lebowitz and Presutti ([7]), who made use of Ruelle's estimates ([12]). Uniqueness however does not hold generally: phase transition is possible even in dimension one and Benfatto, Presutti and Pulvirenti ([1]) determined all DLR measures in the special case of one-dimensional harmonic nearest-neighbour (Markovian) systems.

Nevertheless there may be uniqueness within some class of DLR measures and Cassandro, Olivieri, Pellegrinotti and Presutti ([2]) used Dobrushin's Theorem 1.2 to derive a uniqueness condition for a family of multi-dimensional spin systems. When applied to the system (1) their result implies the following theorem. Assume that, for $\zeta > 0$, $F(\zeta) = \int_0^\zeta G(\eta)d\eta$ where G is a C^1, convex, positive, increasing function, and analogously for $\zeta < 0$, and denote by $V(\zeta_j,\ j\neq 0)$ the variance of the distribution $Q^{\{0\}}(\cdot|\zeta_j, j\neq 0)$.

<u>2.1 Theorem</u> ([2]). If

$$\sup V(\zeta_j, j\neq 0) < (2\beta \sum_{i=1}^{\infty} |J(i)|)^{-1} \tag{2}$$

where the supremum is taken over all $\zeta_j, j\neq 0$, then the specification (1) admits exactly one random field such that $\sup_i E|X_i| < \infty$.

As an illustration, if $F(\zeta) = \alpha|\zeta|^{2+\delta}$ $(\alpha>0,\ \delta>0)$, then (2) holds when the temperature $\frac{1}{\beta}$ is sufficiently high but will fail for low temperatures. The result presented below (Theorem 2.3) asserts that in the present one-dimensional case, if

$$\sum_{i=1}^{\infty} i|J(i)| < \infty \tag{3}$$

then there is only one DLR measure satisfying the condition $\sup_i E|X_i| < \infty$, whatever the value of the temperature $\frac{1}{\beta}$. Before formulating this result we link the condition $\sup_i E|X_i| < \infty$ with Ruelle's notion of a tempered DLR measure and the notion of a tame measure to be introduced in §3 below.

<u>2.2 Theorem.</u> If Π is a DLR measure admitted by the specification (1) and X_i, $i \in \mathbb{Z}$ is a random field with distribution Π, then the following statements

are equivalent.

(i) Π is tempered ([11],[7]) in the sense that the sequence $\frac{1}{2n+1}\sum_{i=-n}^{n} X_i^2$, $n = 1,2,\ldots$ is almost surely bounded (by a random bound).

(ii) $\sup_i E|X_i| < \infty$.

(iii) Π is tame in a sense to be explained below.

Call DLR measures satisfying the conditions of Theorem 2.2 *regular*. Our main assertion concerning the spin systems (1) is the following

2.3 Theorem. If $\sum_{i=1}^{\infty} i|J(i)| < \infty$, then the specification (1) admits exactly one regular DLR measure. This measure is translation invariant.

This theorem follows from the general result of the next section.

§3. A general uniqueness theorem.

In the one-dimensional case, if $\Lambda = [s,t]$, then the external configuration $\zeta_j, j \notin \Lambda$ consists of a forward sequence $z = (\zeta_{t+1}, \zeta_{t+2}, \ldots)$ and a backward sequence $x = (\ldots, \zeta_{s-2}, \zeta_{s-1})$ and, given a specification, it will be convenient to write $Q^\Lambda(\cdot|x,z)$ for $Q^\Lambda(\cdot|\zeta_j, j\notin\Lambda)$. We will always use z, $\bar{z}$ etc. to denote forward sequences and x, $\bar{x}$ etc. to denote backward sequences.

Suppose that a one-dimensional specification is given, and that the set $\vec{T}$ of forward sequences z and the set $\overleftarrow{T}$ of backward sequences x for which the conditional probabilities $Q^\Lambda(\cdot|x,z)$ are defined are complete, separable metric spaces under appropriate metrics. We will use the notation $x_n \to x$, $z_n \to z$ for metric convergence, which will be assumed to be stronger than term-by-term convergence but to generate the same Borel fields. We will also assume that adding a finite number of terms at the beginning of a forward sequence in $\vec{T}$ produces another forward sequence in $\vec{T}$, and analogously for backward sequences.

We make the following hypotheses I-V.

I. The specification is translation invariant.

II. For each $\Lambda = [s,t]$, if $x_n \to x$ and $z_n \to z$, then $Q^\Lambda(\cdot|x_n,z_n) \to Q^\Lambda(\cdot|x,z)$ in total variation.

III. For any Λ, x, $\bar{x}$, z, $\bar{z}$

$$Q^\Lambda(\cdot|x,z) << r^\Lambda(\cdot|x,z;\ \bar{x},\bar{z})$$

where $<<$ denotes absolute continuity and $r^\Lambda(\cdot|x,z;\ \bar{x},\bar{z})$ is the "overlap" of $Q^\Lambda(\cdot|x,z)$ and $Q^\Lambda(\cdot|\bar{x},\bar{z})$, i.e. their greatest lower bound in the space of measures.

For the remaining two hypotheses IV and V we postulate the existence of certain special sets $M_\nu(\ell) \subset \mathbb{R}^\ell$, $\nu = 1,2,\ldots$; $\ell = 1,2,\ldots$ with the properties described below.

IV. For any $\varepsilon > 0$, any (metrically) compact set $C \subset \vec{T}$, any compact set $A \subset \overleftarrow{T}$ and any $\nu \geq 1$ there exists $\mu \geq 1$ such that

$$Q^{[s,t]}(\mathbb{R}^{i-s} \times M_\mu(j-i+1) \times \mathbb{R}^{t-j} | x,z) > 1 - \varepsilon$$

whenever $s < i \leq j < t$, z is a forward sequence, say $(\zeta_1, \zeta_2, \ldots)$, such that $(\zeta_1, \ldots, \zeta_k) \in M_\nu(k)$ for some $k \geq 0$ and $(\zeta_{k+1}, \zeta_{k+2}, \ldots) \in C$, and x is a backward sequence $(\ldots, \zeta_{-2}, \zeta_{-1})$ with $(\zeta_{-k}, \ldots, \zeta_{-1}) \in M_\nu(k)$ and $(\ldots, \zeta_{-k-2}, \zeta_{-k-1}) \in A$. (The case $k = 0$ simply means $z \in C$, $x \in A$).

V. For any $\varepsilon > 0$, any compact sets $C \subset \vec{T}$ and $A \subset \overleftarrow{T}$ and any $\nu \geq 1$, $\mu \geq 1$ there exists an integer $k_0 \geq 1$ such that the following is true. If $z = (\zeta_1, \zeta_2, \ldots)$, $\bar{z} = (\bar{\zeta}_1, \bar{\zeta}_2, \ldots)$, $x = (\ldots, \zeta_{-2}, \zeta_{-1})$, $\bar{x} = (\ldots, \bar{\zeta}_{-2}, \bar{\zeta}_{-1})$ are such that for some $k \geq k_0$

$$(\zeta_1, \ldots, \zeta_k) = (\bar{\zeta}_1, \ldots, \bar{\zeta}_k) \in M_\nu(k)$$

$$(\zeta_{-k}, \ldots, \zeta_{-1}) = (\bar{\zeta}_{-k}, \ldots, \bar{\zeta}_{-1}) \in M_\nu(k)$$

$$(\zeta_{k+1}, \zeta_{k+2}, \ldots) \in C, \quad (\bar{\zeta}_{k+1}, \bar{\zeta}_{k+2}, \ldots) \in C$$

$$(\ldots, \zeta_{-k-2}, \zeta_{-k-1}) \in A, \quad (\ldots, \bar{\zeta}_{-k-2}, \bar{\zeta}_{-k-1}) \in A$$

then, for arbitrary $\Lambda = [s,t]$,

$$\left\| \frac{Q^{[s,t]}((\cdot) \cap M_\mu(t-s+1) | x,z)}{Q^{[s,t]}(M_\mu(t-s+1) | x,z)} - \frac{Q^{[s,t]}((\cdot) \cap M_\mu(t-s+1) | \bar{x},\bar{z})}{Q^{[s,t]}(M_\mu(t-s+1) | \bar{x},\bar{z})} \right\| < \varepsilon.$$

3.1 Definition. A random field admitted by the specification is *tame* if for every $\varepsilon > 0$ there exist a compact set $C \subset \vec{T}$ and a compact set $A \subset \overleftarrow{T}$ such that

$$P((X_i, X_{i+1}, \ldots) \in C) \geq 1 - \varepsilon, \quad P((\ldots, X_{i-1}, X_i) \in A) \geq 1 - \varepsilon$$

for all i. The distribution of such a random field (which is a probability measure on $\mathcal{B}(\mathbb{R}^{\mathbb{Z}})$) will also be called tame.

The general theorem which lies at the root of Theorem 2.3 is the following.

3.2 Theorem. If a one-dimensional specification satisfies hypotheses I-V, then it admits at most one tame probability measure. Such a measure, if it exists, is translation invariant.

Note that every translation invariant probability measure is necessarily tame and therefore, under hypotheses I-V, the specification admits at most one translation invariant measure. Thus Theorem 3.2 is in a way an extension of Theorem 2 in [8]. (Cf. [6] for the case of a countable state space. See also [1] for a special case).

§4. Outline of proofs.

Complete proofs of Theorems 2.2, 2.3 and 3.2 will appear in [9] and [10]. What follows is an outline designed to spotlight the main points and to indicate the methods used.

At the heart of the proof of Theorem 3.2 is the following assertion. Suppose that $\ldots,X_{-1},X_0,X_1,\ldots$ is a random field admitted by the specification and let Π be its distribution. One can prove that for each $\varepsilon > 0$ there exist a $\theta > 0$ and an integer $k \geq 1$ such that for every pair of integers s, t with $t-s \geq 2k+2$ there is a set $G_{s,t} \subset \overleftarrow{T} \times \overrightarrow{T}$ with

$$P(((\ldots,X_{s-1},X_s),\ (X_t,X_{t+1},\ldots)) \in G_{s,t}) \geq 1-\varepsilon$$

and

$$r^{[s+1,t-1]}(\mathbb{R}^{t-s-1}|x,z;\ \bar{x},\bar{z}) \geq \theta$$

for all $(x,z) \in G_{s,t}$ and $(\bar{x},\bar{z}) \in G_{s,t}$. The important point in this assertion is that θ does not depend on $t-s$. One begins by restricting $(\ldots,X_s)$ and $(X_t,\ldots)$ to appropriate compact subsets $A \subset \overleftarrow{T}$ and $C \subset \overrightarrow{T}$ and noting that, for fixed Λ, the absolute continuity assumed in hypothesis III is uniform in x, $\bar{x}$, z, $\bar{z}$, when the latter are restricted to compact sets. Careful use of the hypotheses enables one then to choose $k \geq 1$ and $G_{s,t}$ in such a manner that, if $(x,z) \in G_{s,t}$ and $(\bar{x},\bar{z}) \in G_{s,t}$, the total overlap between the conditional distribution of $((X_{s+1},\ldots,X_{s+k}),\ (X_{t-k},\ldots,X_{t-1}))$ given $(\ldots,X_s) = x$ and $(X_t,\ldots) = z$ on the one hand, and the corresponding conditional distribution given $(\ldots,X_s) = \bar{x}$, and $(X_t,\ldots) = \bar{z}$ on the other, is bounded away from zero. One then applies hypotheses IV and V to extend the overlap to the coordinates $(X_{s+k+1},\ldots,X_{t-k-1})$.

The other major ingredient of the proof of Theorem 3.2 is a limit theorem obtained by means of a martingale argument. If $D = \mathop{X}_{i=-\infty}^{\infty} D_i$ is a rectangle in $\mathcal{B}(\mathbb{R}^{\mathbb{Z}})$, then the sequence $Q^{[-n,n]}(\mathop{X}_{i=-n}^{n} D_i|X_j,|j| > n)$, $n = 1,2,\ldots$ is a backward submartingale and hence converges almost surely and in L_1. Following arguments which parallel those used in [8] (though it must be remembered that the Markov property is not available here) one can show that, if D is finite-dimensional in the sense that $D_i = \mathbb{R}$ for sufficiently large $|i|$, then

$$\lim_{n\to\infty} \int\!\cdots\!\int |Q^{[-n,n]}(\mathop{X}_{i=-n}^{n} D_i|x,z) - Q^{[-n,n]}(\mathop{X}_{i=-n}^{n} D_i|\bar{x},\bar{z})|\,\Pi_n(dxdz)\Pi_n(d\bar{x}d\bar{z}) = 0$$

where $\Pi_n(dxdz) = P((\ldots,X_{-n-1}) \in dx,\ (X_{n+1},\ldots) \in dz)$. It follows easily that for such D

$$\lim_{n\to\infty} Q^{[-n,n]}(\mathop{X}_{i=-n}^{n} D_i|X_j,\ |j| > n) = \Pi(D)$$

almost surely, and the uniqueness of Π follows from this, once it is shown that two tame probability measures admitted by the specification cannot be mutually singular. The translation invariance of Π is a consequence of its uniqueness.

Turning now to the spin systems considered in §2 above, suppose (3) holds and define $\Phi(i) = \sum_{k=i}^{\infty} |J(k)|$, $i = 1,2,\ldots$, so that $\sum_{i=1}^{\infty} \Phi(i) < \infty$. One can show that it is sufficient to consider the conditional probabilities $Q^{\Lambda}(\cdot|x,z)$ defined only for x and z with $\|x\| < \infty$, $\|z\| < \infty$, where the norm $\|\cdot\|$ is defined for forward sequences $z = (\zeta_1, \zeta_2, \ldots)$ by $\|z\| = \sum_{i=1}^{\infty} \Phi(i)|\zeta_i|$ and for backward sequences $x = (\ldots, \zeta_{-2}, \zeta_{-1})$ by $\|x\| = \sum_{i=1}^{\infty} \Phi(i)|\zeta_{-i}|$. The special sets $M_\nu(\ell)$, which appear in hypotheses IV and V of §3, are in the present case defined as follows:

$$M_\nu(\ell) = \{(\zeta_1, \ldots, \zeta_\ell): \frac{1}{n} \sum_{i=1}^{n} \zeta_i^2 \leq \nu \text{ and } \frac{1}{n} \sum_{i=1}^{n} \zeta_{\ell-i+1}^2 \leq \nu \text{ for } n = 1,2,\ldots,\ell\}.$$

The implication (i) $\Rightarrow$ (ii) in Theorem 2.2 follows from Theorem 4.4 in [7]. If (ii) holds, one can choose d so that the sets of all z with $\|z\| \leq d$ and all x with $\|x\| \leq d$ have high probabilities; as pointed out by DeMasi in [3], Ruelle's estimates ([12]) can then be applied to $Q^\Lambda(\cdot|x,z)$, and will hold uniformly for $\|x\| \leq d$, $\|z\| \leq d$. One can deduce from this that, given $\varepsilon > 0$, there is ν such that

$$P(\frac{1}{n} \sum_{i=1}^{n} X_{s+i}^2 \leq \nu, \quad i = 1,2,\ldots,\ell) \geq 1 - \varepsilon$$

for arbitrary s and ℓ. This implies condition (i) of Theorem 2.2 and can also be shown to imply condition (iii). The implication (iii) $\Rightarrow$ (i) is proved similarly.

It is worth mentioning at this point De Masi's result ([3]) that translation invariant DLR measures for the spin systems considered here are tempered. This assertion is contained in the implication (iii) $\Rightarrow$ (i).

Finally the uniqueness part of Theorem 2.3 is an application of Theorem 3.2 : one can show that the specification (1) satisfies hypotheses I-V.

References

1. Benfatto, G., Presutti, E., Pulvirenti, M.: DLR measures for one-dimensional harmonic systems, Z. Wahrscheinlichkeitstheorie und verw. Gebiete 41, 305-312 (1978).

2. Cassandro, M., Olivieri, E., Pellegrinotti, A., Presutti, E.: Existence and uniqueness of DLR measures for unbounded spin systems, Z. Wahrscheinlichkeitstheorie und verw. Gebiete 41, 313-334 (1978).

3. De Masi, A.: One-dimensional DLR invariant measures are regular, Comm. Math. Phys. 67, 43-50 (1979).

4. Dobrushin, R.L.: The description of a random field by means of conditional probabilities and conditions of its regularity (in Russian), Teor. Verojatnost. i Primenen. 13, 201-229 (1968). (English transl.: Theor. Probability Appl. 13, 197-224 (1968)).

5. Dobrushin, R.L.: Prescribing a system of random variables by conditional distributions (in Russian), Teor. Verojatnost. i Premenen. 15, 469-497 (1970). (English transl.: Theor. Probability Appl. 15, 458-486 (1970)).

6. Kesten, H.: Existence and uniqueness of countable one-dimensional Markov random fields, Ann. Probability 4, 557-569 (1976).

7. Lebowitz, J.L., Presutti, E.: Statistical mechanics of systems of unbounded spins, Comm. Math. Phys. 50, 195-218 (1976). Erratum: ibid. 78, 151 (1980).

8. Papangelou, F.: Stationary one-dimensional Markov random fields with a continuous state space. In "Probability, Statistics and Analysis" (ed. J.F.C. Kingman and G.E.H. Reuter), London Math. Soc. Lecture Note Series 79, 199-218. Cambridge: Cambridge University Press 1983.

9. Papangelou, F.: On the absence of phase transition in one-dimensional random fields. (I) Sufficient conditions. Submitted.

10. Papangelou, F.: On the absence of phase transition in one-dimensional random fields. (II) Superstable spin systems. Submitted.

11. Ruelle, D.: Superstable interactions in classical statistical mechanics, Comm. Math. Phys. 13, 127-159 (1970).

12. Ruelle, D.: Probability estimates for continuous spin systems, Comm. Math. Phys. 50, 189-194 (1976).

GENERALISED WEYL OPERATORS

by

R L Hudson
Mathematics Department
University of Nottingham
University Park
Nottingham NG7 2RD

and

K R Parthasarathy
Indian Statistical Institute
7, S.J.S. Sansanwal Marg
New Delhi 110016
India

Abstract Using the quantum Itô's formula of [5] we construct operators satisfying a generalisation of the Weyl commutation relations, in which scalar-valued test functions are replaced by operator-valued ones.

§1. Introduction

Let $\mathcal{H}$ denote the Boson Fock space $\Gamma(L^2[0,\infty))$ over $L^2[0,\infty)$ [2] and for each $f \in L^2[0,\infty)$ let $\psi(f)$ be the corresponding exponential vector [2],

$$\psi(f) = (1, f, (2!)^{-\frac{1}{2}} f \otimes f, (3!)^{-\frac{1}{2}} f \otimes f \otimes f, \dots)$$

in $\mathcal{H}$. The Weyl operators $W(f)$, $f \in L^2[0,\infty)$ are the unitary operators in $\mathcal{H}$ whose actions on exponential vectors are

$$W(f)\psi(g) = \exp\{-\tfrac{1}{2}\|f\|^2 - \langle f,g\rangle\}\psi(f+g)(f,g \in L^2[0,\infty)). \tag{1.1}$$

They satisfy the Weyl relation

$$W(f)W(g) = \exp\{-i \operatorname{Im}\langle f,g\rangle\}W(f+g)(f,g \in L^2[0,\infty)). \tag{1.2}$$

Introducing the mutually adjoint annihilation and creation operators $a(f)$, $a^\dagger(f)$ by means of their actions on exponential vectors

$$a(f)\psi(g) = \langle f,g\rangle\psi(g)$$
$$a^\dagger(f)\psi(g) = \frac{d}{dt}\psi(g+tf)\Big|_{t=0}$$

and noticing that $a^\dagger(f) - a(f)$ is essentially skew-self-adjoint, we may write

$$W(f) = \exp\{a^\dagger(f) - a(f)\}.$$

Quantum Brownian motion [1,5,6] is the family of operators $(A_t, A_t^\dagger, t \geq 0)$

$$A_t = a(\chi_{[0,t]}), \quad A_t^\dagger = a^\dagger(\chi_{[0,t]}).$$

The duality transformation [8] is a Hilbert space isomorphism, which we may use to identify the two spaces from $\mathcal{H}$ onto the Hilbert space $L^2(w)$, where w is Wiener measure, under conjugation by which the self-adjoint operators

$$Q_t = A_t + A_t^\dagger, \quad t \geq 0$$

become multiplications by the canonical realisation X_t, $t \geq 0$ of Brownian motion.

In [5] (see also [3,4,7) a stochastic calculus is developed for quantum Brownian motion generalising the classical Itô calculus in which integration against dX is replaced by integration against the noncommuting independent stochastic differentials dA and dA^{+}, and in which the integrands are operator valued processes $(F(t): t \geq 0)$ which, in the bounded case which concerns us here, are adapted, in the sense that $F(t) \in B(\mathcal{H}_t) \otimes I$ for $t \geq 0$. Here we follow the notation of [5], setting

$$\mathcal{H}_t = \Gamma(L^2[0,t]), \quad \mathcal{H}^t = \Gamma(L^2[t,\infty)),$$

and making the canonical identification $\mathcal{H} = \mathcal{H}_t \otimes \mathcal{H}^t$. When it exists as a bounded operator, the stochastic integral

$$M(t) = \int_0^t (dA^{+}F + GdA + Hds)$$

is determined by the formula

$$\langle\psi(f), M(t)\psi(g)\rangle = \int_0^t \langle\psi(f), (\bar{f}(s)F(s) + G(s)g(s) + H(s))\psi(g)\rangle\, ds. \tag{1.3}$$

Now let $f:[0,\infty) \to \mathbb{C}$ be locally square integrable and, for $t \geq 0$, write

$$f_t = \chi_{[0,t]}f,$$

$$W_t(f) = W(f_t).$$

Combining the relation

$$\langle\psi(g), \psi(h)\rangle = \exp\langle g,h\rangle$$

with (1.2), we have

$$\langle\psi(g), W_t(f)\psi(h)\rangle = \exp\{-\tfrac{1}{2}\|f_t\|^2 - \langle f_t,h\rangle + \langle g, h + f_t\rangle\},$$

whence

$$\frac{d}{dt}\langle\psi(g), W_t(f)\psi(h)\rangle = \langle\psi(g), (\overline{g(t)}f(t) - \overline{f(t)}h(t) - \tfrac{1}{2}|f(t)|^2)W_t(f)\psi(h)\rangle.$$

Comparing with (1.3) we see that the Weyl operators $W_t(f)$, $t \geq 0$ satisfy the stochastic differential equation

$$W_0(f) = I, \quad W_t(f) = (dA^{+}f - \bar{f}dA - \tfrac{1}{2}|f(t)|^2 dt)W_t(f). \tag{1.4}$$

Let $t \to H(t)$ be a strongly continuous self-adjoint valued map from $[0,\infty)$ into $B(\hat{\mathcal{H}})$. The Dyson expansion [7, Theorem X.69] permits the construction of a family of unitary operators $(W_t(H), t \geq 0)$ in $\mathcal{H}$ satisfying the (strong sense) ordinary differential equation

$$W_0(H) = I, \quad \frac{d}{dt}W_t(H) = iH(t)W_t(H). \tag{1.5}$$

If $H(.)$ is adapted, so too is $(W_t(H): t \geq 0)$. $W_t(H)$ is strongly continuous in t. Given two such maps H_1 and H_2, the map

$$t \to \tilde{H}_2(t) = W_t(H_1)H_2(t)W_t(H_1)^{+}$$

is also strongly continuous, and we have

$$W_t(H_1)W_t(H_2) = W_t(H_1 + \tilde{H}_2). \tag{1.6}$$

Our purpose can now be stated; we shall combine and generalise the constructions of the families $W_t(f)$, $W_t(H)$, establishing the existence, for non-anticipating operator valued functions F and H, with H self-adjoint valued, of an operator valued process $(W_t(F,H),\ t \geq 0)$ satisfying the generalisation of (1.4) and of (1.6)

$$W_0(F,H) = I, \quad dW_t(F,H) = (dA^\dagger F - F^\dagger dA + (iH - \tfrac{1}{2}F^\dagger F)dt)W_t(F,H), \tag{1.7}$$

together with the generalisation of (1.2) and (1.6)

$$W_t(F_1,H_1)W_t(F_2,H_2) = W_t(F_1 + \tilde{F}_2,\ H_1 + \tilde{H}_2 - \frac{1}{2i}(F_1^\dagger\tilde{F}_2 - \tilde{F}_2^\dagger F_1)). \tag{1.8}$$

§2. Construction of $W_t(F,H)$

Let h be a Hilbert space and let $F_0, H_0 \in B(h)$ with $H_0 = H_0^\dagger$. We identify F_0 and H_0 with the operators $L_0 \otimes I$ and $H_0 \otimes I$ in $B(h \otimes \mathcal{H})$. In [5] it is proved that there exists a unique unitary adapted process $U = (U(t): t \geq 0)$ in $B(h \otimes \mathcal{H})$ satisfying

$$U(0) = I, \quad dU = U(-dA^\dagger F_0 + F_0^\dagger dA - (iH_0 + \tfrac{1}{2}L_0^\dagger L_0)dt), \tag{2.1}$$

so that the adjoint process satisfies

$$U^\dagger(0) = I, \quad dU^\dagger = (dA^\dagger F_0 - F_0^\dagger dA + (iH_0 - \tfrac{1}{2}L_0^\dagger L_0)dt)U^\dagger. \tag{2.2}$$

We say that the $B(\mathcal{H})$-valued adapted process F is <u>simple</u> if there exists an increasing sequence

$$0 = t_0 < t_1 < \ldots < t_n \underset{n}{\to} \infty$$

such that

$$F = \sum_{j=0}^{\infty} F_j \chi_{[t_j, t_{j+1})}. \tag{2.3}$$

<u>Theorem 2.1</u> Let F, H be simple $B(\mathcal{H})$-valued adapted processes, with H self-adjoint valued. Then there exists a unique unitary-valued adapted process $(W_t(F,H),\ t \geq 0)$ satisfying (1.7).

<u>Proof</u> Assume without loss of generality that, in addition to (2.3),

$$H = \sum_{j=0}^{\infty} H_j \chi_{[t_j, t_{j+1})}.$$

Set $W_0(F,H) = I$, let $t_j \leq t < t_{j+1}$ and assume inductively that $W_{t_j}(F,H)$ has been defined and is unitary. Let T be the shift isomorphism from $L^2[0,\infty)$ onto $L^2[t_j,\infty)$ and let $\Gamma(T)$ be its second quantisation [2] which is an isomorphism from $\mathcal{H}$ onto $\mathcal{H}^{t_j}$. Let S be the isomorphism $I \otimes \Gamma(T)$ from $\mathcal{H}_{t_j} \otimes \mathcal{H}$ onto $\mathcal{H}_{t_j} \otimes \mathcal{H}^{t_j} = \mathcal{H}$. Taking $h = \mathcal{H}_{t_j}$, $F_0 = F_j$ and $H_0 = H_j$, construct the unitary process U in $B(h \otimes \mathcal{H})$ satisfying (2.1). Finally, set

$$W_t(F,H) = SU^\dagger(t-t_j)S^\dagger W_{t_j}(F,H).$$

Then $W_t(F,H)$ is unitary and satisfies (1.7). The uniqueness is a consequence of the quantum Itô's formula [5], if $W' = (W'_t(F,H))$ also satisfies (1.7) then we have

$$d[W^\dagger W'] = dW^\dagger . W' + W^\dagger dW' + dW^\dagger dW' = 0$$

whence $W_t^\dagger W'_t = W_0^\dagger W'_0 = I$ and so $W' = W$. □

We now extend the definition of $W_t(F,H)$ to non-simple F and H. We say that the pair (F,H) of adapted $B(\mathcal{H})$-valued processes, with H self-adjoint valued, is <u>allowed</u> if there exists a sequence (F_n,H_n), $n = 1,2,\ldots$ of pairs of simple processes with H_n self-adjoint valued such that on each finite interval $[0,t]$

$$\sup\{\|F_n(s) - F(s)\|, \|H_n(s) - H(s)\|: s \in [0,t]\} \underset{n}{\to} 0.$$

Note that if F and H are strongly continuous they are allowed.

For an allowed pair (F,H) approximated by a sequence (F_n,H_n), $n = 1,2,\ldots$ and for fixed $t > 0$, we set $W_n = W_t(F_n,H_n)$, and prove

<u>Theorem 2.2</u> The sequence W_n, $n = 1,2,\ldots$ converges strongly to an operator independent of choice of approximating sequence. Denoting the limit by $W_t(F,H)$, the process $(W_t(F,H))$ is adapted and isometric-valued, and satisfies (1.7).

<u>Proof</u> We prove $(W_n\psi(f))$ is Cauchy for each locally bounded $f \in L^2[0,\infty)$. By Theorems 3.1 and 3.3 of [5],

$$\begin{aligned}
&\|(W_n - W_m)\psi(f)\|^2 \\
&= 2\|\psi(f)\|^2 - 2\,\mathrm{Re}\langle W_n\psi(f), W_m\psi(f)\rangle \\
&= -2\,\mathrm{Re}\{\langle (W_n - I)\psi(f), (W_m - I)\psi(f)\rangle + \langle \psi(f), (W_m - I)\psi(f)\rangle \\
&\quad + \langle (W_n - I)\psi(f), \psi(f)\rangle\} \\
&= -2\,\mathrm{Re}\int_0^t \{\langle (W_n - I)\psi(f), (\bar{f}F_m - F_m^\dagger f + (iH_m - \tfrac{1}{2}F_m^\dagger F_m))W_m\psi(f)\rangle \\
&\quad + \langle (\bar{f}F_n - F_n^\dagger f + (iH_n - \tfrac{1}{2}F_n^\dagger F_n))W_n\psi(f), W_m\psi(f)\rangle \\
&\quad + \langle F_nW_n\psi(f), F_mW_m\psi(f)\rangle \\
&\quad + \langle \psi(f), (\bar{f}F_m - F_m^\dagger f + iH_m - \tfrac{1}{2}F_m^\dagger F_m)W_m\psi(f)\rangle \\
&\quad + \langle (\bar{f}F_n - F_n^\dagger f + iH_n - \tfrac{1}{2}F_n^\dagger F_n)W_m\psi(f), \psi(f)\rangle\}\,ds \\
&= -2\,\mathrm{Re}\int_0^t \{\langle W_n\psi(f), (\bar{f}(F_m - F_n) - (F_m^\dagger - F_n^\dagger)f)W_m\psi(f)\rangle \\
&\quad + \langle W_n\psi(f), i(H_m - H_n)W_m\psi(f)\rangle \\
&\quad - \tfrac{1}{2}\langle W_n\psi(f), (F_m^\dagger F_m + F_n^\dagger F_n - 2F_n^\dagger F_m)W_m\psi(f)\rangle\}\,ds.
\end{aligned}$$

Writing $F_m^\dagger F_m + F_n^\dagger F_n - 2F_n^\dagger F_m = (F_m^\dagger - F_n^\dagger)F_m + F_m^\dagger(F_n - F_m)$, it is clear that this $\underset{m,n}{\to} 0$. Since the $\psi(f)$ with f locally bounded are total, (W_n) converges strongly; a modification of the argument given shows that the limit does not depend on choice of (F_n,H_n). Since each W_n is isometric, so too is the limit. That the limit process inherits adaptedness is clear; that it satisfies (1.7) follows by passage to the limit in the corresponding equation for W_n. □

§3. Generalised Weyl relations

Let F_i, H_i, $i = 1,2$ be strongly continuous adapted $B(H)$-valued processes, with H_i self-adjoint valued. Then (F_i,H_i) is allowed and according to Theorem 2.1 we can construct isometric adapted processes $(W_t(F_i,H_i),\ i = 1,2)$ satisfying

$$W_0(F_i,H_i) = I, \quad dW_t(F_i,H_i) = (dA^\dagger F_i - F_i^\dagger dA + (iH_i - \tfrac{1}{2}F_i^\dagger F_i)dt)W_t(F_i,H_i).$$

From Corollary 3 to Theorem 3.1 of [5] it follows that $W_t(F_1,H_1)$ is strongly continuous and hence, since $\|W_t(F_1,H_1)\| \le 1$, that the maps

$$t \to \tilde{F}_2(t) = W_t(F_1,H_1)F_2(t)W_t(F_1,H_1)^\dagger$$

$$t \to \tilde{H}_2(t) = W_t(F_1,H_1)H_2(t)W_t(F_1,H_1)^\dagger$$

are strongly continuous. Hence the pair $(\tilde{F}_2,\tilde{H}_2)$, and with it the pair $(F_1 + \tilde{F}_2,\ H_1 + \tilde{H}_2 - \frac{1}{2i}(F_1^\dagger\tilde{F}_2 - \tilde{F}_2^\dagger F_1))$ is admissible. Hence $W_t(F_1 + \tilde{F}_2,\ H_1 + \tilde{H}_2 - \frac{1}{2i}(F_1^\dagger\tilde{F}_2 - \tilde{F}_2^\dagger F_1))$ is defined.

<u>Theorem 3.1</u> For strongly continuous adapted $B(H)$-valued processes F_i, H_i, $i = 1,2$, with H_i self-adjoint,

$$W_t(F_1 + \tilde{F}_2,\ H_1 + \tilde{H}_2 - \tfrac{1}{2i}(F_1^\dagger\tilde{F}_2 - \tilde{F}_2^\dagger F_1))^\dagger W_t(F_1,H_1)W_t(F_2,H_2) = I. \tag{3.1}$$

<u>Proof</u> By the Itô product formula of [5]

$$\begin{aligned}
&d[W_t(F_1,H_1)W_t(F_2,H_2)] \\
&= (dW_t(F_1,H_1))W_t(F_2,H_2) + W_t(F_1,H_1)dW_t(F_2,H_2) + dW_t(F_1,H_1)dW_t(F_2,H_2) \\
&= (dA^\dagger F_1 - F_1^\dagger dA + (iH_1 - \tfrac{1}{2}F_1^\dagger F_1)dt)W_t(F_1,H_1)W_t(F_2,H_2) \\
&\quad + W_t(F_1,H_1)(dA^\dagger F_2 - F_2^\dagger dA + (iH_2 - \tfrac{1}{2}F_2^\dagger F_2)dt)W_t(F_2,H_2) \\
&\quad + F_1^\dagger W_t(F_1,H_1)F_2W_t(F_2,H_2)dt \\
&= \{dA^\dagger(F_1 + \tilde{F}_2) - (F_1 + \tilde{F}_2)^\dagger dA \\
&\quad + [i(H_1 + \tilde{H}_2 - \tfrac{1}{2i}(F_1^\dagger\tilde{F}_2 - \tilde{F}_2^\dagger F_1)) - \tfrac{1}{2}(F_1 + \tilde{F}_2)^\dagger(F_1 + \tilde{F}_2)]dt\}W_t(F_1,H_1)W_t(F_2,H_2)
\end{aligned}$$

using the isometry of $W_t(F_1,H_1)$. On the other hand

$$\begin{aligned}
&dW_t(F_1 + \tilde{F}_2,\ H_1 + \tilde{H}_2 - \tfrac{1}{2i}(F_1^\dagger\tilde{F}_2 - \tilde{F}_2^\dagger F_1)) \\
&= \{dA^\dagger(F_1 + \tilde{F}_2) - (F_1 + \tilde{F}_2)^\dagger dA \\
&\quad + [i(H_1 + \tilde{H}_2 - \tfrac{1}{2i}(F_1^\dagger\tilde{F}_2 - \tilde{F}_2^\dagger F_1)) - \tfrac{1}{2}(F_1 + F_2)^\dagger(F_1 + \tilde{F}_2)]dt\} \\
&\qquad W_t(F_1 + \tilde{F}_2,\ H_1 + \tilde{H}_2 - \tfrac{1}{2i}(F_1^\dagger F_2 - F_2^\dagger F_1))
\end{aligned}$$

and hence, again using Itô's formula

$$d[W_t(F_1 + \tilde{F}_2,\ H_1 + \tilde{H}_2 - \tfrac{1}{2i}(F_1^\dagger\tilde{F}_2 - \tilde{F}_2^\dagger F_1))^\dagger W_t(F_1,H_1)W_t(F_2,H_2)] = 0.$$

Since the initial value is I we obtain (3.1). □

If $W_t(F_1,H_1)$ and $W_t(F_2,H_2)$ are unitary, then, by multiplying on the right by their inverses and taking adjoints, we deduce (1.8) from (3.1). In this case the product $W_t(F_1 + \tilde{F}_2, H_1 + \tilde{H}_2 - \frac{1}{2i}(F_1^\dagger \tilde{F}_2 - \tilde{F}_2^\dagger F_1))$ is clearly also unitary.

References

[1] Cockroft, AM and Hudson, RL, Quantum mechanical Wiener processes, J. Multivariate Anal. 7, 107-24 (1978).

[2] Guichardet, A, Symmetric Hilbert spaces and related topics, LNM 261, Springer, Berlin (1972).

[3] Hudson, RL, Karandikar, RL and Parthasarathy, KR, Towards a theory of noncommutative semimartingales adapted to Brownian motion and a quantum Itô's formula, in Theory and application of random fields, ed. Kallianpur, LN in Control and Information Sciences 49, Springer, Berlin (1983).

[4] Hudson, RL and Parthasarathy, KR, Quantum diffusions, in Theory and application of random fields, ed. Kallianpur, LN in Control and Information Sciences 49 Springer, Berlin (1983).

[5] Hudson, RL and Parthasarathy, KR, Quantum Itô's formula and stochastic evolutions, submitted to CMP.

[6] Hudson, RL and Streater, RF, Noncommutative martingales and stochastic integrals in Fock space, in Stochastic processes in quantum theory and statistical physics, ed. Albeverio et al., LN in Physics 173, Springer, Berlin (1982).

[7] Reed, M and Simon, B, Methods of modern mathematical physics, Fourier analysis and self-adjointness, Academic Press, New York (1975).

[8] Segal, IE, Tensor algebras over Hilbert space I, Trans. Amer. Math. Soc. 81, 106-34 (1956).

One - dimensional stochastic differential equations involving the local times of the unknown process.

by J.F Le Gall.

Laboratoire de Probabilités ; Tour 56, 4 Place Jussieu. F. 75230 Paris Cédex 05.

0. *Introduction* :

Let B be a one-dimensional brownian motion, $f : \mathbb{R} \to \mathbb{R}$ a bounded measurable function and ν a bounded measure on $\mathbb{R}$. We consider the following stochastic equation :

(0.1) $$X_t = X_0 + \int_0^t f(X_s)dB_s + \int_{\mathbb{R}} \nu(da)\, L_t^a(X)$$

where $L_t^a(X)$ denotes the local time at a for the time t of the semimartingale X.

Two cases of (0.1) are of special interest. Firstly, whenever ν is absolutely continuous with respect to the Lebesgue measure on $\mathbb{R}$ (i.e. $\nu(da)=g(a)da$), (0.1) becomes the usual Ito equation :

(0.2) $$X_t = X_0 + \int_0^t f(X_s)dB_s + \int_0^t (gf^2)(X_s)ds$$

When $\nu = \alpha\, \delta_{(0)}$ ($\delta_{(0)}$ denotes the Dirac measure at 0) and f=1, we get :

(0.3) $$X_t = X_0 + B_t + \alpha\, L_t^0(X)$$

The solution of (0.3) is the well-known process called the skew brownian motion, which has been studied by Walsh ([10]) and Harrison and Shepp ([2]) in particular.

Equations of the type (0.1) were first considered by Stroock and Yor ([9]).

Their goal was to prove the "purity" of certain martingales. In this paper we propose to show that (0.1) is a good extension of (0.2), in the following sense : under a few regularity assumptions, any solution of (0.1) is the strong limit of a sequence of solutions of equations of the type (0.2), and conversely any limit of a sequence of solutions of equations of the type (0.2) is a solution of an equation of the type (0.1). Thus the set of solutions of (0.1) is obtained from the set of solutions of (0.2) through a compactification process.

In section 1 we recall a few results about one-dimensional stochastic equations which we shall use in our study of (0.1). Section 2 is devoted to the proof of basic results about equation (0.1). In section 3 we prove the main limit theorems which enable us to relate equation (0.1) to equation (0.2). In section 4 we investigate the relationship between solutions of equations of the type (0.1) and random walks on the integers.

Throughout this work, $L_t^a(X)$ will denote the symmetric local time (at a and for the time t) of the semi-martingale X. We will use basic results about local times of a continuous semi-martingale, namely the Tanaka's formula, the generalized Ito formula and the density of occupation time formula (see [3] for instance). We will also use the notions of pathwise uniqueness and weak uniqueness of solutions for a stochastic differential equation, as defined by Yamada and Watanabe in their fundamental paper ([11]).

1. *One-dimensional stochastic differential equations and local times :*

In this section we recall several results from [4] that will be of use in our study of (0.1). Let us consider the classical Ito equation :

$$dX_t = \sigma(t,X_t)dB_t + b(t,X_t)dt, \tag{1.1}$$

where B is a one-dimensional brownian motion and $\sigma, b : \mathbb{R}_+ \times \mathbb{R} \to \mathbb{R}$ are bounded measurable functions.

In [4] we used the notion of local time of a semi-martingale to prove pathwise uniqueness results and also comparison theorems and limit theorems for solutions of equation (1.1). Our methods relied on the following simple lemma :

Lemma 1.1 :

Assume that X is a continuous semi-martingale and that $\lambda : [0 ; \infty[\to [0 ; \infty[$ is an increasing function s.t :

$$\int_{0^+} \frac{du}{\lambda(u)} = \infty$$

Assume further that for any $t \geqslant 0$:

$$\int_0^t \frac{d\langle X\rangle_s}{\lambda(X_s)} 1_{(X_s>0)} < +\infty \qquad \text{a.s.}$$

Then, $L_t^{0+}(X) = 0$ for all $t \geqslant 0$ a.s.

Proof :

It is an immediate consequence of the density of occupation time formula :

$$\int_0^t \frac{d\langle X\rangle_s}{\lambda(X_s)} 1_{(X_s>0)} = \int_0^\infty \frac{da}{\lambda(a)} L_t^a(X).$$

Since the latter integral is finite, the well-known continuity properties of local times (see [12]) yield the result of the lemma.

□

Corollary 1.2 :

Let $\sigma, b : \mathbb{R}_+ \times \mathbb{R} \to \mathbb{R}$ be bounded measurable functions. Suppose that σ satisfies one of the two following assumptions :

(A) : There exists a strictly increasing function $\rho : [0 ; \infty[\to [0 ; \infty[$

$$\text{s.t. :} \quad \int_{0^+} \frac{du}{\rho(u)} = \infty,$$

$$\text{and} \qquad (\sigma(t,x)-\sigma(t,y))^2 \leq \rho(|x-y|) \quad \text{for all} \quad (t,x,y)$$

(B) : There exists $\varepsilon > 0$ s.t. : $\sigma(t,x) \geq \varepsilon$ for all (t,x),

and there exists a strictly increasing function $f : \mathbb{R} \to \mathbb{R}$

$$\text{s.t :} \quad (\sigma(t,x)-\sigma(t,y))^2 \leq |f(x)-f(y)| \text{ for all } (t,x,y).$$

Then, whenever X^1 and X^2 are two solutions of (1.1) (on the same filtered probability space, with the same brownian motion B), we have :

$$L_t^0 (X^1 - X^2) = 0 \quad \text{for all} \quad t \geq 0 \quad \text{a.s.}$$

Proof :

We apply *lemma 1.1* to $X = X^1 - X^2$.

Case (A) :

We take $\lambda=\rho$. Then :

$$\int_0^t \frac{d\langle X\rangle_s}{\lambda(X_s)} 1_{(X_s>0)} = \int_0^t \frac{(\sigma(s,X_s^1)-\sigma(s,X_s^2))^2}{\rho(X_s^1 - X_s^2)} ds\, 1_{(X_s^1-X_s^2>0)} \leq t$$

Case (B) :

We take $\lambda(x) = x$.

$$\int_0^t \frac{d\langle X\rangle_s}{X_s} 1_{(X_s>0)} = \int_0^t \frac{(\sigma(s,X_s^1)-\sigma(s,X_s^2))^2}{X_s^1 - X_s^2} ds\, 1_{(X_s^1-X_s^2>0)}$$

$$\leq \int_0^t \frac{f(X_s^1)-f(X_s^2)}{X_s^1-X_s^2} ds\, 1_{(X_s^1-X_s^2>0)}$$

The latter integral can easily be proved to be finite using an approximation of f by C^1 functions (see [4] for the details of the proof).

□

Remark :

The result of the corollary remains valid if X^1 and X^2 are not supposed to be solutions of (1.1) but are simply assumed to satisfy :

$$dX_t^1 = \sigma(t,X_t^1)dB_t + dV_t^1$$
$$dX_t^2 = \sigma(t,X_t^2)dB_t + dV_t^2$$

where V^1,V^2 are continuous and of finite variation on compact sets.

Theorem 1.3 :

Suppose that σ and b satisfy one of the three following assumptions :

a) σ satisfies (A) and b is Lipschitz.

b) σ satisfies (A) and $\sigma \geqslant \varepsilon$ for some $\varepsilon > 0$.

c) σ satisfies (B).

Then pathwise uniqueness holds for (1.1).

Remark :

Each of the three assumptions a), b), c) also implies the weak existence of solutions for (1.1). Thus the well-known results of Yamada and Watanabe ([11]) imply that there exists a unique solution of (1.1) with a given initial value, on any filtered probability space carrying a brownian motion B. Moreover this solution is a strong one.

Proof of *theorem 1.3* :

Cases b) and c).

Since $|\sigma| \geqslant \varepsilon$, we know that weak uniqueness holds for (1.1) ([8] p. 192).

Let X^1,X^2 be two solutions of (1.1) with the same initial value. From *corollary 1.2*, we deduce that $L_t^o(X^1-X^2) = 0$. An easy application of the Tanaka's formula yields that $X^1 \vee X^2$ and $X^1 \wedge X^2$ are also solutions of (1.1) with the same initial value.

Then weak uniqueness implies that $X^1 \vee X^2 = X^1 \wedge X^2$ and thus $X^1 = X^2$.

Case a) :

Let K be a Lipschitz constant for b and X^1, X^2 be two solutions of (1.1) with the same initial value. Tanaka's formula (and *corollary 1.2*) implies that :

$$|X_t^1 - X_t^2| = \int_0^t \text{sgn}\,(X_s^1 - X_s^2)\,d(X_s^1 - X_s^2)$$

$$\text{Thus : } E[\,|X_t^1 - X_t^2|\,] = E[\int_0^t \text{sgn}\,(X_s^1 - X_s^2)(b(s,X_s^1) - b(s,X_s^2))ds\,]$$

$$\leq K \int_0^t E[\,|X_s^1 - X_s^2|\,]\,ds$$

Hence, invoking Gromwall's lemma :

$$E[\,|X_t^1 - X_t^2|\,] = 0 \quad \text{for all } t$$

□

Remarks :

Case a) was proved by Yamada and Watanabe in [11]. Case b) is due to Okabe and Shimizu ([6]). Case c) is a generalization of a result due to Nakao ([5]). Nakao proved that pathwise uniqueness holds whenever $\sigma(t,x)$ does not depend on t and is uniformly positive and of bounded variation on compact sets. In the time-homogeneous case, assumption c) translates the boundedness of the quadratic variation of σ on compact sets. We refer the reader to [1] for many extensions of the above results.

Theorem 1.4 :

Suppose that for $i=1,2$, X^i satisfies :

$$dX_t^i = \sigma(t,X_t^i)dB_t + b_i(t,X_t^i)dt$$

where σ, b_1, b_2 are bounded measurable functions. Assume that :

(i) σ satisfies (A) or (B)

(ii) One of the two functions b_1, b_2 is Lipschitz.

Assume further that :

1) $b_1 \geqslant b_2$

2) $X_0^1 \geqslant X_0^2$ a.s.

Then : $X_t^1 \geqslant X_t^2$ for all t a.s.

Proof :

Suppose for instance that b_1 is Lipschitz. Then :

$$E[(X_t^2 - X_t^1)_+] = E\left[\int_0^t 1_{(X_s^1 < X_s^2)} (b_2(s,X_s^2) - b_1(s,X_s^1))ds\right]$$

$$\leqslant E\left[\int_0^t 1_{(X_s^1 < X_s^2)} (b_1(s,X_s^2) - b_1(s,X_s^1))ds\right]$$

$$\leqslant K \int_0^t E[(X_s^2 - X_s^1)_+]ds.$$

Hence Gromwall's lemma yields :

$$E[(X_t^2 - X_t^1)_+] = 0 \quad \text{for all } t$$

□

We finally state a "strong" limit theorem for solutions of equations of the type (1.1), which we will use in Section 3. The proof of this result, which will not be given here (see [4]), involves techniques similar to those used in the proof of *theorems 1.3* and *1.4.*

Theorem 1.5 :

Suppose that X^n $(n=1,2,...)$ and X are continuous semi-martingales such that :

$$dX_t^n = \sigma_n(X_t^n)dB_t + b_n(X_t^n)dt \qquad n=1,2,\ldots$$
$$dX_t = \sigma(X_t)dB_t + b(X_t)dt$$

where σ_n, b_n $(n=1,2,\ldots)$, σ, b are bounded measurable functions from $\mathbb{R}$ to $\mathbb{R}$.

Assume that :

(i) σ satisfies (A) or (B).

(ii) There exists positive constants ε and K s.t :

$$\varepsilon \leqslant \sigma_n \leqslant K \quad \text{for all } n$$
$$|b_n| \leqslant K$$

Assume further that :

1) $X_0^n \xrightarrow[n\to\infty]{L^1} X_0$

2) $\int_{-M}^{M} |\sigma_n(u)-\sigma(u)|du \xrightarrow[n\to\infty]{} 0$ for any $M > 0$

$\int_{-M}^{M} |b_n(u)-b(u)|du \xrightarrow[n\to\infty]{} 0$ for any $M > 0$

Then : $E[\sup_{0\leqslant s\leqslant t} |X_s^n - X_s|] \xrightarrow[n\to\infty]{} 0$ for all t.

2. *Stochastic differential equations involving the local times of the unknown process* :

We first introduce a few notations. $BV(\mathbb{R})$ will denote the space of all functions $f : \mathbb{R} \to \mathbb{R}$ of bounded variation on $\mathbb{R}$ such that :

1) f is right continuous

2) There exists an $\varepsilon > 0$ s.t. : $f(x) \geqslant \varepsilon$ for all x. If f is in $BV(\mathbb{R})$, $f(x-)$ will denote the left-limit of f at a point x and $f'(dx)$ will be the bounded measure associated with f.

$M(\mathbb{R})$ will denote the space of all bounded measures ν on $\mathbb{R}$ such that :

$$|\nu(\{x\})| < 1 \quad \text{for all } x \text{ in } \mathbb{R}.$$

If ν is in $M(\mathbb{R})$, ν^c will denote the continuous part of ν.

We shall need the following elementary lemma whose proof is left to the reader :

Lemma 2.1 :

Let ν be in $M(\mathbb{R})$.

There exists a function f in $BV(\mathbb{R})$, unique up to a multiplicative constant, such that :

$$f'(dx)+(f(x)+f(x-))\nu(dx) = 0. \tag{2.1}$$

If we require that $f(x)\xrightarrow[x\to-\infty]{} 1$, then f is unique and is given by :

$$f(x)=f_\nu(x) = \exp(-2\nu^c(]-\infty\ ;\ x\ ])) \prod_{y\leq x} \left(\frac{1-\nu(\{y\})}{1+\nu(\{y\})}\right) \tag{2.2}$$

We are interested in the following singular stochastic differential equation :

$$dX_t = \varphi(X_t)dB_t + \int_{\mathbb{R}} \nu(da)d_t\, L_t^a(X) \tag{2.3}$$

where φ is in $BV(\mathbb{R})$ and ν is in $M(\mathbb{R})$. The next proposition allows us to relate equation (2.3) to a classical Ito equation without drift.

Proposition 2.2 : ([9])

Let φ be in $BV(\mathbb{R})$ and ν be in $M(\mathbb{R})$.

Let f_ν be defined by formula (2.2) and set :

$$F_\nu(x) = \int_0^x f_\nu(y)dy$$

Then X is a solution of equation (2.3) if and only if $Y \overset{\text{def}}{=} F_\nu(X)$ is a solution of :

(2.4) $$dY_t = (\varphi\, f_\nu) \circ F_\nu^{-1}\,(Y_t)dB_t$$

Proof :

The proof is an easy application of the generalized Ito formula. We first assume that X satisfies (2.3). Then :

$$Y_t = Y_0 + \int_0^t \frac{1}{2}\,(f_\nu(X_s)+f_\nu(X_s-))dX_s + \frac{1}{2}\int_{\mathbb{R}} L_t^a\,(X)f_\nu'\,(da)$$

$$= Y_0 + \int_0^t \frac{1}{2}\,(f_\nu(X_s)+f_\nu(X_s-))\varphi(X_s)dB_s + \int_0^t \frac{1}{2}\,(f_\nu(X_s)+f_\nu(X_s-))\,\cdot\cdot$$

$$\cdot\cdot \int_{\mathbb{R}} \nu(da)d_s\,L_s^a\,(X)) + \frac{1}{2}\int_{\mathbb{R}} L_t^a\,(X)f_\nu'\,(da)$$

$$= Y_0 + \int_0^t \frac{1}{2}(f_\nu(X_s)+f_\nu(X_s-))\varphi(X_s)dB_s + \int_{\mathbb{R}} \frac{1}{2}\,(f(a)+f(a-))\,\cdot\cdot$$

$$\cdot\cdot\ \nu(da)\,L_t^a\,(X) + \frac{1}{2}\int_{\mathbb{R}} L_t^a\,(X)f_\nu'\,(da)$$

$$= Y_0 + \int_0^t \frac{1}{2}\,(f_\nu(X_s)+f_\nu(X_s-))\varphi(X_s)dB_s$$

The latter equality easily implies that Y satisfies (2.4). We have used (2.1) together with the fact that $L_s^a\,(X)$ only increases when $X_s = a$.

The converse is proved in a similar manner.

□

We note that $(\varphi\, f_\nu) \circ F_\nu^{-1}$ belongs to $BV(\mathbb{R})$. We have thus proved that equation (2.3) is equivalent to a similar one where $\nu=0$.

Theorem 2.3 :

Let φ be in $BV(\mathbb{R})$ and ν be in $M(\mathbb{R})$.

Then existence and pathwise uniqueness of solutions hold for (2.3).

Proof :

In view of *proposition 2.2*, it is enough to examine the case $\nu=0$. In this case however the result of the theorem is an immediate consequence of *theorem 1.3* (case c)).

□

The choice of the additional condition $|\nu(\{x\})| < 1$ for all x, in the definition of $M(\mathbb{R})$ is motivated by the following result :

suppose that φ is in $BV(\mathbb{R})$ and that ν is a bounded measure such that $|\nu(\{x_0\})| > 1$ for some x_0 in $\mathbb{R}$;

then (2.3) has no solution with the initial value x_0.

This was proved by Harrison and Shepp ([2]) in the case where $\varphi=1$ and $\nu=\alpha\,\delta_{(0)}$. The above result derives from an obvious extension of that case.

When ν only satisfies $|\nu(\{x\})| \leqslant 1$ for all x, the result of *theorem 2.3* remains valid. For instance, if $\varphi=1$ and $\nu=\delta_{(0)}$, the solution of (2.3) is a reflecting brownian motion. More generally, a point x such that $|\nu(\{x\})| = 1$ corresponds to a reflection of the process over or below this point depending on the sign of $\nu(\{x\})$, in which case it is no longer possible to transform equation (2.3) into an equation of the form (2.4). We shall thus limit ourselves to measures ν belonging to $M(\mathbb{R})$.

Throughout the rest of this section, K will denote an arbitrary positive constant.

Let φ be in $BV(\mathbb{R})$ and ν be in $M(\mathbb{R})$. $P_{\varphi,\nu}$ will denote the law of the solution of (2.3) with initial value 0, considered on the time interval $[0\,;K]$

Proposition 2.2 implies that the support of $P_{\varphi,\nu}$ is :

$$\text{supp}\,(P_{\varphi,\nu}) = \{u : [0;K] \to \mathbb{R} / \quad u \text{ continuous, } u(0) = 0\}$$
$$\overset{\text{def}}{=} C_0\,([0;K],\mathbb{R})$$

Indeed, *proposition 2.2* shows that we can restrict our attention to the case $\nu=0$. The solution of (2.3) is then obtained by random time change of a brownian motion. Our result follows from the corresponding result for brownian motion.

Of course this result does not remain valid when ν only satisfies the weaker assumption $|\nu(\{x\})| \leqslant 1$ for all x. (Consider for instance the reflecting brownian motion).

We now investigate necessary and sufficient conditions for $P_{\varphi,\nu}$ to be equivalent to another measure $P_{\psi,\mu}$.

Theorem 2.4 :

Let φ,ψ be in $BV(\mathbb{R})$ and ν,μ be in $M(\mathbb{R})$.

The two following assumptions are equivalent :

(i) $P_{\varphi,\nu}$ is equivalent to $P_{\psi,\mu}$

(ii) $$\begin{cases} \varphi = \psi \\ \mu(dx) = \nu(dx) + \theta(x)dx \quad \text{for some } \theta \text{ in } L^2_{loc}(\mathbb{R}) \end{cases}$$

Proof :

(ii) ⇒ (i) is a direct application of Girsanov's formula.

We assume that (i) holds. In the following, ω will denote an element of $C_0([0,K],\mathbb{R})$. The quadratic variation of a semi-martingale remains the same when the probability is replaced by an equivalent one. Thus,

$$\int_0^t \varphi^2(\omega(s))ds = \int_0^t \psi^2(\omega(s))ds \quad \text{for all } t \leqslant K,\ P_{\varphi,\nu} \text{ - a.s.}$$

Hence $\varphi = \psi$.

Using *proposition 2.2* together with a suitable time change we may assume that :

$$\varphi = \psi = 1 \quad \text{and} \quad \nu = 0.$$

Set $\quad W = P_{\varphi,\nu} = P_{1,0}$ and $R = P_{1,\mu}$

W is Wiener measure on $C_0([0;K],\mathbb{R})$ while R is the law of the unique solution of

$$Y_t = B_t + \int_{\mathbb{R}} \mu(da)\, L_t^a\, (Y)$$

Let $(\mathcal{F}_t,\ 0 \leq t \leq K)$ be the canonical filtration and $(X_t,\ 0 \leq t \leq K)$ be the canonical process on $C_0([0;K],\mathbb{R})$

Set :

$$H_t = \frac{dR}{dW}\Big|\mathcal{F}_t \quad \text{for all} \quad t \text{ in } [0;K]$$

H_t is a $(\mathcal{F}_t,W)$ martingale. Since H_t is positive, we know that there exists a martingale U_t such that :

$$H_t = \exp\,(U_t - \tfrac{1}{2} < U >_t)$$

U_t is a $(\mathcal{F}_t,W)$ martingale and thus it admits a representation of the form :

$$U_t = \int_0^t k_s\, dX_s$$

where k_s is a predictable process s.t :

$$\int_0^K k_s^{\,2}(\omega)ds < +\infty \qquad \text{W-a.s.}$$

Girsanov's formula implies that :

$$X_t - \int_0^t k_s\, ds \quad \text{is a } (\mathcal{F}_t,R) \text{ martingale.}$$

But we can also deduce from the definition of R that :

$$X_t - \int_R \mu(da)\, L_t^a(X) \quad \text{is a} \quad (\mathcal{F}_t, R) \quad \text{martingale.}$$

Hence : $$\int_C^t k_s\, ds = \int_{\mathbb{R}} \mu(da)\, L_t^a(X) \quad \text{for all} \quad t \leq K, \quad W\text{-a.s.}$$

This equality cannot hold unless :

$$\mu(da) = \theta(a)\, da \quad \text{for some} \quad \theta \quad \text{in} \quad L^1(\mathbb{R})$$

$$\text{and} \quad k_s(\omega) = \theta(X_s(\omega)) \quad W\text{-a.s.} \quad \text{for all} \quad s \leq K.$$

We also have : $$\int_0^K \theta^2(X_s(\omega))\, ds < \infty \qquad W\text{-a.s.}$$

Hence : $$\int_R L_t^a(X)\theta^2(a)\, da < \infty \qquad W\text{-a.s.}$$

Thus θ belongs to $L^2_{loc}(\mathbb{R})$.

□

3. *Limit theorems* :

In this section we use *theorem 1.5* to prove limit theorems for solutions of stochastic equations of the form (2.3). Those limit theorems will permit us to deduce that any solution of (2.3) is the limit in a strong sense of a sequence of semi-martingales which are solutions of classical Ito equations. The converse holds provided that we restrict our attention to stochastic differential equations whose coefficients satisfy integrability conditions.

Theorem 3.1 :

Let ν_n, n=1,2,... be in $M(\mathbb{R})$ and φ_n, n=1,2,... be in $BV(\mathbb{R})$. Suppose that there exist two positive constants ε, M such that :

$$|\nu_n| \ (R) \leq M \qquad \text{for all} \quad n$$

$$\varepsilon \leq \varphi_n(x) \leq M \qquad \text{for all} \quad n,x$$

$$|\nu_n(\{x\})| \leq 1-\varepsilon \quad \text{for all} \quad n,x$$

Let $(\Omega, \mathcal{F}, \mathcal{F}_t, P)$ be a filtered probability space carrying a brownian motion B. For $n=1,2,\ldots$ let X^n be a continuous semi-martingale such that :

$$X_t^n = X_0^n + \int_0^t \varphi_n(X_s^n)\, dB_s + \int_{\mathbb{R}} \nu_n(da)\, L_t^a(X^n)$$

Assume that $X_0^n \xrightarrow{L^1(\Omega)} X_0$

Assume further that there exist two functions φ, f in $BV(\mathbb{R})$ such that :

$$\int_{-K}^{K} |\varphi_n - \varphi|\,(x)\, dx \xrightarrow[n\to\infty]{} 0 \qquad \text{for all} \quad K > 0$$

$$\int_{-K}^{K} |f_{\nu_n} - f|(x)\, dx \xrightarrow[n\to\infty]{} 0 \qquad \text{for all} \quad K > 0$$

Set : $\nu(dx) = -\dfrac{f'(dx)}{f(x)+f(x-)}$

Then : $E[\sup_{0\leq s\leq t} |X_s^n - X_s|\,] \xrightarrow[n\to\infty]{} 0 \quad$ for all $t \geq 0$

where X is uniquely defined by :

$$X_t = X_0 + \int_0^t \varphi(X_s)\, dB_s + \int_{\mathbb{R}} \nu(da)\, L_t^a(X).$$

Remark :

The limit of the sequence (ν_n), if it exists, does not necessarily coïncide with ν.

For instance, set $X_0^n = 0$ for all n

$$\varphi_n = 1$$

$$\nu_n(dx) = f_n(x)dx$$

where : $f_n \geqslant 0$

$$\int_R f_n(x)\,dx = A > 0$$

$\text{supp}\,(f_n) \subset [-\varepsilon_n\,;\,\varepsilon_n]$ and $\varepsilon_n \xrightarrow[n\to\infty]{} 0$

Then $\nu_n \xrightarrow{(\text{weakly})} A\,\delta_{(0)}$

But the theorem proves that

$$X^n \xrightarrow[n\to\infty]{} X$$

where X is defined by

$$X_t = B_t + \left(\frac{1-\exp(-2A)}{1+\exp(-2A)}\right) L_t^o\,(X)$$

<u>Proof of *theorem 3.1*</u> :

We shall use the following notations :

$$f_n(x) = f_{\nu_n}(x) = \exp(-2\nu_n^c\,]-\infty;x\,])) \prod_{y\leqslant x}\left(\frac{1-\nu(\{y\})}{1+\nu(\{y\})}\right)$$

$$F_n(x) = \int_0^x f_n(y)dy$$

$$h_n(x) = (\varphi_n\,f_n)\circ F_n^{-1}\,(x)$$

Set $Y_t^n = F_n\,(X_t^n)$. By *proposition 2.2* :

$$Y_t^n = Y_0^n + \int_0^t h_n\,(Y_s^n)\,dB_s.$$

We know that :

$$\int_{-K}^K |f_n-f|(x)dx \xrightarrow[n\to\infty]{} 0 \qquad \text{for all } K > 0$$

$$\int_{-K}^K |\varphi_n-\varphi|(x)dx \xrightarrow[n\to\infty]{} 0 \qquad \text{for all } K > 0$$

It follows that :

$$\int_{-K}^{K} |h_n - h|(x)dx \xrightarrow[n\to\infty]{} 0 \quad \text{for all} \quad K > 0$$

Where $h = (f.\varphi) \circ F^{-1}$ and $F(x) = \int_0^x f(y)dy.$

There exist two positive constants ε', M' such that :

$$\varepsilon' \leqslant h_n(x) \leqslant M' \quad \text{for all} \quad n,x.$$

Let Y be the unique solution of :

$$Y_t = F(X_0) + \int_0^t h(Y_s)\, dB_s.$$

We have :

$$E\,[\,|Y_0^n - Y_0|\,] \leqslant E\,[\,|F_n(X_0^n) - F_n(X_0)|\,] + E\,[\,|F_n(X_0) - F(X_0)|\,]$$

Thus : $E\,[\,|Y_0^n - Y_0|\,] \xrightarrow[n\to\infty]{} 0$

Theorem 1.5 implies that :

$$E\,[\sup_{0\leqslant s\leqslant t} |Y_s^n - Y_s|\,] \xrightarrow[n\to\infty]{} 0 \qquad \text{for all} \quad t \geqslant 0$$

which in turn implies that

$$E\,[\sup_{0\leqslant s\leqslant t} |X_s^n - F^{-1}(Y_s)|\,] \xrightarrow[n\to\infty]{} 0 \qquad \text{for all} \quad t \geqslant 0.$$

A similar argument to the one used in the proof of *proposition 2.2* shows that $X_t \overset{\text{déf}}{=} F^{-1}(Y_t)$ satisfies :

$$X_t = X_0 + \int_0^t \varphi(X_s)\, dB_s + \int_{\mathbb{R}} \nu(da)\, L_t^a\,(X)$$

where $\nu = -\dfrac{f'(da)}{f(a)+f(a-)}$

□

Let ν be in $M(\mathbb{R})$. We define

$$N(\nu) = |\nu^c| \, (\mathbb{R}) + \frac{1}{2} \sum_y \left|\log \left(\frac{1-\nu(\{y\})}{1+\nu(\{y\})}\right)\right|$$

If φ belongs to $BV(\mathbb{R})$, $V(\varphi)$ will denote the total variation of φ. We note that

$$N(\nu) = V\left(\frac{1}{2} \operatorname{Log}(f_\nu)\right)$$

Let $(\Omega, \mathcal{F}, \mathcal{F}_t, P)$ be a filtered probability space carrying a brownian motion B. If ν is in $M(\mathbb{R})$ and φ is in $BV(\mathbb{R})$, $X^{\nu,\varphi}$ will denote the unique solution of :

$$X_t = \int_0^t \varphi(X_s)\, dB_s + \int_{\mathbb{R}} \nu(da)\, L_t^a\, (X)$$

Let T be a fixed positive constant. If X,X' are two continuous semimartingales, we set :

$$d(X,X') = E\,[\sup_{0\leq s\leq T} |X_s - X'_s|\,]$$

provided that this expectation is finite.

With these definitions at hand, we can state the following corollary of *theorem 3.1* :

Corollary 3.2 :

Let K,ε,M,C be positive constants.

Then, $H \overset{\text{def}}{=} \{X^{\nu,\varphi} / N(\nu) \leq K,\ \varepsilon \leq \varphi \leq M,\ V(\varphi) \leq C\}$ is a compact set for the topology induced by d.

The set of all $X^{\nu,\varphi}$ belonging to H such that ν is absolutely continuous with respect to Lebesgue measure is dense in H.

Proof :

Let (ν_n, φ_n) be a sequence in $M(\mathbb{R}) \times BV(\mathbb{R})$ s.t. :

$$N(\nu_n) \leq K \qquad \text{for all} \quad n$$
$$\varepsilon \leq \varphi_n \leq M$$
$$V(\varphi_n) \leq C$$

We have to prove that there exist a subsequence $(\nu_{n_k}, \varphi_{n_k})$ and a (ν,φ) in $M(\mathbb{R}) \times BV(\mathbb{R})$ s.t. :

$$d(X^{\nu_{n_k}, \varphi_{n_k}}, X^{\nu,\varphi}) \xrightarrow[k\to\infty]{} 0$$

Since the total variations of the φ_n's are uniformly bounded, we can find a φ in $BV(\mathbb{R})$ and a subsequence (φ_{n_k}) such that :

$$\varphi_{n_k}(x) \xrightarrow[k\to\infty]{} \varphi(x)$$

for all x in $\mathbb{R}$-D where D is at most countable.

φ clearly satisfies :

$$\varepsilon \leq \varphi \leq M \quad \text{and} \quad V(\varphi) \leq C.$$

The same reasoning applies to the sequence (f_{ν_n}). We may thus assume that there exists a f in $BV(\mathbb{R})$ s.t :

$$f_{\nu_{n_k}}(x) \xrightarrow[k\to\infty]{} f(x) \quad \text{for all} \quad x \quad \text{in} \quad \mathbb{R}\text{-D}$$

Set $\qquad \nu(dx) = - \dfrac{f'(dx)}{f(x)+f(x-)}$

Theorem 3.1 implies that :

$$d(X^{\nu_{n_k}, \varphi_{n_k}}, X^{\nu,\varphi}) \xrightarrow[k\to\infty]{} 0$$

It remains to prove that : $N(\nu) \leq K$.

From *lemma 2.1* we deduce that :

$$f = \lambda\, f_\nu \quad \text{for some} \quad \lambda > 0.$$

Hence : $$N(\nu) = V\left(\frac{1}{2}\operatorname{Log}(f_\nu)\right)$$
$$= V\left(\frac{1}{2}\operatorname{Log}(f)\right)$$
$$\leqslant \overline{\lim_{n\to\infty}}\, V\left(\frac{1}{2}\operatorname{Log}(f_{\nu_n}\right)$$
$$\leqslant K.$$

The second assertion of the corollary is easy and left to the reader.

□

<u>*Remark*</u> :

a) Consider the two following kinds of stochastic equations :

(3.1) $$X_t = B_t + \int_0^t g(X_s)ds \qquad (g \text{ is in } L^1(\mathbb{R}))$$

(3.2) $$X_t = B_t + \int_{\mathbb{R}} \nu(da)\, L_t^a\,(X) \qquad (\nu \text{ is in } M(\mathbb{R})).$$

Corollary 3.2 implies the following results. Any solution of (3.2) is the strong limit of a sequence of solutions of equations of the form (3.1). Conversely let (g_n) be a bounded sequence in $L^1(\mathbb{R})$ and let X^n , n=1,2,... be the solutions of the corresponding equations (3.1). Then there exist a measure ν and a subsequence (X^{n_k}) such that X^{n_k} converges strongly to the process X solution of (3.2).

b) As shown in the following simple example, there is no analogue of *Corollary 3.2* for measures ν which only satisfy the weaker assumption $|\nu(\{x\})|\leqslant 1$ for all x.

Set : $$\nu_n = \delta_{(\frac{1}{2^n})} - \delta_{(\frac{-1}{2^n})} \qquad \text{for all } n$$

Let X^n be defined by :

$$X_t^n = B_t + \int_{\mathbb{R}} \nu_n(da)\, L_t^a\,(X^n)$$

X^n is well-defined : see the remarks after *theorem 2.3*. We can easily prove that :

$$X^n \xrightarrow{\text{(weakly)}} X$$

where $X_t = UB_t$, U being a random variable independent of B and such that :

$$P(U=1) = P(U=-1) = \frac{1}{2} .$$

But X is not even a Markov process and thus it cannot be solution of an equation of the form (3.2).

We now use *theorem 3.1* to obtain a new proof and also a slight generalization of a result due to Rosenkrantz ([7]).

Corollary 3.3 :

Let ν be in $M(\mathbb{R})$ and X be a continuous semimartingale such that :

$$X_t = X_0 + B_t + \int_{\mathbb{R}} \nu(da)\, L_t^a (X).$$

Assume that X_0 is in $L^1(\Omega)$ and set $X_t^n = \frac{1}{n} X_{n^2 t}$.

Then X^n converges weakly towards the skew brownian motion of parameter α given by :

$$\frac{1-\alpha}{1+\alpha} = \exp(-2\nu^c(\mathbb{R})) \prod_y \left(\frac{1-\nu(\{y\})}{1+\nu(\{y\})} \right)$$

(recall that the skew brownian motion with parameter α is the process uniquely defined by :

$$X_t = B_t + \alpha\, L_t^0 (X))$$

Remark :

Rosenkrantz treated the case $\nu(da) = g(a)da$. In this case, we have :

$$\frac{1-\alpha}{1+\alpha} = \exp \left(-2 \int_{\mathbb{R}} g(x)dx\right)$$

Proof :

Define, for each integer n and for any Borel set A :

$$\nu_n(A) = \nu(nA)$$

Then :

$$X_t^n = \frac{1}{n} X_0 + \frac{1}{n} B_{n^2 t} + \int_{\mathbb{R}} \nu_n(da)\, L_t^a(X^n)$$

so that X^n has the same law as the process Y^n defined by :

$$Y_t^n = \frac{1}{n} X_0 + B_t + \int_{\mathbb{R}} \nu_n(da)\, L_t^a(Y^n)$$

Theorem 3.1 together with the relations

$$f_{\nu_n}(x) \xrightarrow[n\to\infty]{} 1 \quad \text{for all} \quad x < 0$$

$$f_{\nu_n}(x) \xrightarrow[n\to\infty]{} \exp(-2\nu^c(\mathbb{R})) \prod_y \left(\frac{1-\nu(\{y\})}{1+\nu(\{y\})}\right) \qquad \text{for all} \quad x > 0$$

complete the proof of the corollary.

□

Remark :

Corollary 3.3 provides a result of convergence towards the skew brownian motion with parameter α, where α is in the interval $]-1;1[$. The extreme values of α (α=1 or α=-1) which correspond to reflecting brownian motion, cannot be obtained through these methods. However it is possible to state a similar result of convergence towards a reflecting brownian motion.

Let (f_n) be a sequence in $L^1(\mathbb{R})$ s.t :

$f_n \geqslant 0$

$\text{supp}\,(f_n) \subset [-\varepsilon_n;\varepsilon_n]$ with $\varepsilon_n \xrightarrow[n\to\infty]{} 0$

$\int_R f_n(x)dx \xrightarrow[n\to\infty]{} +\infty$

Let X_n be defined by :

$$X_t^n = B_t + \int_0^t f_n (X_s^n)\, ds.$$

Then it is easy to prove that :

$$X^n \xrightarrow[n\to\infty]{\text{(weakly)}} |B|$$

As a last application of *theorem 3.1*, we state the following comparison theorem.

Theorem 3.4 :

Let ν,μ be in $M(\mathbb{R})$ and φ be in $BV(\mathbb{R})$

Let X,Y be two continuous semi-martingales such that :

$$X_t = X_o + \int_0^t \varphi(X_s)dB_s + \int_{\mathbb{R}} \nu(da)\, L_t^a\,(X)$$

$$Y_t = Y_o + \int_0^t \varphi(Y_s)dB_s + \int_{\mathbb{R}} \mu(da)\, L_t^a\,(Y)$$

Assume that :

1) $X_o \geqslant Y_o$

2) $\nu \geqslant \mu$

Then : $X_t \geqslant Y_t$ for all t a.s.

Proof :

Whenever $\nu(da) = f(a)da$, $\mu(da) = g(a)da$ and f and g are Lipschitz, the theorem holds as a special case of *theorem 1.4*.

In the general case there exist two sequences of measures of the above form, μ_n,ν_n, such that :

$$f_{\mu_n} \xrightarrow[n\to\infty]{} f_\mu$$

$$f_{\nu_n} \xrightarrow[n\to\infty]{} f_\nu$$

$$\mu_n \geqslant \nu_n \quad \text{for all} \quad n.$$

If X^n, Y^n are the corresponding processes we have

$$X_t^n \geqslant Y_t^n \quad \text{for all} \quad t \quad \text{a.s.}$$

and by *theorem 3.1* $\quad X_t^n \longrightarrow X_t \quad$ for all t.

$$Y_t^n \longrightarrow Y_t$$

Hence $X_t \geqslant Y_t$ for all t a.s.

□

4. *Approximation by random walks* :

In [2], Harrison and Shepp proved that the skew brownian motion is the weak limit of $\frac{1}{n} S_{[n^2 t]}$, where S is a random walk with exceptional behaviour at the origin. In this section, we shall extend this result to the case of a process X solution of :

$$X_t = B_t + \int_{\mathbb{R}} \nu(da)\, L_t^a(X)$$

where ν is in $M(\mathbb{R})$.

Our results will also provide some information about the asymptotic behaviour of a certain class of random walks.

Theorem 4.1 :

Let $(\Omega, \mathcal{F}, \mathcal{F}_t, P)$ be a filtered probability space carrying a brownian motion B.

Let ν be in $M(\mathbb{R})$ and X be the process uniquely defined by :

$$X_t = B_t + \int_{\mathbb{R}} \nu(da)\, L_t^a(X).$$

We define α_k^n, for all integers k and all n=1,2,..., by :

$$\frac{1-\alpha_k^n}{1+\alpha_k^n} = \exp\left(-2\nu^c\left(\left]\frac{k}{n}, \frac{k+1}{n}\right]\right)\right) \prod_{\frac{k}{n} < y \leq \frac{k+1}{n}} \left(\frac{1-\nu(\{y\})}{1+\nu(\{y\})}\right) = \frac{f_\nu\left(\frac{k+1}{n}\right)}{f_\nu\left(\frac{k}{n}\right)}$$

Then there exists a sequence S^n, n=1,2,... of random walks defined on Ω such that :

$$S_0^n = 0 \qquad \text{for all } n$$

$$P[S_{p+1}^n = k+1/S_p^n = k] = \frac{1}{2}(1+\alpha_k^n) \qquad \text{for all } n,p,k$$

$$P[S_{p+1}^n = k-1/S_p^n = k] = \frac{1}{2}(1-\alpha_k^n) \qquad \text{for all } n,p,k$$

and : $\frac{1}{n} S^n_{[n^2 t]} \xrightarrow[n\to\infty]{(\text{probability})} X_t$ for all $t \geq 0$.

Proof :

Set $\nu_n = \sum_{k=-\infty}^{\infty} \alpha_k^n \, \delta_{\left(\frac{k}{n}\right)}$ for n=1,2,...

For each n, let X^n be defined by :

$$X_t^n = B_t + \int_{\mathbb{R}} \nu_n(da)\, L_t^a(X^n)$$

Note that : $f_{\nu_n}(x) \xrightarrow[n\to\infty]{} f_\nu(x)$ for all x.

Theorem 3.1 implies that :

$$E\left[\sup_{0\leq s\leq t} |X_s^n - X_s|\right] \xrightarrow[n\to\infty]{} 0$$

Set : $\tau_0^n = 0$

$$\tau_{p+1}^n = \inf\{t > \tau_p^n / |X_t^n - X_{\tau_p^n}^n| = \frac{1}{n}\}, \qquad \text{for } p=0,1,2,...$$

Note that, conditionally on $\{X_{\tau_p^n}^n = \frac{k}{n}\}$:

$$(X^n_{\tau^n_p+u} - X^n_{\tau^n_p}, \ 0 \leq u \leq \tau^n_{p+1} - \tau^n_p) \quad \text{is}$$

distributed as a skew brownian motion with parameter α^n_k.

Define :

$$S^n_p = n\, X^n_{\tau^n_p} \quad \text{for} \quad n=1,2,\ldots \quad \text{and} \quad p=0,1,\ldots$$

Each S^n is a random walk on the integers. Recalling our previous remarks and using the well-known properties of skew brownian motion, we then deduce that S^n satisfies all the required properties.

It remains to prove that :

$$\frac{1}{n}\, S^n_{[n^2 t]} \xrightarrow[n\to\infty]{\text{(probability)}} X_t \qquad \text{for all} \quad t.$$

Note that : $\frac{1}{n}\, S^n_{[n^2 t]} = X^n_{\tau^n_{[n^2 t]}}$

Lemma : $\left\{ \tau^n_{[n^2 t]} \xrightarrow[n\to\infty]{\text{(probability)}} t \right\}$ holds for all $t \geq 0$.

Proof :

We shall only investigate the case $t=1$.

Set :

$$\sigma^n_p = n^2\, (\tau^n_{p+1} - \tau^n_p) \qquad \text{for} \quad p=0,1,\ldots$$

Since X^n satisfies the strong Markov property and since the absolute value of a skew brownian motion is distributed as a reflecting brownian motion ([10]), the σ^n_p's, $p=0,1,2,\ldots$ are independent and identically distributed. Their common distribution is the law of $T = \inf\,\{t \geq 0 / |B_t| = 1\}$.

The weak law of large numbers implies that :

$$\frac{1}{n^2} \sum_{p=0}^{n^2} \sigma^n_p \xrightarrow[n\to\infty]{\text{(probability)}} 1$$

Thus $\tau^n_{n^2} \xrightarrow[n\to\infty]{\text{(probability)}} 1$

□

Let us complete the proof of *theorem 4.1*. Consider $t \geq 0$ and $\alpha > 0$. The following relation is easily established :

$$\sup_n (P [\sup_{|u|\leq\varepsilon} (|X^n_{t+u}-X^n_t|) > \alpha]) \xrightarrow[\varepsilon\to 0]{} 0.$$

The lemma implies that :

$$P [|\tau^n_{[n^2 t]} - t| > \varepsilon] \xrightarrow[n\to\infty]{} 0 \quad \text{for all} \quad \varepsilon > 0.$$

Since $\qquad P [|X^n_t - X_t| > \alpha] \xrightarrow[n\to\infty]{} 0$

we conclude that :

$$P [|X^n_{\tau^n_{[n^2 t]}} - X_t| > \alpha] \xrightarrow[n\to\infty]{} 0$$

Hence : $\qquad P [|\frac{1}{n} S^n_{[n^2 t]} - X_t| > \alpha] \xrightarrow[n\to\infty]{} 0.$

□

Remark :

Let $Y^n_t = \frac{1}{n} S^n_{[n^2 t]}$. From *theorem 4.1*, X can be shown to be the weak limit of the sequence(Y^n). To this effect, it suffices to establish the tightness of the sequence of the laws of Y^n. Since $Y^n_t = X^n_{\tau^n_{[n^2 t]}}$, this task is easily performed with the help of the well-known tightness moment criterion.

The goal of *theorem 4.1* was to construct a sequence (S^n) of random walks on the integers such that $(\frac{1}{n} S^n_{[n^2 t]}, t \geq 0)$ converges weakly towards the process X defined by :

$$X_t = B_t + \int_{\mathbb{R}} \nu(da)\, L^a_t (X). \tag{4.1}$$

The next theorem is a converse of this result : it shows that solutions of (4.1) are the only processes which can arise as limits of $\left(\frac{1}{n} S^n_{[n^2 t]}\right)$ where the S^n's belong to a certain class of random walks on the integers.

Theorem 4.2 :

Let $(\Omega, \mathcal{F}, \mathcal{F}_t, P)$ be a filtered probability space carrying a brownian motion B.

For each $n=1,2,\dots$, let S^n be a random walk on the integers such that :

$$S^n_0 = 0$$

$$P[S^n_{p+1} = k+1 / S^n_p = k] = \frac{1}{2}(1+\alpha^n_k) \qquad \text{for all } k,p$$

$$P[S^n_{p+1} = k-1 / S^n_p = k] = \frac{1}{2}(1-\alpha^n_k) \qquad \text{for all } k,p$$

Assume that there exist two positive constants ε and K such that :

(i) $\quad |\alpha^n_k| \leqslant 1-\varepsilon \qquad$ for all n,k

(ii) $\quad \sum_{k=-\infty}^{+\infty} |\alpha^n_k| \leqslant K \qquad$ for all n

Then there exist a measure ν in $M(R)$ and a subsequence (S^{n_k}) such that

$$\left(\frac{1}{n_k} S^{n_k}_{[n^2 t]}, \ t \geqslant 0\right) \xrightarrow{\text{(weakly)}} X^\nu$$

where X^ν is defined by

$$X^\nu_t = B_t + \int_{\mathbb{R}} \nu(da) \, L^a_t (X^\nu)$$

Proof :

Set : $\qquad \nu_n = \sum_{k=-\infty}^{+\infty} \alpha^n_k \, \delta_{\left(\frac{k}{n}\right)} \qquad$ for $n=1,2,\dots$

Let X^n be defined by :

$$X_t^n = B_t + \int_{\mathbb{R}} \nu(da)\, L_t^a\,(X^n)$$

We define as in the proof of *theorem 4.1* :

$$\tau_0^n = 0$$

$$\tau_{p+1}^n = \inf\,\{t > \tau_p^n / |X_t^n - X_{\tau_p^n}^n| = \frac{1}{n}\}$$

Then :

$$\left(\frac{1}{n}\, S^n_{[n^2 t]}\,,\ t \geq 0\right) \text{ is distributed as } \left(X^n_{\tau^n_{[n^2 t]}}\,,\ t \geq 0\right)$$

Corollary 3.2 shows that it is possible to find a subsequence (X^{n_k}) and a measure ν in $M(\mathbb{R})$ such that :

$$X^{n_k} \xrightarrow[k\to\infty]{\text{(weakly)}} X^\nu$$

(use (i) and (ii) to deduce that the ν_n' s satisfy all the required properties).

The same arguments as in the proof of *theorem 4.1* can now be used to conclude that :

$$\left(\frac{1}{n_k}\, S^{n_k}_{[n_k^2 t]}\,,\ t \geq 0\right) \xrightarrow[k\to\infty]{\text{(weakly)}} X^\nu$$

□

We finally specialize to the case $S^n = S$ for all n and we obtain a result which is the analogue for a random walk of *corollary 3.3*.

Theorem 4.3 :

Let S be a random walk on the integers such that :

$$S_0 = 0$$

$$P[S_{p+1} = k+1 / S_p = k] = \frac{1}{2}(1+\alpha_k) \qquad \text{for all} \quad p,k$$

$$P[S_{p+1} = k-1 / S_p = k] = \frac{1}{2}(1-\alpha_k) \qquad \text{for all} \quad p,k$$

Assume that :

(i) $|\alpha_k| < 1$ for all k

(ii) $\sum_{k=-\infty}^{+\infty} |\alpha_k| < +\infty$

Set $X_t^n = \frac{1}{n} S_{[n^2 t]}$

Then X^n converges weakly towards the skew brownian motion with parameter α given by :

$$\frac{1-\alpha}{1+\alpha} = \prod_{k=-\infty}^{+\infty} \left(\frac{1-\alpha_k}{1+\alpha_k} \right)$$

Remark :

Harrison and Shepp ([2]) proved this result in the case $\alpha_0 = \alpha$, $\alpha_k = 0$ if $k \neq 0$.

Proof :

Let $\nu_n = \sum_{k=-\infty}^{\infty} \alpha_k \, \delta_{(\frac{k}{n})}$

we have : $f_{\nu_n}(x) \xrightarrow[n\to\infty]{} 1$ for all $x < 0$.

$$f_{\nu_n}(x) \xrightarrow[n\to\infty]{} \prod_{k=-\infty} \frac{1-\alpha_k}{1+\alpha_k} \qquad \text{for all} \quad x > 0$$

Hence if $\nu = \alpha \, \delta_{(0)}$

$$f_{\nu_n}(x) \longrightarrow f_\nu(x) \qquad \text{for all} \quad x \neq 0.$$

The remaining part of the argument duplicates the end of the proof of *theorem 4.2.*

□

REFERENCES :

[1] M.T. Barlow, E. Perkins : One-dimensional stochastic differential equations involving a singular increasing process. Preprint (1983).

[2] J.M. Harrison, L.A. Shepp. On skew brownian motion. Annals of probability 9 (1981) p. 309-313.

[3] J. Jacod. Calcul stochastique et problèmes de martingales. Lecture Notes in Mathematics 714. Springer Verlag Berlin 1979.

[4] J.F. Le Gall. Temps locaux et equations differentielles stochastiques. Seminaire de probabilités XVII. Lecture Notes in Mathematics 986 Springer Verlag Berlin 1983.

[5] S. Nakao. On the pathwise uniqueness of solutions of one-dimensional stochastic differential equations. Osaka J. Math. 9 (1972) p. 513-518.

[6] Y. Okabe, A. Shimizu. On the pathwise uniqueness of solutions of stochastic differential equations. J. Math. Kyoto University 15 (1975) p. 455-466.

[7] W. Rosenkrantz. Limit theorems for solutions to a class of stochastic differential equations. Indiana University Math. J. 24 (1975) p. 613-625.

[8] D.W. Stroock, S.R.S. Varadhan. Multidimensional diffusion processes. Springer Verlag Berlin 1979.

[9] D.W. Stroock, M. Yor. Some remarkable martingales. Seminaire de probabilités XV. Lecture Notes in Mathematics 850. Springer Verlag Berlin (1981).

[10] J.B. Walsh. A diffusion with discontinuous local time. Astérisque 52-53 (1978) p. 37-45.

[11] T. Yamada, S. Watanabe. On the uniqueness of solutions of stochastic differential equations. J. Math. Kyoto University II (1971) p. 155-167.

[12] M. Yor. Sur la continuité des temps locaux associés à certaines semi-martingales. Astérisque 52-53 (1978) p. 23-35.

Time changes of Brownian motion and the conditional excursion theorem

by

Paul McGill

Department of Mathematics
The New University of Ulster
Coleraine BT52 1SA
N. Ireland.

Although excursion theory is accepted as 'well-known' and has often been used for calculations it does seem extraordinary that a simple treatment of the relevant results is not readily available in the literature. We point out that as long ago as 1969 Williams [14] used the conditional excursion theorem in order to derive the Ray-Knight results on local time. And the same theorem has been applied by Walsh [13] to investigate the excursion filtration of a diffusion. The work of Maisonneuve [7], [8] deals extensively with unconditional excursion theory and treats the topic in considerable generality. But conditional excursion theory is not nearly so well documented (see [5] however) and it is for this reason that we give a complete treatment of the excursion process obtained by taking a general time-change of one dimensional Brownian motion. It is easy to see that the results apply, without a great deal of difficulty, to any recurrent one dimensional diffusion.

Consider the real-valued Brownian motion process B_t, whose filtration we denote by $\underline{\underline{B}}_t$, and let $A(t) \geq 0$ be an additive functional of the path. It has been shown in [4] that the most general such functional can be written in the form

$$A(t) = \int L(x,t)\mu(dx)$$

where μ is a Radon measure on the real line R and L is the local time. Hence the support of μ will be a closed subset F of R. Define $\tau(t)$ to be the right continuous inverse of A(t). Because A(t) is adapted we easily check that $\tau(t)$ is a $\underline{\underline{B}}_t$ stopping time for each value of t. Now write $X_t = B_{\tau(t)}$. This corresponds to the process B_t run in a time scale which only increases when the process is in the support of μ. By path continuity X_t takes its values in the set F. Such processes are referred to as gap diffusions since they behave like diffusions except for jumps, of deterministic length, across the open intervals of R not charged by the measure μ. We shall denote the filtration of X_t by $\underline{\underline{X}}_t$ and we use the special convention that $\underline{\underline{F}} = \underline{\underline{X}}_\infty$ represents the excursion σ-field of B_t with respect to the boundary set F and the time-change $\tau(t)$.

We wish to study the excursion process of B_t associated to the time change $\tau(t)$. The excursions exit from the set F and are absorbed when they hit F again. And there are two cases. The first is where we study the unconditional excursion process. This is the content of the first section, and follows a standard argument, implicit in [10], which proves the relevant Ito excursion result. This is already contained in the work of Maisonneuve [8] but our method is simpler in that it only uses the Ito formula together with the strong Markov property. Also we are able to give a formula for computing the entrance law in terms of the resolvent of $\bar{B}_t$, the process B_t killed when it first hits F. The other case is where we wish to describe the behaviour of the excursions, given that we know X_t, the process on the boundary. This is more complicated and is carried out in two stages. First of all we look at the case where the boundary is the negative real line. Here the conditional and unconditional excursion measures are the same and, provided we take care in applying the strong Markov property, the argument of the first section goes through. Next we consider the general case. Since each excursion will now carry some information about the future we must use the conditional excursion measure which we construct by using the theory of grossissement (or filtration enlargement). Our basic reference for this is the monograph of Jeulin [6].

We should point out that there is little in this article which is really new. It is true that our results on conditional excursions go further, in the case of Brownian motion, than those found in [5]. For example we give a more detailed description of the excursion measure, though again we must admit that this is essentially contained in the work of Williams. Theorem 3.2 (b) is closely related to his description of the excursion law by conditioning on the maximum, as can be seen by consulting [12]. But we have been encouraged, somewhat paradoxically, by the number of experts in related areas who have professed ignorance of excursion theory and its applications. And we hope that, by proving the results using the methods of stochastic calculus, we have made them seem less esoteric.

1.THE UNCONDITIONAL EXCURSION THEOREM

We begin with the Brownian motion process B_t, which we will assume starts at zero (this is just a convenient normalisation). Let F be the closed subset of the real line R which supports the Radon measure μ and write $T = \inf\{t > 0: B_t \in F\}$. We wish to look at the structure of the excursions of the process away from F. The behaviour of these is independent of the time change (in that they are only affected by the support of μ). Therefore we introduce the resolvent of the process $\bar{B}_t = B_{t\wedge T}$ which we will write as $R_\lambda f(x)$. This function is the unique bounded solution of

$$u'' = 2\lambda u - 2f \qquad (1.a)$$

which vanishes on the set F and at infinity. Here we will assume that f has compact support in the complement of F, so that in the neighbourhood of the boundary points of F the resolvent can be expressed as the difference of two convex functions. This is important because later on we will apply the generalised Ito formula [11] which needs this condition. Also we need to write the complement of F as $\bigcup(a_n, b_n)$ where these are distinct disjoint open intervals.

The excursion space W associated to the set F is the collection of all paths γ which start at some point of F and are then absorbed when next they hit F. We shall adopt the usual convention that all functions defined on W are zero on the null excursion. The excursion process is defined as follows. Let $D(\omega) = \{t : \Delta\tau(t)(\omega) \neq 0\}$ be the random subset of R which is the domain of our excursion functional. The excursion process is a mapping $\underline{\underline{E}}:\Omega\times R^+ \to W$ defined by

$$\begin{aligned} \underline{\underline{E}}(\omega,s) &= \{\bar{B}_t\circ\theta_{\tau(s)-}(\omega) : t \geq 0\} & s \in D(\omega) \\ &= \Delta & s \notin D(\omega) \end{aligned}$$

Here θ_t is the translation operator on B_t and Δ is the null excursion. Then if W is equipped with the topology of pointwise convergence $\underline{\underline{E}}$ is a measurable mapping.

We now construct the excursion measure Q on the set W. It is enough to show how Q is defined when restricted to each W_n, the set of excursions which take values in the interval (a_n, b_n). And we further divide these into two sets namely W_n^+, the excursions which start at a_n and go upwards, and the downward excursions W_n^-. So let us write the transition density of $\bar{B}_t$ as $p_t(x,y)$. Then we can define the entrance law into W_n^+ at time t to be

$$Q_n^+[t;dy] \quad = \quad dy \frac{\partial}{\partial x} p_t(x,y)\Big|_{x=a_n+} \qquad (a_n < y < b_n) \qquad (1.b)$$

The excursion entrance law into W_n^- is defined similarly to be

$$Q_n^-[t;dy] \quad = \quad -dy \frac{\partial}{\partial x} p_t(x,y)\Big|_{x=b_n-} \qquad (a_n < y < b_n) \qquad (1.c)$$

The terminology 'excursion entrance law' means that if t is positive and Y is a Borel subset of (a_n,b_n) then

$$Q_n^+[\gamma(t) \in Y] \quad = \quad \int_Y Q_n^+[t;dy]$$

The excursion measure on W_n^+ is now specified by declaring that the Q_n^+ conditional distribution of $\{\gamma(t+s) : s \geq 0\}$, given that $\gamma(t) \notin F$, is that of a Brownian motion started at $\gamma(t)$ and absorbed at the hitting time of F. More precisely we can write

$$Q_n^+[f_1(\gamma(t+t_1)).....f_m(\gamma(t+t_m))] \quad =$$

$$\int Q_n^+[t;dy]\ E_y[f_1(\bar{B}_{t_1})\ ......f_m(\bar{B}_{t_m})]$$

The proof that this does specify a measure (i.e. that the above definition is consistent) is deferred to Corollary 1.3 below. Anyway, since the sets W_n are disjoint this provides us with the general description of the excursion measure Q on the set W. Next we wish to justify this description by showing how it is related to the original process. The fundamental idea of Ito [3] is that if the Brownian motion is run in a suitable time scale, and we use a large enough state space, then we obtain a Poisson point process. Classically the time scale is the one which corresponds to the case where μ is a single point mass, so that conditional on the corresponding local time we obtain a Poisson point process with values in the space of excursions from a point. The standard reference for excursion theory is the work of Maisonneuve (see [7] and [8] in particular).

As usual we write L(x,t) to denote the bicontinuous version of the local time of B_t, normalised so that the occupation density formula becomes

$$\int_0^t f(B_s)ds \quad = \quad \int f(a)L(a,t)da$$

Also it is convenient to write $\tilde{L}(x,t) = \frac{1}{2}L(x,\tau(t))$ whenever x is a boundary point of F. Now recall that if N(t) is an integer-valued increasing process with unit jumps then it is Poisson with rate λ if and only if $N(t) - \lambda t$ is a martingale. This remark enables us to interpret the following lemma from the Ito point of view.

Lemma 1.1 Let f be any continuous function whose compact support is contained in the complement of F. Then if $u > 0$

$$\sum_{0<s\leq t} f(\bar{B}_u)\circ\theta_{\tau(s)-} \; - \; \sum_n \{Q_n^+[u;f]\tilde{L}(a_n,t) + Q_n^-[u;f]\tilde{L}(b_n,t)\}$$

is a $\underline{\underline{B}}_{\tau(t)}$ purely discontinuous martingale, whose jumps coincide with those of $\tau(t)$.

Proof: Writing $u(x) = R_\lambda f(x)$ we can, by the remarks made at the beginning of this section, apply the generalised Ito's formula of [11] to $e^{-\lambda t}u(B_t)$. From (1.a) this shows that

$$e^{-\lambda t}R_\lambda f(B_t) \; = \; R_\lambda f(0) \; + \; \int_0^t e^{-\lambda s}u'(B_s)dB_s \; +$$
$$\sum_n \int_0^t e^{-\lambda s}\{u'(a_n+)d_sL(a_n,s) \; - \; u'(b_n-)d_sL(b_n,s)\}$$
$$- \; \int_0^t e^{-\lambda s}f(B_s)ds$$

is a uniformly integrable martingale. Now let us time change by $\tau(t)$. The first term on the l.h.s. will vanish, by the definition of the resolvent kernel, and therefore we find that we obtain the purely discontinuous $\underline{\underline{B}}_{\tau(t)}$ martingale

$$\sum_{0<s\leq t} \int_{\tau(s)-}^{\tau(s)} e^{-\lambda\tau(u)}f(B_u)du \; -$$
$$\sum_n \int_0^t e^{-\lambda\tau(s)}\{u'(a_n+)d_s\tilde{L}(a_n,s) \; - u'(b_n-)d_s\tilde{L}(b_n,s)\}$$

But the singular integrals do not charge the jumps of $\tau(t)$, so we can integrate against the $\underline{\underline{B}}_{\tau(t)}$ predictable process $\exp\{\lambda\tau(t)-\}$ to find that

$$\int_0^\infty e^{-\lambda u} \; [\sum_{0<s\leq t} f(\bar{B}_u)\circ\theta_{\tau(s)-} \; -$$
$$\sum_n \{Q_n^+[u;f]\tilde{L}(a_n,t) + Q_n^-[u;f]\tilde{L}(b_n,t)\}]du$$

is a $\underline{\underline{B}}_{\tau(t)}$ martingale (note how we have used the description of the excursion measure given above). The proof is completed by inverting the Laplace transform.

The above lemma shows how the local time terms arise in the excursion theorem. They are produced by the singular behaviour of the process at each boundary point, where the excursions are born at times which are not stopping times. We are now able to prove the unconditional version of Ito's excursion theorem.

Theorem 1.2 Let K be defined on the excursion space W and suppose that Q[K] < +∞. Then

$$H_t = \sum_{0<s\leq t} K\circ\underline{\underline{E}}(\omega,s) - \sum_n \{Q_n^+[K]\tilde{L}(a_n,t) + Q_n^-[K]\tilde{L}(b_n,t)\}$$

is a $\underline{\underline{B}}_{\tau(t)}$ martingale.

Proof: It is enough to prove the result for K varying in a dense set. The most convenient such collection, for our purposes, is the family of functions K of the form

$$K(\gamma) = f_1(\gamma(u+t_1))\ldots\ldots f_m(\gamma(u+t_m)) \qquad u > 0$$

where the $\{f_i\}$ are bounded, continuous and u is fixed. So let us introduce the $\underline{\underline{B}}_{\tau(t)}$ stopping times $u + \tau(\xi_n)-$ by writing

$$\xi_n = \inf\{t > \xi_{n-1} : \tau(t) - \tau(t)- > u\}$$

with $\xi_0 = 0$ (the fact that this procedure defines stopping times is well-known). If $g(x) = f(x)E_x[f_1(\bar{B}_{t_1}) \ldots\ldots f_m(\bar{B}_{t_m})]$ then by the previous lemma

$$\sum_{\xi_n\leq t} g(\bar{B}_u)\circ\theta_{\tau(\xi_n)-} - \sum_n \{Q_n^+[K]\tilde{L}(a_n,t) + Q_n^-[K]\tilde{L}(b_n,t)\}$$

is a $\underline{\underline{B}}_{\tau(t)}$ martingale. However by the strong Markov property at each of the (finite number of) stopping times $u + \tau(\xi_n)-$ we find that

$$E[g(\bar{B}_u)\circ\theta_{\tau(\xi_n)-}] = E[K\circ\underline{\underline{E}}(\omega,\xi_n)]$$

so that $E[H_t] = 0$. It is now immediate from the strong Markov property of B_t at time $\tau(s)$, that $E[H_t|\underline{\underline{B}}_{\tau(s)}] = H_s$ so the proof is complete.

Probably the most general form of the above theorem can be found in [8] Theorem 6.3. However we have also obtained the following results. Part (b) is sometimes referred to as the 'strong Markov property of the excursion process' though this is not entirely accurate from our point of view.

Corollary 1.3 (a) Q is a σ-finite measure on W.
(b) Let S be a strictly positive stopping time defined on the excursion space (W,Q) with the canonical right continuous completed filtration $\underline{\underline{Q}}_t$. Then, given $\gamma(S)$, the process $\{\gamma(S+t): t \geq 0\}$ is conditionally independent of $\underline{\underline{Q}}_S$, with the same law as $\bar{B}_t$ started at $\gamma(S)$.

Proof: (a) We need to show that the entrance law is consistent, namely that if $Y \subseteq (a_n, b_n)$ is a Borel set then

$$\int_Y Q_n^+[u+t;dx] \quad = \quad \int_{a_n}^{b_n} Q_n^+[u;dx] P_x[\bar{B}_t \in Y]$$

To see this take K to be the indicator of the set $\{\gamma: \gamma(0) = a_n, \gamma(u+t) \in Y\}$ and use the Doob optional stopping theorem on the martingale H at the end of the first excursion into W_n^+ of length greater than u+t. The result follows from our description of Q, since K is also the indicator of $\{\gamma: \gamma(u) \in (a_n, b_n)\} \cap \{\gamma: \gamma(0) = a_n, \gamma(u+t) \in Y\}$ and the local time at a_n is non-zero. To see that Q_n^+ is σ-finite we remark that $\int_0^\infty Q_n^+[\gamma(t) \notin F] e^{-\lambda t} dt$ is the derivative of the resolvent $R_\lambda 1(x)$ at $x = a_n+$, and this is finite. Thus $Q_n^+[\gamma : \gamma(t) \notin F]$ is finite for each $t > 0$, so W_n^+ is a countable union of sets of finite measure $W_n^+ \cap \{\gamma: \gamma(\frac{1}{m}) \notin F\}$.
(b) Our description of the measure Q shows this to be true if S is a simple stopping time. The general case follows by taking limits using discrete approximation of S from above by a sequence of simple stopping times.

2. CONDITIONAL EXCURSIONS FROM $(-\infty, 0)$

We next look at the conditional behaviour of the excursion process given the process on the boundary, in the special case where the boundary is the infinite interval $(-\infty, 0)$ and the measure μ is the corresponding restriction of Lebesgue measure. The first problem is to identify the filtration $\underline{\underline{X}}_t$ in some more tractable way. One method of doing this is to use the following result.

Theorem 2.1 ([2]) Subject to the boundary conditions that

(a) $Y_t \leq 0$

(b) $\bar{L}(0,t)$ is continuous and increasing

(c) $\int Y_s \, d_s\bar{L}(0,s) = 0$

the stochastic differential equation

$$Y_t = Y_0 + \beta_t - \bar{L}(0,t)$$

has a unique solution Y_t which is adapted to the filtration of the given Brownian motion β_t. This solution is defined by the formula $\bar{L}(0,t) = \sup\{(Y_0+\beta_s)^+ : 0\leq s\leq t\}$.

Now let us write the Tanaka formula [11]

$$B_t^- = B_0^- + \int_0^t 1_{\{B_s<0\}} dB_s - \tfrac{1}{2}L(0,t)$$

If $\tilde{\beta}_t = \int_0^{\tau(t)} 1_{\{B_s<0\}} dB_s$, this being a Brownian motion by the P. Levy martingale characterisation, we can time change this equation by $\tau(t)$ to get

$$B_{\tau(t)} = B_0^- + \tilde{\beta}_t - \tilde{L}(0,t)$$

The above theorem now shows that we have proved the following.

Corollary 2.2 (a) $\underline{\underline{X}}_t$ is the filtration of $\tilde{\beta}_t$.
(b) $\tilde{L}(0,t)$ is $\underline{\underline{X}}_t$ adapted.

The next lemma is apparently well-known but its relevance for excursion theory seems to have been first pointed out in [9]. The proof is so easy that we reproduce it here for completeness.

Lemma 2.3 If N_t is a square integrable $\underline{\underline{B}}_{\tau(t)}$ martingale which is orthogonal to $\tilde{\beta}_t$ (i.e. $\langle\tilde{\beta},N\rangle = 0$) then

$$E[N_\infty - N_0 | \underline{\underline{F}}] = 0.$$

Proof: We work directly from the definition. Suppose that f is a bounded $\underline{\underline{F}}$ measurable function. Then $f_t = E[f|\underline{\underline{X}}_t]$ is an $\underline{\underline{X}}_t$ (and a fortiori a $\underline{\underline{B}}_{\tau(t)}$) martingale. Since, by Corollary 2.2, $\underline{\underline{X}}_t$ is the filtration of $\tilde{\beta}_t$ we can apply the well-known Ito martingale representation theorem to see that there exists an $\underline{\underline{X}}_t$ predictable process u_t such that $f_t = f_0 + \int_0^t u_s d\tilde{\beta}_s$. But by the properties of the stochastic integral we have $\langle N,f\rangle = 0$ so that $f_t(N_t$

$- N_0)$ is also a $\underline{\underline{B}}_{\tau(t)}$ uniformly integrable martingale. Using the Doob optional stopping time theorem we obtain

$$E[f(N_\infty - N_0)] \quad = \quad E[f_0(N_0 - N_0)] \quad = \quad 0$$

which is the result we want.

Notice that $\tilde{\beta}_t$ is a martingale which depends only on the original process B_t below the zero. The above 'projection lemma' is applied typically in situations where we have used Ito's formula to construct (discontinuous) $B_{\tau(t)}$ martingales registering only when B_t is above zero. These are therefore orthogonal to $\tilde{\beta}_t$.

At this stage we ought to think of constructing the conditional excursion measure but fortunately, in this special case, the conditional and unconditional measures are the same.

Lemma 2.4 Let f be any continuous function whose compact support is contained in $(0,+\infty)$. Then

$$\sum_{0<s\leq t} f(\bar{B}_u)\circ\theta_{\tau(s)-} \quad - \quad Q[u;f]\tilde{L}(0,t)$$

is a $\underline{\underline{B}}_{\tau(t)}$ martingale which is orthogonal to $\tilde{\beta}_t$.

Proof: This is the same as for Lemma 1.1. Our martingale is again purely discontinuous, hence orthogonal to every continuous $\underline{\underline{B}}_{\tau(t)}$ martingale.

A difficulty now arises if we try to imitate the proof of Theorem 1.2 in the conditional case. This is because the times ξ_n are not measurable w.r.t. the σ-field $\underline{\underline{F}}$ and hence we require a more subtle argument than that used in [9] Lemma 4.3. We begin with a general result whose proof can be found in [1].

Lemma 2.5 The σ-fields $\underline{\underline{F}}^1$ and $\underline{\underline{F}}^2$ are conditionally independent given $\underline{\underline{F}}^3$ if and only if

$$E[Z|\underline{\underline{G}}] \quad = \quad E[Z|\underline{\underline{F}}^3]$$

where $\underline{\underline{G}} \ = \ \sigma\{\underline{\underline{F}}^3,\underline{\underline{F}}^2\}$ and Z is any bounded $\underline{\underline{F}}^1$ measurable random variable.

The following corollary is an obvious consequence of this criterion. It is extremely useful if we wish to enlarge the conditioning σ-field.

Corollary 2.6 If $\underline{\underline{F}}^1$ and $\underline{\underline{F}}^2$ are conditionally independent given $\underline{\underline{F}}^3$ then this remains true if we replace the latter by any σ-field $\underline{\underline{G}}$ such that $\underline{\underline{F}}^3 \subseteq \underline{\underline{G}} \subseteq \sigma\{\underline{\underline{F}}^1, \underline{\underline{F}}^3\}$.

Lemma 2.7 Suppose that S is any $\underline{\underline{B}}_t$ stopping time.

(a) $\underline{\underline{B}}_S$ and $\underline{\underline{F}}$ are conditionally independent given $\underline{\underline{X}}_{A(S)}$.

(b) If f is any $\underline{\underline{B}}_S$ measurable function then for each bounded measurable g

$$E[fg \circ \theta_S | \underline{\underline{F}}] = E[fE_{B_S}[g|\underline{\underline{F}}]|\underline{\underline{F}}]$$

Proof: (a) Let us define the stopping time $U = \inf\{t > S: B_t \in F\}$. The strong Markov property implies that $\underline{\underline{B}}_U$ and $\{B_t \circ \theta_U: t \geq 0\}$ are conditionallly independent given $X_{A(U)} = X_{A(S)}$, this last equality since we have chosen $\tau(t)$ to be right continuous. By the previous corollary we can enlarge the conditioning σ-field to $\underline{\underline{X}}_{A(U)}$. Which completes the proof since $\underline{\underline{F}}$ is contained in the σ-field generated by the $\underline{\underline{X}}_{A(U)}$ and $\{B_t \circ \theta_U: t \geq 0\}$.

(b) By the strong Markov property of B_t at the time S we have

$$E[fg \circ \theta_S | \underline{\underline{B}}_S, \underline{\underline{F}} \circ \theta_{A(S)}] = fE_{B_S}[g|\underline{\underline{F}}]$$

The result follows when we condition by $\underline{\underline{F}}$.

Remark: Notice that the above lemma holds in the general case where μ is an arbitrary Radon measure. This will be important in the next section.

Theorem 2.8 Let K be any function defined on the excursion space W and such that $Q[K] < +\infty$. Then

$$\sum_{0<s\leq t} K \circ \underline{\underline{E}}(\omega, s) - Q[K]\tilde{L}(0,t) \qquad (2.b)$$

is a $\underline{\underline{B}}_{\tau(t)}$ martingale orthogonal to $\tilde{\beta}_t$.

Proof: Here we repeat the argument of Theorem 1.2 only we need to be a little more careful. So as before we introduce the function $g(x) = f(x)E_x[f_1(\bar{B}_{t_1}) \ldots\ldots f_m(\bar{B}_{t_m})]$ where all functions are compactly supported on $(0,+\infty)$. By Lemmas 2.3 and 2.4 we have

$$E[\sum_{0<s\leq t} g(\bar{B}_u) \circ \theta_{\tau(s)-} | \underline{\underline{F}}] = Q[u;g]\tilde{L}(0,t) \qquad (2.c)$$

But if $K(\gamma) = f_1(\gamma(u+t_1)) \ldots\ldots f_m(\gamma(u+t_m))$ then we can apply Lemma 2.7 (b) at each of the times $u + \tau(\xi_n)-$ (as defined in the proof of Theorem 1.2) and argue inductively to get

$$E[\sum_{\xi_n \leq t} g(\bar{B}_u)\circ\theta_{\tau(\xi_n)-} \mid \underline{\underline{F}}] = E[\sum_{\xi_n \leq t} K\circ\underline{\underline{E}}(\omega,\xi_n) \mid \underline{\underline{F}}]$$

But from this and (2.c) we can now finish the argument in the same way as before.

One reason for expressing the above result in terms of random measures (i.e. discontinuous martingales) is because we can use it to generate other martingales orthogonal to $\tilde{\beta}_t$. The following corollary is a conditional version of the Maisonneuve formula (see [8] Theorem 6.3).

Corollary 2.9 If Z_t is a bounded $\underline{\underline{B}}_{\tau(t)}$ predictable process then

$$E[\sum_{0<s\leq t} Z_s K\circ\underline{\underline{E}}(\omega,s) \mid \underline{\underline{F}}] = Q[K]\int_0^t \tilde{Z}_s \, d\tilde{L}(0,s)$$

where $\tilde{Z}_t$ is the $\underline{\underline{X}}_t$ optional projection of Z_t.

Proof: Since Z_s is predictable we can integrate it against the discontinuous martingale (2.b) to get another martingale orthogonal to $\tilde{\beta}_t$. Now apply Lemma 2.3 and use the fact that, by Corollary 2.2, $\tilde{L}(0,t)$ is $\underline{\underline{X}}_t$ adapted.

3. THE CONDITIONAL EXCURSION THEOREM

We resume our discussion of the general case only this time we examine the conditional behaviour of the excursion process, given the process on the boundary. This we do by factoring the time-change $\tau(t)$ via the time change we used in section two, which we shall now relabel as $\eta(t)$. To do this factorisation conveniently we shall work with the normalisation which fixes the excursion interval under consideration to be $(0,a)$, and we will assume that $B_0 \leq 0$. The immediate problem is therefore to describe the conditional behaviour of the excursions of B_t into $(0,a)$ given $\underline{\underline{F}}$, the entire history of the boundary process X_t. However knowing the process on the boundary enables us to identify when, in the $\tau(t)$ time scale, we have excursions which cross the gaps in the support of μ. Therefore $\underline{\underline{X}}_{A(T_a)} \subseteq \underline{\underline{F}}$ and by the strong Markov property, and Lemma 2.7 (a), it is enough to consider only those excursions which take place during the time interval $[0,T_a]$. The difference between the conditional and unconditional excursion measures is due to the fact that the former must necessarily carry some information about the future. In keeping with our martingale approach to these problems we shall use the technique of grossissement (or filtration enlargement) to derive our results. The standard reference for this is the work of Jeulin [6].

So let ρ be the last leaving time of zero before B_t hits a. It is well known that ρ is not a stopping time of the process B_t, since it carries some information about the future. However, in the language of the general theory of processes, it is the end of an optional set and therefore one can use the technique of grossissement. Let V_t^ρ denote the $\underline{\underline{B}}_t$ supermartingale obtained by taking the $\underline{\underline{B}}_t$ optional projection of the process $1_{[0,\rho[}$. We shall write its Doob-Meyer decomposition as

$$V_t^\rho = M_t^\rho - D_t^\rho$$

where M_t^ρ is a martingale and D_t^ρ is a predictable increasing process.

<u>Lemma</u> 3.1 $V_t^\rho = 1_{[0,T_a[}(1-B_t^+/a)$. Therefore we have

$$M_t^\rho = 1 - \frac{1}{a}\int_0^{t\wedge T_a} 1_{\{B_s>0\}} dB_s \quad ; \quad D_t^\rho = \frac{1}{2a}L(0,t\wedge T_a).$$

<u>Proof</u>: Let S be any $\underline{\underline{B}}_t$ stopping time. By the strong Markov property,

$$E[1_{[0,\rho[}(S)] = P[S < \rho] = E[E_{B_S}[T < T_a];S < T_a]$$

And, since B_t is in natural scale, this gives the first part. The rest follows from the Tanaka formula

$$B_t^+ = \int_0^t 1_{\{B_s>0\}} dB_s + \tfrac{1}{2}L(0,t)$$

by the uniqueness of the Doob-Meyer decomposition.

Next let $\underline{\underline{B}}_t^\rho$ denote the (right-continuous, completed) filtration obtained by initial enlargement of the filtration $\underline{\underline{B}}_t$, using the random variable $L(0,\rho)$ (warning: this is not the same as the filtration of the process B_t^ρ defined in the next theorem!). Using the above lemma and [6] p.86 we obtain the following.

<u>Theorem</u> 3.2 The process

$$B_t^\rho = B_t - \int_0^{t\wedge\rho} \frac{1}{B_s-a} 1_{\{B_s>0\}} ds - \int_{t\wedge\rho}^{t\wedge T_a} \frac{1}{B(s)} 1_{\{B_s>0\}} ds$$

is a $\underline{\underline{B}}_t^\rho$ Brownian motion.

Notice how this shows that the process B_t runs, before time ρ, as a Brownian motion conditioned to hit $(-\infty,0]$ before it hits a. And that after time ρ it behaves like a BES(3) process exiting from zero. Also remark that since we have used initial enlargement, B_t is independent of $L(0,\rho)$.

For the construction of the conditional excursion measure, denoted here by $\tilde{Q}$, we use the subset $\tilde{W}$ of W, consisting of all those excursions whose initial and final boundary points are the same. Also we will write $\tilde{W}_n^+$ to denote the intersection of W_n^+ with $\tilde{W}$ so that this is the set of excursions from a_n which end at a_n (recall that W_n^+ is the family of excursions from a_n into the interval (a_n,b_n)). In order to describe the restriction of $\tilde{Q}$ to this set we proceed as in the first section, except that in equations (1.b) and (1.c) we replace the semi-group of $\overline{B}_t$ by the semi-group of Brownian motion conditioned not to hit b_n and absorbed at a_n. Notice that this is exactly as required by Theorem 3.2. As before we normalise by requiring that $\tilde{Q}_0^+$ shall relate to our fixed interval (0,a). And since we are now looking only at $\tilde{W}$ it is convenient to consider the reduced excursion process $\underline{\underline{E}}^*$ which is defined by

$$\underline{\underline{E}}^*(\omega,s) = \underline{\underline{E}}(\omega,s) \qquad X_{s-} = X_s$$

$$= \Delta \qquad X_{s-} \neq X_s$$

Thus far our convention has been that the excursion process is defined with respect to the time change $\tau(t)$. But it is convenient to introduce $\zeta = \int_0^\rho 1_{\{B_s<0\}} ds$, which is a stopping time for the filtration $\underline{\underline{B}}_t^\rho$, and to adopt the notation $\underline{\underline{V}}$ for the σ-field generated by $\{B_{\eta(t)}: 0 \leq t \leq \zeta\}$ and the random variable $L(0,\rho)$. If, as in section two, we introduce the Brownian motion $\overline{\beta}_t = \int_0^{\eta(t)} 1_{\{B_s<0\}} dB_s$ then we can show as before that $\underline{\underline{V}}$ is generated by $L(0,\rho)$ and the process $\{\overline{\beta}_t: 0 \leq t \leq \zeta\}$. The proofs of the following lemmas are omitted since they are similar to the corresponding results of the previous section. See also [9] section five.

Lemma 3.3 Let f be any continuous function whose compact support is contained in (0,a). Then

$$\sum_{0<s\leq t\wedge\zeta} f(\overline{B}_u)\circ\theta_{\eta(s)-} \quad - \quad \tfrac{1}{2}\tilde{Q}_0^+[u;f]L(0,\eta(t)\wedge T_a)$$

is a $\underline{\underline{B}}^{\rho}_{\eta(t)}$ martingale which is orthogonal to $\overline{\beta}_t$.

Lemma 3.4 If N_t is a square integrable $\underline{\underline{B}}^{\rho}_{\eta(t)}$ martingale which is orthogonal to $\overline{\beta}_t$ (i.e. $\langle\overline{\beta},N\rangle = 0$) then

$$E[N_\infty - N_0|\underline{\underline{V}}] \quad = \quad 0.$$

Lemma 3.5 If $S \leq \rho$ is any $\underline{\underline{B}}^{\rho}_t$ stopping time then $\underline{\underline{V}}$ and $\underline{\underline{B}}_S$ are conditionally independent given the σ-field generated by $L(0,\rho)$ and $\{B_{\eta(t)}: 0 \leq t \leq \int_0^S 1_{\{B_s<0\}}ds\}$.

Note that in the previous lemma we have $B^{\rho}_{\eta(t)} = B_{\eta(t)}$ and that we could have replaced $B_{\eta(t)}$ by $\overline{\beta}_t$.

Theorem 3.6 (a) Let K be any function defined on the excursion space $\tilde{W}$ such that $\tilde{Q}[K] < +\infty$. Then

$$E[\sum_{0<s\leq t} K\circ\underline{\underline{E}}^*(\omega,s)\mid \underline{\underline{F}}] \quad = \quad \sum_n \{\tilde{Q}_n^+[K]\tilde{L}(a_n,t) \quad + \quad \tilde{Q}_n^-[K]\tilde{L}(b_n,t)\}$$

(b) The excursions in $W\backslash\tilde{W}$ (i.e. those which cross the complementary intervals of the support of μ) are independent diffusions Y_t satisfying the stochastic differential equation

$$Y_t \quad = \quad x \quad + \quad \beta_t \quad + \int_0^t \frac{ds}{Y_s-x} \qquad (0 < t < T_z)$$

where x is the starting point and z is the absorption point of the excursion.

(c) $$E[\sum_{0<s\leq t} 1_{W\backslash\tilde{W}}\circ\underline{\underline{E}}(\omega,s)|\underline{\underline{F}}] \quad = \quad \sum_n [1/(b_n-a_n)]\{\tilde{L}(a_n,t) \quad + \quad \tilde{L}(b_n,t)\}$$

Proof: Remember that we retain normalisation adopted at the beginning of this section. Therefore we only consider the measure Q_0^+, and we only look at the process up to time T_a.

(a) As in the proof of Theorem 2.8 we will show this for functions of the form $K(\gamma) = f_1(\gamma(u+t_1)).....f_m(\gamma(u+t_m))$, where we will assume that all the functions $\{f_i\}$ are supported on the interval $(0,a)$. Arguing as before, only now using Lemmas 3.4 and 3.5, shows that Lemma 3.3 remains valid when $f(\gamma(u))$ is replaced by any bounded measurable function defined on $\tilde{W}$. Using Lemma 3.4 again we deduce that the result we want holds if $t \leq \zeta$ and the conditioning is on the σ-algebra $\underline{\underline{V}}$, provided we run the process according to the

time change $\eta(t)$ and replace $\tilde{L}(0,t)$ by $\frac{1}{2}L(0,\eta(t))$. The general result, where we run in the time change corresponding to $\tau(t)$ and condition on $\underline{\underline{X}}_{A(T_a)}$, now follows because the extra time change is $\underline{\underline{V}}$ measurable. And an application of Lemma 2.7 (a) enables us to replace $\underline{\underline{X}}_{A(T_a)}$ by $\underline{\underline{F}}$.
(b) This is immediate from Theorem 3.2 and the strong Markov property at the jump times of X_t since ρ is a stopping time of the enlarged filtration.
(c) All that needs to be calculated here is the jump rate. But for this note that, by stopping the martingale $B_t^+ - \frac{1}{2}L(0,t)$ at time T_a, we have $E_x[L(0,T_a)] = 2a$ if $x \leq 0$. And this is independent of the time change.

Of course this is neither the most general nor the most useful form of the conditional excursion theorem. In particular it is advantageous to have a way of handling functions defined on the excursion space which vary predictably relative to the entire process history.

Theorem 3.7 Let $K : \tilde{W}\times\Omega\times R^+ \to R$ be Borel measurable, such that for each fixed $\gamma \in \tilde{W}$ the process $K(\gamma,\omega,t)$ is $\underline{\underline{B}}_{\tau(t)}$ predictable. Then

$$E\Big[\sum_{0<s\leq t} K(\underline{\underline{E}}^*(\omega,s),\omega,s) \quad - \quad \sum_n \int_0^t \{Q_n^+[K(\gamma,\omega,s)]\, d\tilde{L}(a_n,s)$$

$$+ \quad Q_n^-[K(\gamma,\omega,s)]\, d\tilde{L}(b_n,s)\}\,\Big|\,\underline{\underline{F}}\Big] \quad = \quad 0$$

provided $E[\int_0^t |\tilde{Q}[K(\gamma,\omega,s)]|\,d\tilde{L}(0,s)] < +\infty$. Furthermore, under the measure $\tilde{Q}$, the canonical process on $\tilde{W}$ is strongly Markovian in that for each strictly positive excursion space stopping time S the conditional distribution of $\{\gamma(S+t): t \geq 0\}$, given $\tilde{\underline{\underline{Q}}}_S$, is identical in law to $\{\bar{B}_t: t \geq 0\}$ started at $\gamma(S)$ and conditioned to hit the boundary F at the initial point of the excursion γ.

Proof: By the strong Markov property it is enough to look at the situation up to the first jump time of the boundary process. We can then argue as in the proof of Corollary 2.9 to get the result when K has the form $K(\gamma,\omega,t) = Z_t(\omega)K'(\gamma)$ where $K': \tilde{W} \to R$ and Z_t is a $\underline{\underline{B}}_{\tau(t)}$ predictable process. The general case now follows by an approximation argument using the dominated convergence theorem. The final statement follows as for Corollary 1.3 (b).

The next result appears in [14], though it was probably also known to Williams. However the crucial application in [14] is to the transient case, which we do not consider here.

Theorem 3.8 Let α be any $\underline{\underline{F}}$ measurable time. Then $\{\underline{\underline{E}}(\omega,s): 0 < s \leq \alpha\}$ and $\{\underline{\underline{E}}(\omega,\alpha+s): s \geq 0\}$ are conditionally independent given $\underline{\underline{F}}$.

Proof: Since $B_{\tau(\alpha)}$ is $\underline{\underline{F}}$ measurable the result follows from Lemma 2.7 (b).

Throughout all of this article we have tried to ignore the initial excursion. The main reason for this is that calculations involving the initial excursion need to be carried out separately, and are usually fairly straightforward. And anyway, by the strong Markov property, the initial excursion $\{\overline{B}_t: t \geq 0\}$ is independent of the conditioning field $\underline{\underline{F}}$.

REFERENCES

1. C. Dellacherie and P.A. Meyer, Probabilités et potentiel, Hermann, Paris, 1975.

2. N. El-Karoui et M. Chaleyat-Maurel, Un problème de réflexion et ses applications au temps local et aux équations différentielles stochastiques sur R. Cas continu. In Temps Locaux, Astérisque 52-3, Société Mathématique de France, 1978.

3. K. Ito, Poisson point processes attached to Markov processes, Proc. 6th Berkley Symp. Math. Stat. and Prob. Vol. 3 , Univ. of California Press (1971) 225-240.

4. K. Ito and H.P. McKean, Diffusion processes and their sample paths, Springer-Verlag, Berlin and New York, 1965.

5. P.A. Jacobs, Excursions of a Markov process induced by continuous additive functionals, Zeit. fur Wahrscheinlichkeitstheorie 44 (1978) 325-336.

6. T. Jeulin, Semi-martingales et grossissement d'une filtration. Lecture Notes in Mathematics Vol. 833, Springer-Verlag, Berlin and New York, 1980.

7. B.Maisonneuve, Systèmes régénératifs. Astérisque 15. Société Mathématique de France.1974.

8. B. Maisonneuve, Exit systems, Ann. of Prob. Vol.3 No.3 (1975) 399-411.

9. P. McGill, Markov properties of diffusion local time : a martingale approach, Adv. Appl. Prob. 14 (1982) 789-810.

10. P. McGill, Calculation of some conditional excursion formulae, Zeit. fur Wahrscheinlichkeitstheorie 61 (1982) 255-260.

11. P. A. Meyer, Un cours sur les intégrales stochastiques, Séminaire de Probabilités X, Lecture Notes in Mathematics Vol. 510, Springer-Verlag, Berlin and New York (1976) 245-400.

12. L.C.G. Rogers, Williams' characterisation of the Brownian excursion law: proof and applications, Séminaire de Probabilités XV, Lecture Notes in Mathematics Vol.850, Springer-Verlag, Berlin and New York (1981) 227-250.

13. J.B. Walsh, Excursions and local time. In Temps Locaux, Astérisque 52-53, Société Mathématique de France, 1978.

14. D. Williams, Markov properties of Brownian local time, Bull. Amer. Math. Soc. 75 (1969) 1035-36.

ON SQUARE-ROOT BOUNDARIES FOR BESSEL PROCESSES, AND POLE-SEEKING BROWNIAN MOTION.

Marc YOR

Laboratoire de Probabilités ; Université P. et M. Curie ; 75230 Paris Cedex 05

1. Introduction :

Let $(B_t)_{t \geq 0}$ be a real-valued Brownian motion, starting at 0 ; define

$$\tilde{T}_c = \inf\{t : |B_t| = c\sqrt{1+t}\} \qquad (c > 0).$$

This family of stopping times plays a key rôle in the computation by B. Davis [2] of the optimal constants A_p and a_p for the inequalities :

$$(1.a) \qquad E(|B_T|^p) \leq A_p\, E(T^{p/2}), \qquad \text{for } 0 < p \leq 2$$

and

$$(1.a') \qquad a_p\, E(T^{p/2}) \leq E(|B_T|^p), \quad \text{for } p \geq 2, \text{ and } E(T^{p/2}) < \infty.$$

A part of Davis' proof relies upon the following formula, due to L. Shepp [8] :

$$(1.b) \qquad E\left[(1+\tilde{T}_c)^{-\mu}\right] = \frac{1}{M(\mu;\frac{1}{2};\frac{c^2}{2})} \qquad (\mu \geq 0).$$

On the other hand, it was remarked in [2] that, although one has, for $p > 0$, and $k > 1$:

$$(1.c) \qquad E(1/L^{p/2}) \leq C_{p,k} \| 1/|B_L|^p \|_k,$$

for any r.v. $L > 0$, (1.c) does not hold for $k = 1$, even when L is assumed to be a stopping time ; again, this may be done by taking $L = \tilde{T}_c$, and letting $c \to \infty$. These two results clearly show the importance of the stopping times $\{\tilde{T}_c\}$ in connection with the study of reflecting Brownian motion ; in this Note, we take up the next natural step, that is the study of :

$$\tilde{T}_c = \inf\{t : \rho_t = c\sqrt{1+t}\},$$

for (ρ_t) a Bessel process, with dimension $d \geq 2$, and we extend L. Shepp's formula in this set-up.

Moreover, with the help of the mutual (local) absolute continuity of the Bessel laws, for dimensions ≥ 2 , when the processes start at $a > 0$, the Fourier transform of the total winding of complex BM around 0, up to $\tilde{T}_c$, is obtained (see formula (2.b.2) below).

This formula (2.b.2) is very similar to D. Kendall's formula (32) in [5], which gives the Fourier transform of the total winding around 0 for the pole-seeking complex BM stopped when it first hits a circle centered at 0.
In the third paragraph below, a probabilistic explanation is given for this similarity, using the time substitution method, as advocated by D. Williams ([5], p.414) in the discussion following D. Kendall's paper [5] !.

2. An extension of Shepp's formula (1.b) :

(2.1) We consider, on the space $\Omega = C(\mathbb{R}_+,\mathbb{R}_+)$, the process of coordinates $(\rho_t(\omega) \equiv \omega(t)\ ;\ t \geq 0)$, and its natural filtration $(\mathcal{F}_t = \sigma\{\rho_s\ ;\ s \leq t\}\ ;\ 0 \leq t \leq \infty)$
To any couple of numbers $\nu \geq 0$, and $a \geq 0$, we associate the dimension $d = 2(\nu+1)$, and the distribution P_a^ν, on $(\Omega, \mathcal{F}_\infty)$, of the d-dimensional Bessel process, that is the $\mathbb{R}_+$-valued diffusion with infinitesimal generator $A_\nu \equiv \frac{1}{2}\frac{d^2}{dx^2} + \frac{2\nu+1}{2x}\frac{d}{dx}$, starting at a.
Now, fix $a > 0$, $\mu,\nu \geq 0$. According to [13], and [7], one has, for any $(\mathcal{F}_{t+})$ stopping time T :

$$\text{(2.a)} \qquad \frac{dP_a^\lambda}{dP_a^\mu} = \left(\frac{\rho_T}{a}\right)^{\lambda-\mu} \exp\left(-\frac{\nu^2}{2} C_T\right), \quad \text{on } \mathcal{F}_{T+} \cap (T < \infty)$$

where :

$$C_t = \int_0^t ds\Big/\rho_s^2, \text{ and } \quad \lambda = (\mu^2 + \nu^2)^{1/2}.$$

(2.2) Confluent hypergeometric functions appear repeatedly in the main formulae below. We have conformed with the notations and definitions used in Abramowitz - Stegun ([1], p. 504 and following).

(2.3) The main result in this paper is the following

Theorem : For any $\mu \geq 0$, $\nu \geq 0$, $\alpha \geq 0$, and $a,c > 0$, one has :

$$(2.b) \qquad E_a^\mu\left[(1+\tilde{T}_c)^{-\alpha} \exp(-\frac{\nu^2}{2} C_{\tilde{T}_c})\right] = (\frac{a}{c})^{\lambda-\mu} \frac{\Lambda(\alpha + \frac{\lambda-\mu}{2} ; \lambda+1 ; \frac{a^2}{2})}{\Lambda(\alpha + \frac{\lambda-\mu}{2} ; \lambda+1 ; \frac{c^2}{2})},$$

where $\lambda = (\mu^2 + \nu^2)^{1/2}$, and $\Lambda = M$, if $a < c$; $= U$, if $a > c$.

In particular,

$$(2.b.1) \qquad E_a^\mu\left[(1+\tilde{T}_c)^{-\alpha}\right] = \frac{\Lambda(\alpha ; \mu+1 ; \frac{a^2}{2})}{\Lambda(\alpha ; \mu+1 ; \frac{c^2}{2})}$$

$$(2.b.2) \qquad E_a^0\left[\exp(-\frac{\nu^2}{a} C_{\tilde{T}_c})\right] = (\frac{a}{c})^\nu \frac{\Lambda(\frac{\nu}{2} ; \nu+1 ; \frac{a^2}{2})}{\Lambda(\frac{\nu}{2} ; \nu+1 ; \frac{c^2}{2})}$$

Before entering properly into the proof of the theorem, we remark that, if $\pi_{a,c}$ denotes the distribution of the $\mathbb{R}_+^2$-valued r.v. $\left[\log(1+\tilde{T}_c) ; C_{\tilde{T}_c}\right]$, under P_a^μ, then :

$$(2.c) \qquad \pi_{a,c} = \pi_{a,b} * \pi_{b,c}$$

for any b between a and c, proving at once the infinite divisibility of $\pi_{a,c}$. This is a probabilistic proof (and improvement) of Hartman's result ([4], p. 271-2), asserting that the right-hand side of (2.b.1), resp. (2.b.2), is the Laplace transform in α, resp. : $\frac{\nu^2}{2}$, of an infinitely divisible distribution on $\mathbb{R}_+$.
We also note that identity (2.c) is probabilistically easier understood after time-changing the Bessel process (ρ_t), with the inverse (τ_t) of (C_t).
It is well-known (cf. D. Williams [12] ; [7]) that :

$$(2.d) \qquad \rho_{\tau_t} = \exp\{\beta_t\}, \quad \text{under } P_a^\mu,$$

where (β_t) stands here for $BM^\mu_{\log a}$, a real-valued BM, with constant drift μ, starting at $(\log a)$. Using (2.d), one obtains :

$$(2.e) \qquad \begin{cases} C_{\tilde{T}_c} \stackrel{\text{def}}{=} \sigma(c) \equiv \inf\{t : \beta_t = \log c + \frac{1}{2}\log\left[1 + \int_0^t \exp(2\beta_s)ds\right] ; \\ \log(1+\tilde{T}_c) = \log\left[1 + \int_0^{\sigma(c)} ds \exp(2\beta_s)\right] \end{cases}$$

Formula (2.c) now appears as a consequence of the strong Markov property taken at time $\sigma(b)$, for $BM^{\mu}_{(\log a)}$.
We now proceed to the proof of the theorem, via two steps.

Step 1. We first prove formula (2.b.1).
The following notation will be helpful :

$$\tilde{I}_{\mu}(z) = (\frac{z}{2})^{-\mu}\, I_{\mu}(z) \; ; \; \tilde{K}_{\mu}(z) = (\frac{z}{2})^{-\mu}\, K_{\mu}(z).$$

Recall that, for any $\theta > 0$, the processes :

$$(2.f) \qquad (\tilde{I}_{\mu}(\theta\rho_t)\exp(-\frac{\theta^2}{2}t)\; ;\; t \geq 0), \text{ and } (\tilde{K}_{\mu}(\theta\rho_t)\exp(-\frac{\theta^2}{2}t),\; t \geq 0)$$

are two P^{μ}_a-local martingales, an assertion from which the Laplace transform of the law of $T_c \equiv \inf\{t : \rho_t = c\}$ under P^{μ}_a is easily deduced (cf. J. Kent [6]).
We now suppose that $a < c$, and deduce from (2.e) that :

$$E^{\mu}_a\left[\tilde{I}_{\mu}(\theta c\sqrt{1+\tilde{T}_c})\exp(-\frac{\theta^2}{2}\tilde{T}_c)\right] = \tilde{I}_{\mu}(\theta a).$$

Then, following Shepp's method [8], we integrate both sides of the previous equality with respect to $d\theta\, e^{-\theta^2/2}\cdot\theta^p$, and obtain, after the change of variables $\theta' = \theta\sqrt{1+\tilde{T}_c}$:

$$(2.g) \qquad E^{\mu}_a\left[(1+\tilde{T}_c)^{-\frac{1+p}{2}}\right] \quad u_p(c) = u_p(a),$$

where : $u_p(c) = \int_0^{\infty} d\theta\, e^{-\frac{\theta^2}{2}}\cdot\theta^p\cdot\tilde{I}_{\mu}(\theta c)$.

We now use the expansion : $\tilde{I}_{\mu}(z) = \sum_{n=0}^{\infty} (\frac{z}{2})^{2n}\frac{1}{n!}\frac{1}{\Gamma(\mu+n+1)}$, to obtain :

$$(2.h) \qquad u_p(c) = \frac{\Gamma(\alpha)}{\Gamma(\mu+1)} M(\alpha\; ;\; \mu+1\; ;\; \frac{c^2}{2})\, 2^{\alpha-1}, \text{ where : } \alpha = \frac{1+p}{2}.$$

This proves (2.b.1), as a consequence of (2.g) ; in the case $a > c$, we use the same method, with $\tilde{K}_{\mu}$ now replacing $\tilde{I}_{\mu}$.

Step 2. The complete formula (2.b) now follows from (2.b.1), using the explicit Radon-Nikodym density formula (2.a) for $T = \tilde{T}_c$.

Remark : In fact, formula (2.h) has a long history ; it is due to Hankel (cf. Watson [11], p. 384-394) and is a generalisation of formulae due to Lipschitz, Weber, and Gegenbauer ; at the beginning of the century, formula (2.h) has been frequently used by some physicists (again, see Watson [11], p. 385).

3. Another interpretation of the total winding for pole-seeking BM :

(3.1) For any $\delta > 0$, we introduce a new family of distributions $\{{}^{\delta}P_a^{\nu}\}$ on Ω. ${}^{\delta}P_a^{\nu}$ is the distribution of the $d \equiv 2(\nu+1)$ dimensional Bessel process, starting from a, with "naïve drift" δ, that is the distribution of the $\mathbb{R}_+$-valued diffusion with infinitesimal generator :

$$\text{(3.a)} \qquad A_\nu = \frac{1}{2}\frac{d^2}{dx^2} + \left(\frac{2\nu+1}{2x} - \delta\right)\frac{d}{dx}.$$

The introduction of a terminology such as "naïve drift" seems necessary, in order to avoid confusion with the diffusion obtained by taking the radial part of a $\mathbb{R}^d$-valued BM, started at the origin, with $\vec{\delta}$ ($\in \mathbb{R}^d$) ; this latter diffusion is usually called Bessel process with drift $\delta = |\vec{\delta}|$ (cf. Shiga - Watanabe [9] ; Watanabe [10] ; [7]).

(3.2) In the course of his mathematical study of Bird Navigation, D. Kendall [5] obtained some remarkable formulae (see formula (32) and (34), p. 384 of [5]) from which the following is easily deduced : for $c' < a'$,

$$\text{(3.b)} \qquad {}^{\delta}E^{o}_{a'}\left(\exp - \frac{\nu^2}{2} C_{T_{c'}}\right) = \left(\frac{a'}{c'}\right)^{\nu} \frac{U(\nu\,;\,2\nu+1\,;\,2\delta a')}{U(\nu\,;\,2\nu+1\,;\,2\delta c')},$$

where $T_{c'} = \inf\{t : \rho_t = c'\}$.
This formula (3.b), when compared with (2.b.2), immediately shows the distributional identity :

$$\text{(3.c)} \qquad (4C_{\tilde{T}_c}\,;\,P^{o}_{a}) \overset{(d)}{\equiv} (C_{T_{c'}}\,;\,{}^{\delta}P^{o}_{a'}),$$

where :

$$a^2 = 4\delta a' \;;\; c^2 = 4\delta c'.$$

We recall that both the total windings of the complex valued BM, and of the pole-seeking BM introduced by D. Kendall may be written as $(\beta_{C_t}\,;\,t \geq 0)$, where, in each case, (β_t) is a real-valued BM started at 0, independent from the radial part of the process. Therefore, formula (3.c) immediately transforms into a distributional identity between the total windings of these two complex valued processes.

(3.3) We now give a probabilistic interpretation of the identity (3.c). Since we have already given an interpretation of $C_{\tilde{T}_c}$ as a hitting time for BM (cf., formula (2.e) above), it remains to do likewise for $C_{T_{c'}}$, under ${}^{\delta}P^{o}_{a'}$, which finally amounts to studying the process (ρ_{τ_t}), where (τ_t) denotes the inverse of (C_t), under ${}^{\delta}P^{o}_{a'}$. Recall that (ρ_t) satisfies the equation :

$$\rho_t = a' + \beta_t + \frac{1}{2}\int_0^t \frac{ds}{\rho_s} - \delta t,$$

where (β_t) is a real-valued BM.

With the help of Itô's formula, one easily deduces that $Y_t \equiv \frac{1}{\rho_{\tau_t}}$ satisfies :

$$\text{(3.d)} \qquad Y_t = \frac{1}{a'} - \int_0^t Y_s \, d(\gamma_s - \frac{s}{2}) + \delta t,$$

with (γ_t) a new BM_o. The method of variation of constants now gives the explicit formula :

$$Y_t = \exp(-\gamma_t)\{\frac{1}{a'} + \delta\int_0^t \exp(\gamma_s)ds\}.$$

Therefore, under ${}^{\delta}P^{o}_{a'}$, one has :

$$\begin{aligned} C_{T_{c'}} &= \inf\{t : Y_t = \frac{1}{c'}\} = \inf\{t : \log Y_t = -\log c'\} \\ &= \inf\{t : \gamma_t - \log\left[\frac{1}{a'} + \delta\int_0^t \exp(\gamma_s)ds\right] = \log c'\} \\ &= \inf\{t : \gamma_t = \log(\frac{c'}{a'}) + \log\left[1 + (\delta a')\int_0^t \exp(\gamma_s)ds\right]\}. \end{aligned}$$

On the other hand, one has, under P^{o}_{a}, from formula (2.e) :

$$C_{\tilde{T}_c} \overset{(d)}{=} \inf\{t : \gamma_t = \log(\frac{c}{a}) + \frac{1}{2}\log\left[1 + a^2\cdot\int_0^t ds \exp(2\gamma_s)\right]\},$$

and therefore, as $\tilde{\gamma}_t = 2\gamma_{t/4}$ is a new BM :

$$4C_{\tilde{T}_c} \overset{(d)}{=} \inf\{t : \tilde{\gamma}_t = \log\left[(\frac{c}{a})^2\right] + \log\left[1 + \frac{a^2}{4}\int_0^t dv \exp(\tilde{\gamma}_v)\right]\},$$

and the proof of (3.c) is ended.

(3.4) The following extension of D. Kendall's formulae (32) and (34) in [5] has been obtained in ([7] ; section 12) : for any $\nu \geq 0$, $\delta > 0$, $\mu > 0$, $\theta \geq 0$,

$$ {}^{\delta}E_a^{\nu}\left[\exp\left(-\frac{\mu^2}{2} C_{T_b} - \frac{\theta^2}{2} T_b\right)\right] = \left(\frac{b}{a}\right)^{\nu+\frac{1}{2}} \cdot \exp \delta(b-a) \cdot \frac{\mathcal{L}_{k\lambda}(2\hat{\theta}a)}{\mathcal{L}_{k\lambda}(2\hat{\theta}b)}, $$

where :

$$ \lambda = (\mu^2 + \nu^2)^{1/2} \; ; \; \hat{\theta} = (\theta^2 + \delta^2)^{1/2} \; ; \quad k = \frac{2\nu+1}{2} \cdot \frac{\delta}{\hat{\theta}}, $$

and :

$$ \mathcal{L}_{k\lambda} = M_{k\lambda}, \text{ if } a < b \; ; \; = W_{k\lambda}, \text{ if } a > b. $$

In particular, for $\theta = 0$, one gets :

$$ {}^{\delta}E_a^{\nu}\left[\exp - \frac{\mu^2}{2} C_{T_b}\right] = \left(\frac{a}{b}\right)^{\lambda-\nu} \frac{\Lambda(\lambda-\nu \; ; \; 1+2\lambda \; ; \; 2\delta a)}{\Lambda(\lambda-\nu \; ; \; 1+2\lambda \; ; \; 2\delta b)}. $$

The comparison of this formula with (2.b.2) implies the following extension of (3.c):

(3.e) $$ (4C_{\tilde{T}_c} \; ; \; P_a^{2\nu}) \overset{(d)}{\equiv} (C_{T_{c'}} \; ; \; {}^{\delta}P_{a'}^{\nu}) $$

where :

$$ a^2 = 4\delta a' \; ; \; c^2 = 4\delta c'. $$

The proof given in (3.3) for the identity (3.c) is still valid for (3.e), provided the process (γ_t) in (3.d) now stands for BM_0^{ν}, a real-valued Brownian motion, started at 0, with constant drift ν.

REFERENCES :

[1] M. ABRAMOVITZ, I. STEGUN : Handbook of Mathematical Functions. New-York - Dover - 1970.

[2] M.T. BARLOW, S.D. JACKA, M. YOR : Inequalities for a couple of processes stopped at an arbitrary random time. To appear (1983).

[3] B. DAVIS : On the L^p norms of stochastic integrals and other martingales. Duke Math. Journal, vol. 43, n° 4, 697-704 (1976).

[4] P. HARTMAN : Uniqueness of principal values, complete monotonicity of logarithmic derivatives of principal solutions and oscillation theorems. Math. Ann. 241, 257-281 (1979).

[5] D. KENDALL : Pole-seeking Brownian Motion and Bird Navigation. Journal of the Royal Statistical Society. Series B, 36, n° 3, p. 365-417, 1974.

[6] J. KENT : Some probabilistic properties of Bessel functions. Ann. Prob. 6, 760-770 (1978).

[7] J. PITMAN, M. YOR : Bessel processes and Infinitely divisible laws. In : "Stochastic Integrals". Lecture Notes in Maths 851. Springer (1981) (ed. D. Williams).

[8] L. SHEPP : A first passage problem for the Wiener process. Ann. Math. Stat. 38 (1967), p. 1912-1914.

[9] T. SHIGA, S. WATANABE : Bessel diffusions as a one-parameter family of diffusion processes. Z.f.W, 27 (1973), 37-46.

[10] S. WATANABE : On Time Inversion of One-Dimensional Diffusion processes. Zeitschrift für Wahr. 31 (1975), 115-124.

[11] G.N. WATSON : A treatise on the theory of Bessel functions. Second edition. Cambridge University Press (1966).

[12] D. WILLIAMS : Path-decomposition and continuity of local time for one-dimensional diffusions, I Proc. London Math. Soc., Ser. 3, 28, 738-768 (1974).

[13] M. YOR : Loi de l'indice du lacet Brownien, et distribution de Hartman-Watson. Z.f.W, 53, 71-95 (1980).

DISTRIBUTIONAL APPROXIMATIONS FOR NETWORKS OF QUASIREVERSIBLE QUEUES

P.K. Pollett
Department of Mathematical Statistics
and Operational Research
University College
Cardiff CF1 1XL
Great Britain

ABSTRACT.

This paper is concerned with establishing Poisson approximations to flows in general queueing networks. Bounds are provided to assess the departure of a given flow from Poisson and these lead to simple criteria for good Poisson approximations. The class of networks considered here are those with a countable collection of customer classes and where the service requirement of a customer at a given queue has a general distribution which may depend upon the class of the customer.

KEYWORDS.

Queueing networks, Poisson Approximations.

1. INTRODUCTION.

In a recent paper, Brown and Pollett (1982) exhibited a method for approximating customer flow processes in single class queueing networks with exponential service requirements and servers with state-dependent rates. The distance of customer flows from Poisson processes was estimated using formulas derived by Brown (1982) for general point processes.

It is the purpose of the current exposition to extend their results to a class of quasireversible networks with customers of different classes and associated general service requirements. Bounds are provided to assess the degree of deviation of arrival processes from suitably chosen Poisson processes. Although the arithmetic values of these bounds are of doubtful practical significance, they are of some theoretical interest and give rise to simple criteria for good Poisson approximations. These criteria tend to fall into three categories: light traffic, heavy traffic and evenly distributed customer routing. However, in contrast to the situation where service requirements are exponential (Brown and Pollett (1982)), the heavy traffic approximation seems only to be possible if service effort is distributed evenly among all customers present in a given queue (the server sharing discipline).

In section 2 a standard notation is defined and various preliminary results on queueing networks are collected. Sections 3 and 4 are devoted to discussing Poisson approximations to arrival processes in both open and closed networks of symmetric queues.

2. NOTATION AND PRELIMINARY RESULTS.

Let $\mathcal{N}$ denote a multiclass network consisting of J queues $\{1,2,\ldots,J\}$ (with J possibly infinite) and a countable set of customer classes, $\mathcal{C}$. If customers are allowed to enter or leave the network it is said to be *open*;

otherwise, there is a fixed number of customers of each class and the network is said to be *closed*. In the open case we suppose that arrivals from outside the network occur as independent Poisson streams, the class c arrival stream at queue j having a bounded rate of $\nu_j(c)$. Define for each c in C a *routing matrix* $\Lambda(c) = (\lambda_{jk}(c))$ to be the collection of probabilities that govern internal transitions from queues j to k for customers of class c, and let $\lambda_{j0}(c) = 1-\sum_{k=1}^{J}\lambda_{jk}(c)$ be the probability that after completion of service at queue j a class c customer leaves the network. If N is closed, $\lambda_{j0}(c)$ is taken to be zero for each j and c.

In the open case define $\underset{\sim}{\alpha}(c) = (\alpha_1(c),\alpha_2(c),\ldots.,\alpha_J(c))$ to be a vector with non-negative entries that satisfies

(1) $$\underset{\sim}{\alpha}(c) = \underset{\sim}{\nu}(c) + \underset{\sim}{\alpha}(c)\Lambda(c).$$

In order that this vector be unique, we assume that it is possible for any class c customer to eventually leave the network either directly or indirectly via some sequence of queues. The quantity $\alpha_j(c)$ may be interpreted as the equilibrium arrival rate for class c customers at queue j and will be positive if it is possible for such customers to visit the queue.

In the closed case we suppose that $\Lambda(c)$ is irreducible and non-null persistent. This ensures that there exists a unique (up to a constant multiple) vector with positive entries that satisfies

(2) $$\underset{\sim}{\alpha}(c) = \underset{\sim}{\alpha}(c)\Lambda(c)$$

and it will be of no loss in generality to assume that $\sum_{j=1}^{J}\alpha_j(c) = 1$. By Chang and Laverberg (1974) the quantity $\alpha_j(c)/\alpha_k(c)$ is the ratio of the class c arrival rates at queues j and k.

We suppose that each queue in the network is *symmetric* (Kelly (1976)), that is, each queue j in N operates as follows:

(i) A total service effort is offered at a rate $\phi_j(n_j)$ (units per second) when there are n_j customers present;

(ii) A proportion $\gamma_j(\ell,n_j)$ of this effort is directed to the customer occupying queue position ℓ; when this customer leaves the queue, customers in positions $\ell+1$, $\ell+2,\ldots.,n_j$ move into positions ℓ, $\ell+1,\ldots.,n_j-1$ respectively;

(iii) When a customer arrives he chooses to occupy position ℓ in the queue with probability $\gamma_j(\ell,n_j+1)$; customers previously in positions ℓ, $\ell+1,\ldots.,n_j$ move into positions $\ell+1,\ell+2,\ldots.,n_j+1$ respectively.

For each j in $\{1,2,\ldots.,J\}$ we assume that $\phi_j(0) = 0$ and for $n>0$, $\phi_j(n)>0$ and $\sum_{\ell=1}^{n} \gamma_j(\ell,n) = 1$. The fact that the same function γ_j is used in both (ii) and (iii) places some restriction upon the types of possible service discipline. However, it allows service requirements to take a quite general form without making equilibrium analysis unmanagable. We suppose that successive service requirements for customers of class c at queue j are i.i.d. random variables with distribution function $F_{jc}(x)$ and mean $\mu_j^{-1}(c)$. Thus, when there are n_j customers present at queue j the rate at which class c customers are served is $\mu_j(c)\phi_j(n_j)$ (customers per second).

Let $\underset{\sim}{x}(t) = (x_1(t),x_2(t),\ldots.,x_J(t))$ be a Markov process that describes the network N and that contains enough information for one to deduce the number of customers in each queue and the classes of each of them. In particular, when queue j is symmetric we let $x_j = (n_j; x_j(1), x_j(2), \ldots, x_j(n_j))$ where $x_j(\ell) = (c_j(\ell), z_j(\ell), u_j(\ell))$ describes the customer in queue position ℓ. Here $c_j(\ell)$ is the class of the customer $z_j(\ell)$ is his service requirement and $u_j(\ell)$ is the amount of service so far received. In general $\underset{\sim}{x}(t)$ will have a continuous state space. However, if each of the F_{jc}, $c \in \mathcal{C}$, admit a *Cox-phase representation* (Cox (1955)), for example if F_{jc} is Hyperexponential or a mixture of Gamma distributions, it is sometimes convenient to let $z_j(\ell)$ and $u_j(\ell)$ determine, respectively, the total number of (fictitious) stages of service and the number of stages reached. In this case the state space will be countable.

For each j in $\{1,2,\ldots,J\}$ and c in $\mathcal{C}$ let $a_j(c) = \alpha_j(c)\mu_j^{-1}(c)$, the average amount of service required by class c customers arriving in queue j, and let $a_j = \sum_{c \in \mathcal{C}} a_j(c)$, the total average requirement. For the closed network let N(c) be the total number of class c customers and define $\underset{\sim}{N} = (\ldots, N(c), \ldots)$ to be the vector which determines the number of customers of each class in the network. Denote the vector with m fewer customers of class c as $\underset{\sim}{N}_c^m$. Define $n_j(c)$ to be the number of class c customers at queue j and let

$$\zeta_N^J = \{(x_1, x_2, \ldots, x_J) : \sum_{j=1}^{J} n_j(c) = N(c),\ c \in \mathcal{C}\}$$

denote the state space of $\underset{\sim}{x}(t)$.

The following results summarise some of the important equilibrium properties of the network consisting of symmetric queues. Lemma 1 is a direct consequence of Theorems 3.7(i) and 3.10 of Kelly (1979).

Lemma 1.

For the open multiclass network N consisting of symmetric queues, an equilibrium distribution exists for $\underset{\sim}{x}$ *if and only if* for all j in {1,2,...,J},

$$b_j^{-1} = \sum_{n=0}^{\infty} a_j^n / \{ \prod_{r=1}^{n} \phi_j(r) \} < \infty.$$

In equilibrium the states of the individual queues are *independent* with queue j having the following properties:

(i) The probability that queue j contains n customers is

$$\pi_j(n) = b_j a_j^n / \{ \prod_{r=1}^{n} \phi_j(r) \}$$

(ii) Given the number of customers in the queue, the classes of customers are *independent* and the probability that the customer in a given position is of class c is

$$a_j(c)/a_j$$

(iii) Given the number of customers in the queue and the classes of each of them together with their service requirements, the amounts of service already received are *independent* with $u_j(\ell)$ uniformly distributed on $(0, z_j(\ell))$

(iv) Given the number of customers in the queue and the classes of each of them, the amounts of service already received are *independent* and the probability a customer of class c has received an amount of service effort not greater than x is

$$(F_{jc})_e(x) = \mu_j(c) \int_0^x (1 - F_{jc}(y))\,dy,$$

the *residual life distribution* corresponding to F_{jc}.

<u>Lemma 2.</u>

The closed multiclass network N consisting of symmetric queues has the following equilibrium properties:

(i) The joint distribution for the numbers of customers in each queue together with the classes of each of them is proportional to

$$\prod_{j=1}^{J} \prod_{\ell=1}^{n_j} \frac{a_j(c_j(\ell))}{\phi_j(\ell)}$$

(ii) Given the numbers of customers in each queue and the classes of each of them, the service requirements are *independent* and if the customer in position ℓ at queue j is of class c, the probability that his service requirement does not exceed x is given by

$$\mu_j(c) \int_0^x z dF_{jc}(z)$$

(iii) Given the numbers of customers in each queue and the classes of each of them together with their service requirements, the amounts of service already received are *independent* with $u_j(\ell)$ uniformly distribution on $(0, z_j(\ell))$

(iv) Given the numbers of customers in each queue and the classes of each of them, the amounts of service already received are *independent* and if the customer in position ℓ at queue j is of class c, the probability that the amount of service he has already received does not exceed x is $(F_{jc})_e(x)$.

Lemma 2 is proved by showing that the equilibrium distribution for $\underset{\sim}{x}$ has p.d.f. given by

$$\pi^{(\underset{\sim}{N})}(\underset{\sim}{x}) = B_{\underset{\sim}{N}} \prod_{j=1}^{J} \prod_{\ell=1}^{n_j} \frac{\alpha_j(c_j(\ell))}{\phi_j(\ell)} du_j(\ell) dF_{jc_j(\ell)}(z_j(\ell))$$

where $B_{\underset{\sim}{N}}$ is chosen so that the $\pi^{(\underset{\sim}{N})}(\underset{\sim}{x})$, $\underset{\sim}{x} \in \zeta^J_{\underset{\sim}{N}}$, sum to unity. This follows from Theorem 3.12(i) of Kelly (1979) and the result of Barbour (1976). Properties (i), (ii) and (iii) follow from integrating over $u_j(\ell)$ and $z_j(\ell)$, $\ell=1,2,\ldots,n_j$, for each j, and then using the appropriate conditioning. Property (iv) may be deduced from (ii) and (iii) by showing that given the numbers of customers in each queue and the classes of each of them, the joint distribution for the amounts of service already received as a function of $v_j(\ell)$, $\ell=1,2,\ldots,n_j$, $j=1,2,\ldots,J$, is

$$\prod_{j=1}^{J} \prod_{\ell=1}^{n_j} \mu_j(c_j(\ell)) \int_{v_j(\ell)}^{\infty} \int_{v_j(\ell)}^{z_j(\ell)} du_j(\ell)\, dF_{jc_j(\ell)}(z_j(\ell)).$$

The result then follows after reversing the order of integration.

Remarks.

If a phase representation is used for the state of each queue then parts (i), (ii) and (iii) of lemma 1 and parts (i) and (iii) of lemma 2 hold good with $z_j(\ell)$ being the total number of stages corresponding to the class of the customer occupying position ℓ and $u_j(\ell)$ counting the number of stages reached.

Observe that in the closed case we may not deduce the independence of customer classes given the numbers of customers in each queue. However, part (i) of lemma 2 may be used to establish the identity,

$$(3) \qquad P\{n_j = n; c_j(\ell) = c\} = \frac{a_j^{(\underset{\sim}{N})}(c)}{\phi_j(n)} \pi_j^{(\underset{\sim}{N}^1_c)}(n-1)$$

for all ℓ in $\{1,2,\ldots,n\}$, where $\pi_j^{(\underset{\sim}{N})}(n)$ is the marginal distribution for n_j in the network whose customer numbers are determined by $\underset{\sim}{N}$ and $a_j^{(\underset{\sim}{N})}(c)=a_j(c)B_{\underset{\sim}{N}}/B_{\underset{\sim}{N}^1_c}$ is the average amount of service requirement arriving at queue j. By using the same method we obtain a slightly more general identity,

(4) $$P\{n_j = n,\ c_j(\ell) = c;\ E\} = \frac{a_j^{(\underset{\sim}{N})}(c)}{\phi_j(n)}\ P^{(\underset{\sim}{N}_c^1)}\{n_j = n-1; E\}$$

where E is any event which does not depend explicitly on n_j or $x_j(\ell)$, and the probability on the right hand side pertains to a network whose customer numbers are determined by $\underset{\sim}{N}_c^1$.

Observe that the amounts of service effort already received are distributed in accordance with the equilibrium age distributions for independent renewal processes, each being constructed from successive i.i.d. service times for customers of a given class at a particular queue. A partial generalisation of this result may be found in Pollett (1983).

The basis for approximation results presented here is the result of Brown (1982) which states that if (ξ, F) is a point process with conditional intensity η and Π is a Poisson process with rate λ then for all $t>0$,

(5) $$d(\xi^t, \Pi^t) \leq \int_0^t E|\eta(s) - \lambda(s)|\,ds$$

where $d(\cdot,\cdot)$ is the *total variation distance* between two probability distributions, in this case between ξ^t and Π^t, the distributions of ξ and Π respectively on the interval $[0,t]$.

When applying this result to establish approximations to flows in queueing networks it is often convenient to choose $\lambda = E\eta$ and further, to simplify computation and interpretation of bounds, we will use the slightly weaker bound,

(6) $$d(\xi^t, \Pi^t) \leq \int_0^t (\mathrm{Var}\ \eta(s))^{\frac{1}{2}}\,ds$$

which follows since the L_p norm increases with p. We call bounds given by (5) L_1 bounds and those given by (6) L_2 bounds.

3. OPEN NETWORKS.

In this section we consider the open multiclass network that consists of symmetric queues and initially we assume general service requirements. We are concerned with the point processes that govern the circulation of customers within the network. Consider specifically the aggregate flow into queue j. Define $\xi_{jc}(t)$ to be the number of customers of class c to arrive at queue j on (0,t]. Let $F(t)$ be the completion of the σ-algebra generated by the path of $\underset{\sim}{x}$ on [0,t]. It is easy to see that the conditional intensity, η_{jc}, of the point process (ξ_{jc}, F) is given by

$$(7) \qquad \eta_{jc}(t) = \nu_j(c) + \sum_{k=1}^{J} \lambda_{kj}(c) d_{kc}(t),$$

where

$$(8) \qquad d_{kc}(t) = \sum_{n=1}^{\infty} \phi_k(n) \sum_{\ell=1}^{n} \gamma_k(\ell,n) \int_0^{\infty} g_{kc}(x) I[x<u_k(\ell) \leqq x+dx, c_k(\ell)=c, n_k(t)=n]$$

$$= \phi_k(n_k(t)) \sum_{\ell=1}^{n_k(t)} \gamma_k(\ell,n_k(t)) g_{kc}(u_k(\ell)) I[c_k(\ell)=c]$$

is the conditional intensity of the departure process from queue k and $g_{kc}(x)$ is the so-called *age specific failure rate* corresponding to $F_{kc}(x)$ defined by

$$g_{kc}(x) = f_{kc}(x)/(1-F_{kc}(x)),$$

where f_{kc} is the density of F_{kc}. Roughly speaking, $g_{kc}(x)$ gives the probability of the almost immediate completion of service of a class c customer at queue k who has already received x units of service effort.

The equilibrium expectation of η_{jc} may be calculated by applying lemma 1 and expression (1) and by observing that in equilibrium

$$Ed_{kc}(t) = \sum_{n=1}^{\infty} \phi_k(n)\pi_k(n) \sum_{\ell=1}^{n} \gamma_k(\ell,n) P\{c_k(\ell) = c | n_k(t) = n\}$$

$$= \int_0^{\infty} g_{kc}(x) P\{x<u_k(\ell) \leq x+dx | c_k(\ell) = c, n_k(t) = n\}.$$

We obtain $E\eta_{jc}(t) = \alpha_j(c)$ and thus we may approximate ξ_{jc} by a Poisson process, Π_{jc}, with rate $\alpha_j(c)$. The next theorem establishes an L_2 bound for the approximation.

Theorem 1.

Consider the open network N in equilibrium. Suppose that the arrival process of customers of class c at queue j is ξ_{jc} and that Π_{jc} is a Poisson process with rate $\alpha_j(c)$. Then for all $t \geq 0$, c in C and j in $\{1,2,\ldots,J\}$,

$$d(\xi_{jc}^t, \Pi_{jc}^t) \leq t\left(\sum_{k=1}^{J} \mu_k^2(c)\lambda_{kj}^2(c) a_k(c)\left(\sum_{n=1}^{\infty} \phi_k(n)\pi_k(n-1)\right.\right.$$

(9)

$$\left.\left.\left(\frac{a_k(c)}{a_k} + \left(I_k(c) - \frac{a_k(c)}{a_k}\right)\sum_{\ell=1}^{n} \gamma_k^2(\ell,n)\right) - a_k(c)\right)\right)^{\frac{1}{2}}$$

where

$$I_k(c) = \mu_k^{-1}(c) \int_0^{\infty} g_{kc}(x)\, dF_{kc}(x)$$

Proof.

Fix j in $\{1,2,\ldots,J\}$ and c in C. Using an argument similar to that used to calculate $Ed_{kc}(t)$ we can show that

$$Ed_{kc}^2(t) = \alpha_k(c)\left(\sum_{n=1}^{\infty} \phi_k(n)\pi_k(n-1)\left(\frac{\alpha_k(c)}{a_k} + \left(\mu_k(c)I_k(c) - \frac{\alpha_k(c)}{a_k}\right)\sum_{\ell=1}^{n} \gamma_k^2(\ell,n)\right)\right.$$

The bound follows from expression (6) and the fact that

$$\operatorname{Var} \eta_{jc}(t) = \sum_{k=1}^{J} \lambda_{kj}^2(c) \operatorname{Var} d_{kc}(t)$$

is independent of t in equilibrium.

The bound in theorem 1 is not explicit, but depends upon the choice of ϕ_k and γ_k. For example, if $\gamma_k(n,n) = 1$ and $\phi_k(n) = 1$, $n>0$, for all k in $\{1,2,\ldots,J\}$, each queue has a single server operating at unit rate with a *pre-emptive resume last come first served* discipline and we have that

$$(10) \qquad d(\xi_{jc}^t, \Pi_{jc}^t) \leq t\left(\sum_{k=1}^{J} \mu_k^2(c)\lambda_{kj}^2(c)a_k(c)(I_k(c)-a_k(c))\right)^{\frac{1}{2}}.$$

This is very similar to the bound obtained in theorem 1 of Brown and Pollett (1982) and is quite easy to interpret. Define $q_{kj}(c) = \mu_k(c)\lambda_{kj}(c)$, the rate at which class c customers flow from queues k to j when a class c service is in progress at queue k. The bound will be small if each term in the sum is small compared with J^{-1} and thus good Poisson approximations will obtain if $q_{kj}(c)$ is $o(J^{-\frac{1}{2}})$ or if $a_k(c)$ is $o(Jq_{kj}^2(c))^{-1})$. However, if class c traffic is large $(a_k(c) \to 1)$ we would not expect a good approximation unless $q_{kj}(c)$ is quite small. This is borne out by considering the nature of $I_k(c)$. By the Cauchy-Schwartz inequality we have

$$I_k(c) = \int_0^\infty 1-F_{kc}(x)dx \int_0^\infty \frac{f_{kc}^2(x)dx}{1-F_{kc}(x)} \geq \int_0^\infty f_{kc}(x)dx = 1,$$

with equality obtained *if and only if* F_{kc} is exponential. Thus $a_k(c) \sim 1$ for each k does not necessarily imply that the bound is small and a heavy traffic approximation is possible only if class c service requirements are exponential.

The above discussion is based on arrival rates, $\alpha_j(c)$, which are only implicitly defined. However, if we observe that $\nu_k(c) \leqq \alpha_k(c) < \mu_k(c)$ then by virtue of expression (1) we may write

$$\nu_j(c) + \sum_{k=1}^{J} \nu_k(c)\lambda_{kj}(c) \leqq \alpha_j(c) < \nu_j(c) + \sum_{k=1}^{J} \mu_k(c)\lambda_{kj}(c)$$

which leads to bounds that are defined explicitly in terms of the parameters defining the network;

<u>Corollory 1</u>.

For the open network consisting of queues with single servers operating at unit rate with a pre-emptive resume last come first served discipline we may delimit three cases in which the class c arrival process at queue j is approximated well by a Poisson process. They are

(i) *Small transition rates* - for example, in relatively large networks with even routing and moderate service requirements. A simple measure of the quality of the approximation is

$$d(\xi_{jc}^t, \Pi_{jc}^t) \leqq t\left(\sum_{k=1}^{J} \mu_k^2(c)\lambda_{kj}^2(c)\right)^{\frac{1}{2}}$$

(ii) *Light traffic* - for example, in networks with sparse exogenous input and small service requirements. A simple bound is

$$d(\xi_{jc}^t, \Pi_{jc}^t) \leqq t\left(\sum_{k=1}^{J} \mu_k(c)\lambda_{kj}^2(c)\left(\nu_k(c) + \sum_{\ell=1}^{J} \mu_\ell(c)\lambda_{\ell k}(c)\right)\right)^{\frac{1}{2}}$$

(iii) *Heavy traffic* - when class c service requirements are exponential and, for example, when exogenous input is heavy. An appropriate bound is

$$d(\xi_{jc}^t, \Pi_{jc}^t) \leqq t\left(\sum_{k=1}^{J} \mu_k(c)\lambda_{kj}^2(c)\left(\mu_k(c) - \nu_k(c) - \sum_{\ell=1}^{J} \nu_\ell(c)\lambda_{\ell k}(c)\right)\right)^{\frac{1}{2}}.$$

Remark.

It should be pointed out that light traffic at the given queue j is only possible if $\alpha_j(c)$, the class c arrival rate there, is small. Hence, although theorem and corollary remain true in this special case, they may reduce to the trivial statement that a stationary point process with a very small rate is almost Poissonian, an immediate property of the metric $d(\cdot,\cdot)$.

It has already been pointed out that the bound in theorem 1 is not explicit. However, by using maximum service efforts $\overline{\phi}_k = \max_{n\geq 1} \phi_k(n)$ and by observing that $\sum_{\ell=1}^{n} \gamma_k^2(\ell,n) \leq 1$ we obtain

$$d(\xi_{jc}^t, \Pi_{jc}^t) \leq t \left(\sum_{k=1}^{J} \overline{\mu}_k^{-2}(c)\lambda_{kj}^2(c)\overline{a}_k(c)(I_k(c)-\overline{a}_k(c)) \right)^{\frac{1}{2}}$$

where $\overline{\mu}_k(c) = \overline{\phi}_k\ \mu_k(c)$ is the maximal rate at which class c customers are served and $\overline{a}_k(c) = \alpha_k(c)\overline{\mu}_k^{-1}(c)$ is the corresponding minimal class c traffic intensity. Thus all the preceeding criteria will apply in the general case provided that we use extremal parameter values.

On the other hand, if service efforts are unbounded none of the criteria will apply. As an example of this consider the network consisting of infinite server queues obtained by setting $\phi_k(n) = n$ for each k. In this case we have

$$d(\xi_{jc}^t, \Pi_{jc}^t) \leq t \left(\sum_{k=1}^{J} \mu_k^2(c)\lambda_{kj}^2(c)a_k(c)I_k(c) \right)^{\frac{1}{2}}$$

Notice that the choice $\gamma_k(\ell,n) = 1/n$, $\ell=1,2,\ldots,n$, minimizes $\sum_{\ell=1}^{n}\gamma_k^2(\ell,n)$ subject to $\sum_{\ell=1}^{n}\gamma_k(\ell,n) = 1$. Therefore, since $I_k(c) - a_k(c)/a_k$ is positive, this choice also minimises the bound in theorem 1. If in addition we have

$\phi_k(n) = 1$, $n>0$, for each k in $\{1,2,\ldots,J\}$, all the queues are *server sharing* and the service effort is distributed evenly among all customers in the queue with a customer's remaining service requirement decreasing at a rate of 1/n (units per second). In this case the bound reduces to

$$d(\xi^t_{jc},\Pi^t_{jc}) \leq t\left(\sum_{k=1}^{J} \mu_k^2(c)\lambda_{kj}^2(c)a_k(c)\left[a_k(c)\left(\frac{1}{a_k}-1\right)\right.\right.$$
$$\left.\left.+\left(I_k(c) - \frac{a_k(c)}{a_k}\right) \log\left(\frac{1}{1-a_k}\right)^{(1-a_k)}\right]\right)^{\frac{1}{2}},$$

which unlike (10) depends through a_k, the net traffic intensity, upon parameters pertaining to other customer classes. Now as $a_k \to 0$ or 1, the *log* term tends to zero. Thus, since $a_k(c) \leq a_k$, the bound tends to zero as $a_k \to 1$ for all $c \in \mathcal{C}$ and a heavy traffic approximation is possible as long as $I_k(c)$ is finite for all k and c.

All the preceeding bounds are given in terms of the $I_k(c)$, $k=1,2,\ldots,J$, quantities that depend solely upon the class c service requirement distributions. In certain cases it is possible to calculate $I_k(c)$ explicitly. For example, if F_{kc} is the *Weibull* distribution with r-th moment $\lambda^{-r}\Gamma(1+rq^{-1})$ then $I_k(c)$ has value $\Gamma(q^{-1})\Gamma(2-q^{-1})$. In most cases, however, it is necessary to resort to numerical means or alternatively obtain bounds for $I_k(c)$. One case in point is the Gamma distribution. Suppose F_{kc} is a Gamma distribution with mean $K\lambda^{-1}$ and variance $K\lambda^{-2}$ so that a class c customer visiting queue k requires K independent exponentially distributed stages of service, each with mean λ^{-1}. In this case $I_k(c)$ depends only upon K and if K is an integer we have

$$I_k(c) = \int_0^\infty \frac{Ky^{2(K-1)}e^{-y}}{[(K-1)!]^2}\left(\sum_{\ell=0}^{K-1} \frac{y^\ell}{\ell!}\right)^{-1} dy$$

which may conveniently be calculated using Gauss-Laguerre quadrature. It is possible to usefully bound this expression by replacing the summation term by $y^{(K-1)}/(K-1)!$ and e^y to obtain respectively the upper and lower bounds

$$(11) \qquad \binom{2K-2}{K-1} \frac{K}{2^{2K-1}} \leq I_k(c) \leq K .$$

Using Stirling's approximation the left-hand side is, for large K, $\frac{1}{2}(K/\pi)^{\frac{1}{2}}$ which tends to infinity with K. Thus, when F_{kc} is deterministic $I_k(c)$ is finite.

Since the Gamma distribution admits a Cox-phase representation, it might be more appropriate to establish a bound using the state description which records the number of stages reached. To do this let $z_j(\ell) = K_j(c_j(\ell))$ where $K_j(c)$ is the number of exponential stages of service required by class c customers at queue j. In this case the conditional intensity, $\eta_{jc}(t)$, is given by (7) but with

$$d_{kc}(t) = \phi_k(n_k(t)) \sum_{\ell=1}^{n_k(t)} \gamma_k(\ell, n_k(t)) \mu_k(c) K_k(c) I[c_k(\ell)=c, u_k(\ell)=K_k(c)].$$

Using lemma 1 and expression (1) we find that $E\eta_{jc}(t) = \alpha_j(c)$ and upon application of expression (6) we obtain the bound given by (9) but with $I_k(c)$ replaced by $K_k(c)$. Consideration of the right-hand inequality of (11) shows that this bound is weaker than that obtained using theorem 1 with equality *if and only if* service requirements are exponential.

4. CLOSED NETWORKS.

Again we restrict our attention to the case where N consists of symmetric queues and we will assume general service requirements. Suppose that $\underset{\sim}{x}(t)$ has a continuous state space description and that $\xi_{jc}(t)$ counts class c arrivals at queue j on (0,t]. Using the same history as before the conditional intensity of (ξ_{jc}, F) is

$$\eta_{jc}(t) = \sum_{k=1}^{J} \lambda_{kj}(c) d_{kc}(t)$$

where $d_{kc}(t)$, the conditional intensity of the departure process from queue k, is given by expression (8).

The equilibrium expectation of η_{jc} may be calculated using the identity (3), together with lemma 2 and expression (2). We obtain $E\eta_{jc}(t) = \mu_j(c) a_j^{(\underset{\sim}{N})}(c)$, a quantity which we will denote as $\alpha_j^{(\underset{\sim}{N})}(c)$.

The next theorem assesses the deviation of ξ_{jc} from a Poisson process with rate $\alpha_j^{(\underset{\sim}{N})}(c)$.

Theorem 2.

Consider the closed network N in equilibrium and let Π_{jc} be a Poisson process with rate $\alpha_j^{(\underset{\sim}{N})}(c)$. Then for all $t \geq 0$, c in C and j in $\{1,2,...,J\}$,

$$d(\xi_{jc}^t, \Pi_{jc}^t) \leq t\left(\alpha_j^{(\underset{\sim}{N})}(c)\{\alpha_j^{(\underset{\sim}{N}_c^1)}(c) - \alpha_j^{(\underset{\sim}{N})}(c)\} + \sum_{k=1}^{J} \mu_k^2(c)\lambda_{kj}^2(c) a_k^{(\underset{\sim}{N})}(c)\left[a_k^{(\underset{\sim}{N}_c^1)}(c)\{\Sigma_{kc}^2 - 1\} + I_k(c)\Sigma_{kc}^1\right]\right)^{\frac{1}{2}}$$

where

$$\Sigma_{kc}^1 = \sum_{n=1}^{N} \phi_k(n)\pi_k^{(\underset{\sim}{N}_c^1)}(n-1)\sum_{\ell=1}^{n} \gamma_k^2(\ell,n)$$

and

$$\Sigma_{kc}^2 = \sum_{n=2}^{N} \frac{\phi_k(n)}{\phi_k(n-1)} \pi_k^{(\underset{\sim}{N}_c^2)}(n-2)\left[1 - \sum_{\ell=1}^{n} \gamma_k^2(\ell,n)\right].$$

Proof.

Fix j in $\{1,2,...,J\}$ and c in C and consider the right-hand side of expression (6) with $\eta = \eta_{jc}$. Since the states of the individual queues are not independent we have

$$\text{Var } \eta_{jc}(t) = \sum_{k=1}^{J} \lambda_{kj}^2(c) \text{ Var } d_{kc}(t)$$

$$+ \sum_{k=1}^{J} \sum_{\substack{\ell=1 \\ \ell\neq k}}^{J} \lambda_{kj}(c)\lambda_{\ell j}(c)\text{Cov}\{d_{kc}(t), d_{\ell c}(t)\}.$$

Now using (3) and (4) together we can show that for $r_2 \neq r_1$,

$$P^{(\underset{\sim}{N})}\{n_j=n, c_j(r_2)=c_j(r_1)=c\} = \frac{a_j^{(\underset{\sim}{N})}(c)\, a_j^{(\underset{\sim}{N}_c^1)}(c)}{\phi_j(n)\phi_j(n-1)} \; \pi_j^{(\underset{\sim}{N}_c^2)}(n-2),$$

which in turn yields

$$E\, d_{kc}(t) d_{\ell c}(t) = \mu_k(c)\mu_\ell(c) a_k^{(\underset{\sim}{N})}(c)\, a_\ell^{(\underset{\sim}{N}_c^1)}(c) \qquad \ell \neq k$$

and

$$E\, d_{kc}^2(t) = \mu_k^2(c) a_k^{(\underset{\sim}{N})}(c)\{a_k^{(\underset{\sim}{N}_c^1)}(c)\textstyle\sum_{kc}^2 + I_k(c)\sum_{kc}^1\}.$$

The above expressions combine to give the required bound.

Although the bound in thoerem 2 is not explicit, it simplifies considerably in a number of special cases. For example, if each queue j has a single server operating at unit rate with a pre-emptive resume last come first served discipline, it is easy to see that $\sum_{kc}^2$ and $\sum_{kc}^1$ are 0 and 1 respectively and that the bound reduces to

$$d(\xi_{jc}^t, \Pi_{jc}^t) \leq t\left(\alpha_j^{(\underset{\sim}{N})}(c)\{\alpha_j^{(\underset{\sim}{N}_c^1)}(c) - \alpha_j^{(\underset{\sim}{N})}(c)\}\right.$$

$$\left. + \sum_{k=1}^{J} \mu_k^2(c)\lambda_{kj}^2(c) a_k^{(\underset{\sim}{N})}(c)\{I_k(c) - a_k^{(\underset{\sim}{N}_c^1)}(c)\}\right)^{\frac{1}{2}} .$$

This is clearly analogous to (10) in the same way that the bounds in theorems 1 and 2 of Brown and Pollett (1982) are analogous. The first term under the square root is clearly negative, so we may draw the same conclusions as those made following expression (10). Moreover, these conclusions are valid in the general case provided we use maximal service rates. To see this it is necessary to use the inequality

$$a_k^{(\underset{\sim}{N}_c^1)} \sum_{kc}^2 + \sum_{kc}^1 \leq \sum_{n=1}^{N} \phi_k(n)\pi_k^{(\underset{\sim}{N}_c^1)}(n-1)$$

and the reasoning used in the corresponding open case. Notice that in contrast to the open case the maximal service rate is never infinite.

Bounds in terms of explicit parameters are also possible. To deal with an analogue of corollary 1 we introduce quantities m(c) and M(c) which are respectively the minima and maxima over j of $\mu_j(c)$ and quantities g(c) and G(c) being respectively the minimum and maximum elements of $\Lambda(c)$. From (2) it is clear that the latter quantities bound the elements of $\underset{\sim}{\alpha}(c)$ and thus for all j in {1,2,....,J} and c in C we have

$$g(c)/M(c) \leq a_j(c) \leq G(c)/m(c).$$

This in turn, by virtue of lemma 2 (i), gives a bound for the normalising constant $B_{\underset{\sim}{N}}$ as

$$f_{\underset{\sim}{N}} \prod_{c\in C} \left(\frac{g(c)}{M(c)}\right)^{N(c)} \leq B_{\underset{\sim}{N}}^{-1} \leq f_{\underset{\sim}{N}} \prod_{c\in C} \left(\frac{G(c)}{m(c)}\right)^{N(c)}$$

where $f_{\underset{\sim}{N}}$ is the multinomial coefficient given by

$$f_{\underset{\sim}{N}} = (N+J-1)!/\{(J-1)! \prod_{c\in C} (N(c))!\}$$

and $N = \sum_{c\in C} N(c)$ is the total number of customers in the network. From the definition of $a_j^{(\underset{\sim}{N})}(c)$ we have, therefore,

$$\frac{N(c)r(c)^{-N(c)}}{N+J-1} \leq a_j^{(\underset{\sim}{N})}(c) \leq \frac{N(c)r(c)^{N(c)}}{N+J-1}$$

where $r(c) = G(c)M(c)/\{g(c)m(c)\}$. This ratio is a simple measure of the discrepancies among the rates of transition between queues for class c customers, since $q_{kj}(c) = \mu_k(c)\lambda_{kj}(c)$ lies between $g(c)m(c)$ and $G(c)M(c)$. Thus, we have

Corollary 2

For the closed network consisting of queues with single servers operating at unit rate with a pre-emptive resume last come first served discipline, we may delimit three cases where the class c arrival process at queue j is approximated well by a Poisson process. They are

(i) *Small transition rates* - for example, in large networks with even routing and moderate service requirements. In this case a simple bound is

$$d(\xi_{jc}^t, \Pi_{jc}^t) \le t\left(\sum_{k=1}^{J} \mu_k^2(c)\lambda_{kj}^2(c)\right)^{\frac{1}{2}}$$

(ii) *Light traffic* - for example, in networks with a moderate number of class c customers circulating and with moderate discrepancies among transition rates. Here an appropriate bound is

$$d(\xi_{jc}^t, \Pi_{jc}^t) \le t\left(\frac{N(c)r(c)^{N(c)}}{N+J-1} \sum_{k=1}^{J} \mu_k^2(c)\lambda_{kj}^2(c)\right)^{\frac{1}{2}}$$

(iii) *Heavy traffic* - when class c service requirements are exponential and, for example, when discrepancies among transition rates are moderate and the number of class c customers is relatively large. In this case a simple measure of the quality of the approximation is

$$d(\xi_{jc}^t, \Pi_{jc}^t) \le t\left(\left(1 - \frac{N(c)r(c)^{-N(c)}}{N+J-1}\right) \sum_{k=1}^{J} \mu_k^2(c)\lambda_{kj}^2(c)\right)^{\frac{1}{2}} .$$

To illustrate how all three of the above criteria may be satisfied, consider the closed multiclass network which exhibits *maximum similarity* (Brown and Pollett (1982)) with respect to each customer class. That is, we suppose that for each c in C $a_j(c) = a(c)$, a constant, for each j. This will certainly be true if, for each c, $\Lambda(c)$ is doubly stochastic and $\mu_j(c)$ is constant with respect to j. Upon application of lemma 2(i) we find that $a_j^{(\underset{\sim}{N})}(c) = N(c)/(N+J-1)$ for all j and assuming exponential service requirements for class c customers we obtain

$$d(\xi^t_{jc},\Pi^t_{jc}) \leq \frac{t\,\mu_j(c)N(c)}{N+J-1}\left(\frac{(N-N(c)+J-1)(N+J-1)}{N(c)(N+J-2)}\left(\sum_{k=1}^{J}\lambda^2_{kj}(c) - \frac{1}{N+J-1}\right)\right)^{\frac{1}{2}}.$$

This bound will be small if $t\,\mu_j(c)$ is moderate and we have either J large with N being o(J) corresponding to light traffic or, N large with both J and N(c) being o(J) corresponding to heavy traffic or finally, $\sum\lambda^2_{jk}(c)$ small, with N and J both being large, corresponding to balanced traffic.

Another case for which the bound of theorem 2 may be simplified is obtained by insisting that all queues have ample servers, that is, at least N. Setting $\phi_k(n) = n$ and $\gamma_k(\ell,n) = 1/n$ for each ℓ and n, it is clear that both $\sum^1_{kc}$ and $\sum^2_{kc}$ are unity and the bound reduces to

$$d(\xi^t_{jc},\Pi^t_{jc}) \leq t\left(\alpha_j^{(\underset{\sim}{N})}\{\alpha_j^{(\underset{\sim}{N}^1_c)}(c) - \alpha_j^{(\underset{\sim}{N})}(c)\} + \sum_{k=1}^{J}\mu^2_k(c)\lambda^2_{kj}(c)a_k^{(\underset{\sim}{N})}(c)I_k(c)\right)^{\frac{1}{2}}.$$

If c is the only customer class then it can be shown that

$$d(\xi^t_{jc},\Pi^t_{jc}) \leq t\left(\frac{N(c)a_j(c)}{\sum_{k=1}^{J}a_k(c)}\left(\sum_{k=1}^{J}\mu^2_k(c)\lambda^2_{kj}(c)I_k(c)\,\frac{a_k(c)}{a_j(c)} - \frac{\mu^2_j(c)a_j(c)}{\sum_{k=1}^{J}a_k(c)}\right)\right)^{\frac{1}{2}}$$

which, in the maximum similarity case, reduces to

$$d(\xi^t_{jc},\Pi^t_{jc}) \leq t\,\mu_j(c)\left(\frac{N(c)}{J}\left(I_j(c)\sum_{k=1}^{J}\lambda^2_{kj}(c) - \frac{1}{J}\right)\right)^{\frac{1}{2}}.$$

This bound is clearly minimised by the choice $\lambda_{kj}(c) = 1/J$, but it cannot be reduced to zero unless service requirements are exponential.

ACKNOWLEDGEMENTS.

The author wishes to thank Tim Brown and Saul Jacka for valuable conversations and Frank Kelly for his continued and constructive criticism.

REFERENCES.

BARBOUR, A.D. (1976), Networks of queues and a method of stages, *Adv. Appl. Prob.*, *8*, 584-591.

BROWN, T.C. (1982) Some Poisson Approximations, Statistics Research Report, Dept. of Maths. Monash University.

BROWN, T.C. and POLLETT, P.K. (1982) Some Distributional Approximations in Markovian Queueing Networks, *Adv. Appl. Prob.*, *14*, 654-671.

CHANG, A. and LAVENBERG, S.S. (1974) Work Rates in Closed Queueing Networks with General Independent Servers. *Opns. Res.*, *22*, 883-847.

COX, D.R. (1955) A use of complex probabilities in the theory of stochastic processes. *Proc. Camb. Phil. Soc.*, *51*, 313-319.

KELLY, F.P. (1976) Networks of queues, *Adv. Appl. Prob.*, *8*. 416-432.

KELLY, F.P. (1979) *Reversibility and Stochastic Networks*, Wiley and Sons, New York.

POLLETT, P.K. (1983) Some Poisson approximations for departure processes in general queueing networks. Submitted for publication.

SOME GEOMETRIC ASPECTS OF POTENTIAL THEORY

John Hawkes
Department of Mathematics and Computer Science
University College of Swansea
Singleton Park
Swansea SA2 8PP
Great Britain

§1. INTRODUCTION.

Let $\{\mu_t\}_{t \geq 0}$ be a weakly continuous semigroup of probability measures on $\mathbb{R}^d$ (d-dimensional euclidean space) such that μ_0 is the unit mass at the origin. A Lévy process is a Markov process, X_t, whose semigroup and resolvent operators are given by

$$E^x f(X_t) = (P_t f)(x) = (f*\overline{\mu}_t)(x)$$

and

$$E^x \int_0^\infty e^{-\lambda t} f(X_t)\, dt = (U^\lambda f)(x)$$

respectively. It follows that the transition function of the process satisfies the geometric condition

$$P_t(x,A) \equiv P_t(y + x, y + A). \qquad (1)$$

Since X_t has a Feller semigroup the general theory of Markov processes can be applied to study the sample path properties of X_t and its potential theory. Chung ([9], pp. 137-144) provides a suitable introduction, and terms not defined here can usually be found in Chung's book. The property (1) ensures that the process inherits a strong geometric structure. This has been exploited to the full in the analysis of sample path properties, see for example Pruitt ([48]). Our object here is to describe some of those aspects of the potential theory of Lévy processes that illuminate this geometric structure. We shall describe some new and not so new results that emphasize these relationships.

In section 2 we bring in a few definitions, so as to establish our notation. In section 3 we give the briefest of accounts of the classical potential theory of brownian motion with a special emphasis on those facets that are peculiar to Lévy processes, which we describe in section 4. In later sections we show how our geometric stance leads quickly to new results and to new proofs on Lévy processes.

Mention should be made of some other approaches to the potential theory of Lévy processes. In particular there is the group theoretic approach by Port and Stone ([43] and [44]) and Berg and Forst ([1]); for the approach using Dirichlet spaces see Silverstein ([50] and [51]) and Fukushima ([13]); Zabczyck ([60]) adopts the integral equation, eigenfunction approach.

For purely analytic treatments of modern potential theory there are the books by Helms ([23]), Wermer ([57]) and Landkof ([39]), whilst the classical book by Kellogg ([34]) should be read for an insight into the underlying problems in mathematical physics. The survey article by Brelot ([5]) is also strongly recommended.

For a full account of the potential theory of Markov processes see Hunt ([24] and [25]) or Blumenthal and Getoor ([3]).

§2. PRELIMINARIES.

Let X_t be a Lévy process in $\mathbb{R}^d$. We shall suppose that X_t has all the usual properties of a strong Markov process. The process $\tilde{X}_t = -X_t$ is also a Lévy process, called the dual of X_t, and its transition and resolvent operators are denoted by $\tilde{P}_t$ and $\tilde{U}^\lambda$ respectively. Technical expressions, where not defined here, can be found in any standard book on processes. (See, for example, Chung ([9]).) We mention here only a few topics in order to establish notation.

Let K be a generic compact set. Then $\mathcal{M}(K)$ will be the set of Radon measures supported by K. The first entry time and last exit time of K are denoted by

$$F = F_K = \inf\{t > 0 : X_t \in K\}$$

and

$$L = L_K = \sup\{t > 0 : X_{t-} \in K\}$$

respectively.

The objects of fundamental interest are the quantities

$$h(x) = P^x(F < \infty) = P^x(L > 0)$$

and

$$\phi^\lambda(x) = E^x e^{-\lambda F} \qquad (\lambda \geq 0).$$

A point x is regular for K if $P^x(F = 0) = 1$. The set of regular points of K is denoted by $K^{(r)}$.

The polar and essentially polar compact sets are defined by

$$\mathcal{P} = \{K : \phi^\lambda(x) \equiv 0\}$$

and

$$\mathcal{EP} = \{K : \phi^\lambda(x) = 0, \ \Lambda \text{ almost everywhere}\}.$$

Here Λ denotes Lebesgue measure. If $\Lambda(A) > 0$ then almost all points of A are regular for A.

The characteristic function of $X_t - X_0$ has the form

$$E\, e^{i(z, X_t - X_0)} = \hat{\mu}_t(z) = e^{-t\psi(z)}$$

where $\psi(z)$, the exponent of the process, has the familiar Lévy-Hinčin representation. We merely note that this representation ensures that

$$\text{Re } \psi(z) \geq 0 \quad \text{and} \quad \text{Re } \left(\frac{1}{\lambda + \psi(z)}\right) \geq 0 \quad \text{if} \quad \lambda \geq 0.$$

It is shown by Port and Stone ([43], Theorem 16.2) that the process is transient if and only if

$$\int_{\|z\|\le 1} \operatorname{Re}\left(\frac{1}{\psi(z)}\right) dz < \infty .$$

Kingman ([36]) had an earlier version of this result based on the Chung-Fuchs criteria for the recurrence of random walks.

§3. CLASSICAL POTENTIAL THEORY (BROWNIAN MOTION).

Brownian motion, B_t, in $\mathbb{R}^d$ is characterized by the transition function

$$P_t(x,A) = \int_A p_t(y-x)dy$$

where $p_t(z) = (2\pi t)^{-\frac{1}{2}d} \exp(-\|z\|^2/2t)$, the Gauss-Weierstrass kernel, also satisfies

$$p_t * p_s \equiv p_{t+s} .$$

The process B_t is transient if and only if $d \ge 3$. In this case one obtains the potential theory by considering the occupation-time measure

$$U(x,A) = \int_0^\infty P_t(x,A)\, dt$$

$$= \int_A u(y-x)\, dy$$

where

$$u(z) = \int_0^\infty p_t(z)\, dt$$

$$= \frac{\Gamma(\frac{1}{2}d-1)}{2\pi^{\frac{1}{2}d}} \frac{1}{\|z\|^{d-2}}$$

is, up to a constant of proportionality, the classical Newtonian potential kernel.

It can be shown that h is a harmonic function in the classical sense. Thus, given any compact set K, there is a unique measure, the equilibrium measure, $\mu \in \mathcal{M}(K)$, such that

$$h(x) = \int_K u(y-x)\, d\mu(y). \tag{2}$$

We now discuss the equilibrium measure from a probabilistic point of view.

Properties of the equilibrium measure.

(a) Capacity. The capacity of K is defined by $\text{cap}(K) = \mu(K)$. The set function $K \to \text{cap}(K)$ extends in the usual way to the analytic sets and the analytic sets are capacitable.

(b) Polar sets. It follows immediately from (2) that $K \in \mathcal{P}$ if and only if $\text{cap}(K) = 0$.

(c) Probabilistic interpretation. The equilibrium distribution admits two interpretations in terms of the last exit time and last exit distribution. Firstly

$$d\mu(x) = w \lim_{\varepsilon \to 0} dx\, P^x(0 < L < \varepsilon)/\varepsilon\ . \tag{3}$$

If $L(x,A) = P^x(B(L-) \in A,\ L > 0)$ is the last exit kernel then

$$L(x,A) = \int_A u(y-x)\, d\mu(y). \tag{4}$$

These results were proved by McKean in ([40]) and later, in a more general context, by Chung ([7]).

(d) The Kelvin principle. The energy of a measure ν is defined by

$$E(\nu) = \iint u(y-x)\, d\nu(y)\, d\nu(x).$$

Let

$$I(K) = \inf\{E(\nu) : \nu(K) = 1,\ \nu \in \mathcal{M}(K)\}. \tag{5}$$

Then one has $I(K) = [\text{cap } K]^{-1}$. The infimum in (5) is uniquely attained by the probability measure $d\nu(x) = d\mu(x)/\mu(K)$, where μ is the equilibrium measure.

(e) The Spitzer-Whitman property. The capacity of a set K has the following geometric interpretation:

$$\lim_{t\to\infty} \frac{\Lambda\{\cup_{s\le t} (B_s + K)\}}{t} = \text{cap }(K)$$

almost surely. This property was noted by Spitzer in ([53], p.121) and Whitman ([58]).

The connection between the Dirichlet problem and random walk had been known for a long time (see Courant, Friedrichs and Lewy ([11]) and Philips and Wiener ([46])) but the connection with brownian motion seems to be due to Kakutani ([29] and [30]). For a detailed probabilistic computation of nearly all the classical results see the book by Port and Stone ([45]). For an alternative probabilistic approach see Kac ([27]).

§4. POTENTIAL THEORY (LÉVY PROCESSES).

We now indicate how the classical theory generalizes. The potential theory is arrived at as follows. Choose $\lambda \ge 0$, $\lambda > 0$ for recurrent processes, and consider the discounted occupation-time

$$U^\lambda(x,A) = \int_0^\infty e^{-\lambda t}\, P_t(x,A)\, dt.$$

Unlike the classical case the measure U^λ need not be absolutely continuous. However the result on hitting probabilities is that there is a unique measure, the equilibrium measure, $\mu^\lambda \in \mathcal{M}(K)$ such that

$$\int_A \phi^\lambda(x)\, dx = \int U^\lambda(0,y-A)\, d\mu^\lambda(y). \tag{6}$$

See Port and Stone ([43], Theorem 6.2). Note that (6) has the alternative form

$$(\tilde{U}^{\lambda} f, \mu^{\lambda}) = (f, \phi^{\lambda}) \tag{7}$$

for measurable functions f.

Again one can investigate the properties of μ.

Properties of the equilibrium measure.

(a) Capacity. The λ-capacity of K is defined by $\operatorname{cap}^{\lambda}(K) = \mu^{\lambda}(K)$. The set function $K \to \operatorname{cap}^{\lambda}(K)$ again extends to a Choquet capacity on the analytic sets.

(b) Essentially polar sets. It follows immediately from (6) that $K \in \mathcal{EP}$ if and only if $\operatorname{cap}^{\lambda}(K) = 0$ for some and hence all $\lambda > 0$.

(c) Probabilistic interpretation. For transient processes there are direct analogues of the brownian motion case. The identity (3) is essentially proved by Port and Stone ([43], Corollary 8.2). The analogue of (4) is

$$\int f(x)\, L(x,A)\, dx = \int_A (\tilde{U} f)(z)\, d\mu(z),$$

as is shown by Port and Stone ([43], Proposition 11.2).

(d) Energy. Here the generalization is not so straightforward. We shall return to this topic later.

(e) The Spitzer result. The analogue of this does not seem to have been investigated except partially by Getoor ([14]). He notes that

$$E^0 \Lambda\{\cup_{s \le t} (X_s + K)\} = \int P^x(F \le t)\, dx,$$

whilst Port and Stone ([43], Theorem 11.1) show that for transient processes the latter is asymptotically $t \operatorname{cap}(K)$. One can then apply

Kingman's subergodic theorem ([37], see the remarks on pp.904-905) to show that

$$\lim_{t\to\infty} \frac{\Lambda\{\cup_{s\le t}(X_s + K)\}}{t} = \text{cap}(K)$$

almost surely.

We remark that it follows automatically that $\text{cap}(K) = 0$ if and only if $\Lambda\{\cup_{s\le t}(X_s + K)\} = 0$ almost surely for some and hence all $t > 0$. Kahane ([28]) gives an analytic proof of this fact, based on the Itô representation.

Finally we mention some special problems that arise in the potential theory of Lévy processes.

Problem 1. Given a set K and a process X_t when can one say that $K \in \mathcal{P}$? For a long time this was a problem even for the special case $K = \{0\}$. See Orey ([41]) and Port and Stone ([42]) for the very special case of the asymmetric Cauchy process on the line.

Problem 2. Given two processes X_1 and X_2 when can one say that $\mathcal{P}(X_2) \subseteq \mathcal{P}(X_1)$?

To obtain some insight into this problem we mention a simple example. For brownian motion in $\mathbb{R}^d$ one has $\psi_2(z) = \frac{1}{2}\|z\|^2$. If X_1 is a spherically symmetric process with a resolvent density then (Blumenthal ([2]) as noted by Orey ([41])) we have $\mathcal{P}_2 \subseteq \mathcal{P}_1$. However one always has $\psi_1(z) = O(1)\psi_2(z)$ $(\|z\| \to \infty)$. This suggests that some kind of domination of exponents yields an inclusion relation between the polar sets.

§5. REGULARITY CONDITIONS.

In this section we review the little-known results on the criteria for the resolvent and semigroup operators to be well behaved.

First let $\mathcal{C}$ be the class of continuous functions bounded on $\mathbb{R}^d$, and M_K the class of bounded measurable functions having compact support. A linear operator L is said to be a strong Feller operator if

$$Lf \in \mathcal{C} \quad \text{whenever} \quad f \in M_K.$$

The following is the main result on the regularity of the resolvent operator.

PROPOSITION 1. (Hawkes ([18])). The following statements are equivalent:

(i) for each $\lambda \geq 0$, and each x, $A \to U^\lambda(x,A)$ is absolutely continuous with respect to Lebesgue measure;

(ii) X has a strong Feller resolvent;

(iii) all λ-excessive functions are lower semicontinuous; and

(iv) $\mathcal{E}\mathcal{P} = \mathcal{P}$.

In the circumstances of the above result there is a version, $u^\lambda(x)$, of the density such that

$$U^\lambda(x,A) = \int_A u^\lambda(y-x)\,dy$$

and, for each y,

$$x \to u^\lambda(y-x)$$

is λ-excessive. The function is lower semicontinuous and positive on the interior of its closed support. We call this version the canonical version of the resolvent density. In this case (7) takes the special form

$$\phi^\lambda(x) = \int u^\lambda(y-x)\,d\mu^\lambda(y), \tag{8}$$

and essentially polar sets are polar. The identity (8) is originally due to Hunt ([25], Theorem 18.4).

PROPOSITION 2. (Hawkes ([18])). The semigroup $\{P_t\}_{t>0}$ is a strong Feller semigroup if and only if for each x and each $t > 0$ the measure

$$A \to P_t(x,A)$$

is absolutely continuous with respect to Lebesgue measure. In this case there exists a family $\{p_t(x)\}_{t>0}$ of densities such that:

(i) $(t,x) \to p_t(x)$ is jointly measurable;

(ii) $x \to p_t(x)$ is lower semicontinuous;

(iii) $p_t * p_s \equiv p_{t+s}$; and

(iv) $\int_A p_t(y-x)\,dy = P_t(x,A)$ for all measurable sets A.

The version of the density satisfying (i) to (iv) is called the canonical transition density. This is related to the canonical resolvent density by the following result.

PROPOSITION 3. If the semigroup of X is a strong Feller semigroup with a canonical density, $p_t(x)$, the resolvent of X is a strong Feller resolvent with canonical density, $u^\lambda(x)$, given by

$$u^\lambda(x) = \int_0^\infty e^{-\lambda t}\, p_t(x)\, dt.$$

There are examples of processes having strong Feller resolvents but not strong Feller semigroups, Hawkes ([18]). On the other hand if the distributions are spherically symmetric or only locally symmetric the converse does indeed hold, see Zabczyck ([60], p.245) and Fukushima ([12], Theorem 4) respectively. Furthermore there are examples where P_t is strong Feller for some <u>but not all</u> positive t.

See Stratton ([55]) for some of the remarkable pathological possibilities for the Hausdorff dimension of the support of μ_t. For examples on even more pathological behaviour of densities see Hawkes ([19]).

§6. MAIN RESULT.

Let A and B be compact sets and let $\mu \in M(A)$ and $\nu \in M(B)$. The main idea is to attempt to define a measure $\mu \circ \nu$ in $M(A \cap B)$ which has the interpretation

$$d(\mu \circ \nu)(x) = \frac{\mu(dx)\nu(dx)}{dx}$$

or $(\tau_t \mu) \circ \nu$ on $(t + A) \cap B$ with

$$d(\tau_t \mu \circ \nu)(x) = \frac{(\tau_t \mu)(dx)\nu(dx)}{dx} \tag{9}$$

where τ_t denotes the translation by distance t. In general this cannot be done. However it is possible to derive the following sufficient condition.

THEOREM 1. With μ and ν as above, having Fourier transforms $\mu^\wedge$ and $\nu^\wedge$, if

$$\int |\mu^\wedge(z)|^2 \, |\nu^\wedge(z)|^2 \, dz < \infty$$

then $(\tau_t \mu) \circ \nu$ can be defined for almost all t, is non-trivial and has the interpretation (9). In particular $(t + A) \cap B$ is non-empty on a set of positive measure.

This is proved in Hawkes ([20]).

The following lemma from harmonic analysis, essentially proved in Stein and Weiss ([54], p.15), also proves useful.

LEMMA 1. Suppose that $f \in L^1(\mathbb{R}^d) \cap L^\infty(\mathbb{R}^d)$. Then if the Fourier transform of f is real and non-negative we have $f^\wedge \in L^1(\mathbb{R}^d)$.

Theorem 1 can be applied to potential theory by noting that if T is the discounted occupation-time measure

$$T(A) = \int_0^\infty e^{-\lambda t} I_A(X_t) \, dt$$

then, see for example Hawkes ([22]),

$$E|T^{\wedge}(z)|^2 = \lambda^{-1}\,\mathrm{Re}\,\left(\frac{1}{\lambda + \psi(z)}\right) \geq 0. \tag{10}$$

This leads to the following criterion for polarity.

THEOREM 2. We have $K \notin \mathcal{P}$ if and only if there exists $\mu \in M(K)$ such that, for some $\lambda > 0$,

$$J^{\lambda}(\mu) = \frac{1}{(2\pi)^d}\int \mathrm{Re}\,\left(\frac{1}{\lambda + \psi(z)}\right)|\mu^{\wedge}(z)|^2\,dz < \infty.$$

Proof. If $J^{\lambda}(\mu) < \infty$ then

$$\int E|T^{\wedge}(z)|^2|\mu^{\wedge}(z)|^2\,dz < \infty$$

and

$$\int |T^{\wedge}(z)|^2|\mu^{\wedge}(z)|^2\,dz < \infty$$

almost surely. By Theorem 1 $(\tau_t\mu)\circ T$ is defined almost everywhere. It thus follows that $\{(t,\omega)\}$ such that

$$\{t + R(\omega)\} \cap K \neq \emptyset$$

has positive product measure and that $K \notin \mathcal{P}$.

Now suppose that $K \notin \mathcal{P}$, let $\mu = \mu_K$ and define

$$f(y) = \int_K \phi^{\lambda}(y + x)\,d\mu(x).$$

Then we can see that $f \in L^1(\mathbb{R}^d) \cap L^{\infty}(\mathbb{R}^d)$ and that

$$[f(y) + f(-y)]^{\wedge}(z) = 2\,\mathrm{Re}\left(\frac{1}{\lambda + \psi(z)}\right)|\mu^{\wedge}(z)|^2 \geq 0.$$

Thus, by Lemma 1, $J^{\lambda}(\mu)$ is finite and the theorem is proved. For fuller details of this proof see Hawkes ([21]).

The quantity $J^{\lambda}(\mu)$ of the theorem is related to the energy of μ in the following way. Suppose that U^{λ} admits a canonical resolvent density, u^λ, and define the λ-energy

$$E^\lambda(\mu) = \iint u^\lambda(y-x)\,d\mu(y)\,d\mu(x).$$

Then a purely formal application of Parseval's theorem yields $E^\lambda(\mu) = J^\lambda(\mu)$. However this cannot in general be justified and all one can actually prove is that $E^\lambda(\mu) \le J^\lambda(\mu)$.

§7. APPLICATIONS.

In this section we mention some of the immediate applications of Theorem 2.

(a) Singletons.

If $K = \{0\}$ and $\nu \in \mathcal{M}(K)$ it is automatic that $\hat{\nu}(z) \equiv c$. The following result is then an immediate consequence of Theorem 2 and the fact (used only when $d \ge 2$) that $\psi(z) = O(1)\|z\|^2$ $(\|z\| \to \infty)$.

THEOREM 3. Suppose that $d = 1$. Then $\{0\} \notin \mathcal{ER}$ if and only if, for some $\lambda > 0$,

$$\int \mathrm{Re}\left(\frac{1}{\lambda + \psi(z)}\right) dz < \infty. \tag{11}$$

If $d \ge 2$ then $\{0\} \in \mathcal{ER}$.

Kesten ([35], Theorem 2) gave the first proof of this fact, a proof that was later shortened by Bretagnolle ([6]). Kesten and Bretagnolle also discuss the more difficult problem of when does $\{0\} \in \mathcal{ER}$ imply $\{0\} \in \mathcal{R}$.

We also have the following alternative characterization of when $\{0\} \in \mathcal{ER}$.

THEOREM 4. Suppose that $d = 1$ and that $\lambda > 0$. We have $\{0\} \notin \mathcal{ER}$ if and only if X has a strong Feller resolvent whose canonical density, $u^\lambda(x)$, is bounded. In this case there is a positive constant c^λ such that

$$\phi^\lambda(x) = c^\lambda u^\lambda(-x). \tag{12}$$

Proof. If there is a bounded resolvent density $u^{\lambda}(x)$ we have

$$[u^{\lambda}(x) + u^{\lambda}(-x)]^{\wedge}(z) = 2\ \mathrm{Re}\left(\frac{1}{\lambda + \psi(z)}\right) \geq 0.$$

Thus, by Lemma 1 the latter is integrable and so $\{0\} \notin \mathcal{E}p$.

If $\{0\} \notin \mathcal{E}p$ then (11) holds and so the measure $U^{\lambda}(0,dx) + U^{\lambda}(0, -dx)$ has an integrable Fourier transform. By the Fourier inversion theorem, this measure has a bounded density. Thus $U^{\lambda}(0,dx)$ has a bounded density and there is a bounded canonical resolvent density.

The final statement of the theorem follows from (8).

Now Orey's elegant argument ([41], p.122, lines 3-10) can be applied to yield the following criterion for 0 to be regular for $\{0\}$. (It is necessary to remark that one does not need the full force of Orey's assumptions, but only those in operation here.)

THEOREM 5. Suppose that $d = 1$ and that $\lambda > 0$. Then 0 is regular for $\{0\}$ if and only if X has a canonical resolvent density, u^{λ}, that is bounded, and continuous at the origin. In this case

$$\phi^{\lambda}(x) = u^{\lambda}(-x)/u^{\lambda}(0).$$

Bretagnolle classifies the circumstances under which this situation occurs.

Theorems 4 and 5 are given, with quite different proofs, in Port and Stone ([43], pp.207-210). Note that in the circumstances of Theorem 4 one sees that $u^{\lambda}(x) + u^{\lambda}(-x)$ is almost everywhere equal to a continuous function. But this is <u>not</u> sufficient to ensure that the lower semicontinuous function $u^{\lambda}(x)$ is continuous at the origin. Consider, for example, a Poisson process with unit drift.

(b) The comparison problem.

Our main result (Theorem 2) yields an immediate solution to the comparison problem.

THEOREM 6. Suppose that $\lambda > 0$. Let X_1 and X_2 be two Lévy processes having exponents ψ_1 and ψ_2 respectively. If

$$\mathrm{Re}\left(\frac{1}{\lambda + \psi_2(z)}\right) = O(1)\ \mathrm{Re}\left(\frac{1}{\lambda + \psi_1(z)}\right) \qquad (\|z\| \to \infty)$$

then $\mathcal{P}(X_2) \subseteq \mathcal{P}(X_1)$.

Proof. The finiteness of $J_1(\mu)$ implies that of $J_2(\mu)$ so the result follows from Theorem 2.

The result has been obtained, with progressive weakening of the assumptions, by Orey ([41]), Kanda ([31]) and Hawkes ([18]). The point here is the very simplicity of the proof that results from our geometric approach.

In fact in [18] we proved slightly more, namely that if

$$\mathrm{Re}\left(\frac{1}{\lambda + \psi_2(z)}\right) \le M\ \mathrm{Re}\left(\frac{1}{\lambda + \psi_1(z)}\right)$$

then $\frac{1}{4}\ \mathrm{cap}_1^{\lambda}(A) \le M\ \mathrm{cap}_2^{\lambda}(A)$ for all analytic sets A. This has the consequence that if X_1 and X_2 are two linear stable processes of index α, $0 < \alpha < 1$, so that the exponents take the form

$\psi(z) = |z|^{\alpha}\{1 - i\beta \mathrm{sgn}(z) \tan \frac{1}{2}\pi\alpha\}$ with $-1 \le \beta \le 1$, then there are constants M_1 and M_2 such that

$$M_1\ \mathrm{cap}_1(B) \le \mathrm{cap}_2(B) \le M_2\ \mathrm{cap}_1(B) \tag{13}$$

for all analytic sets B. This answers a question due to Taylor ([56]).

(c) <u>The symmetrization problem</u> (Orey).

Let X be a Lévy process and let Z_1 and Z_2 be independent copies of X. Then $X^{(s)} = Z_1 - Z_2$ is called the *symmetrization* of X. Orey conjectured that

$$\mathcal{P}(X) \subseteq \mathcal{P}(X^{(s)}). \tag{14}$$

There are examples of varying degrees of sophistication to show that this inclusion can be strict.

Example 1. Let $X_t = P_t - t$ where P_t is a Poisson process of rate one, so that $X^{(s)} = P^{(s)}$. One can see that

$$\mathcal{P}(X) = \{\emptyset\} \quad \text{whilst} \quad \mathcal{P}(X^{(s)}) = \{B:\Lambda(B) = 0\}.$$

Thus the one class of sets is smallest possible whilst the other is largest possible.

Example 2. Pruitt ([47]) has given an example of a subordinator X which has $\gamma(X) = \frac{2}{3}$ and $\gamma(X^{(s)}) = \frac{3}{5}$, γ denoting Pruitt's index. If we choose α so that $\frac{1}{3} < \alpha < \frac{2}{5}$, then an intersection argument of the type used in Hawkes ([17], the argument deducing Theorem 4 from Theorem 3) can be applied to show that almost all realizations of the range of a linear stable process of index α are in $\mathcal{EP}(X^{(s)})$ but not in $\mathcal{EP}(X)$.

The inclusion relation (14) follows from Theorem 6 and the observation that

$$\mathrm{Re}\left(\frac{1}{\lambda + \psi}\right) \le \left(\frac{1}{\lambda + \mathrm{Re}\,\psi}\right) = \mathrm{Re}\left(\frac{1}{\lambda + \frac{1}{2}\psi_s}\right)$$

where ψ is the exponent of X and ψ_s that of $X^{(s)}$.

§8. ENERGY AND CAPACITY

We now return to the problem that was left unanswered in §4(d). In ([18]) we showed that for Lévy processes, and open sets D, one has

$$[4 \operatorname{cap}(D)]^{-1} \le I(D) \le [\operatorname{cap}(D)]^{-1}. \tag{15}$$

The upper inequality is, as is well known, in fact equality when the process is symmetric. In ([10]) Chung and Rao ask whether this is true in general. The answer is no! This is seen by taking $0 < \alpha < 1$ and considering the symmetric and increasing stable processes of index α

and taking D to be the unit interval. In [16] we showed that as α varies I(D) can be arbitrarily close to $[2 \text{ cap } (D)]^{-1}$. Thus the number 4 in (15) cannot be replaced by any number θ, $\theta < 2$. Thus the Kelvin principle even fails for linear stable processes. It would be interesting to know the best constants in (15).

For more information on energy see Chung ([8]) and Chung and Rao ([10]).

§9. WIENER TESTS.

In this section we mention a criterion for a point x to be regular for a set. This criterion is essentially in keeping with our geometric theme. We also indicate how the comparison results for capacity can be applied to yield comparison results for regular points.

(a) The classical result. In [59] Wiener gave a criterion for a point x to be a regular boundary point for the Dirichlet problem associated with a given domain. We shall discuss a probabilistic version of this.

(b) Brownian motion. Let B_t be the brownian motion in $\mathbb{R}^d$. Then x is regular for K if and only if

$$\sum_{n\geq 0} 2^{n(d-2)} \text{cap } [K_n(x)] = \infty \quad (d \geq 3) \tag{16}$$

or

$$\sum_{n\geq 0} n \text{ cap } [K_n(x)] = \infty \quad (d = 2), \tag{17}$$

where $K_n(x) = \{z \in K: 2^{-(n+1)} \leq \|z - x\| < 2^{-n}\}$.

This can be proved by analytic methods (see Landkof [39], Chapter 5) for details and further references). On the other hand purely elementary probabilistic proofs of these statements, essentially due to Lamperti ([38]), can be found in Itô and McKean ([26], p.255) and Port and Stone ([45], p.66).

(c) Stable processes. If $0 < \alpha < \min(2,d)$ and X_t is a spherically symmetric stable process of index α, whence $\psi(z) = \|z\|^{\alpha}$, then the analogue of (16) is that x is X_t-regular for K if and only if

$$\sum_{n\geq 0} 2^{n(d-\alpha)} \operatorname{cap}[K_n(x)] = \infty .$$

A version of this can be established for all linear stable processes. The inequalities (13) can then be applied, as in Hawkes ([16]), to yield

THEOREM 7. Suppose that $d = 1$ and that $0 < \alpha < 1$. Let X be the symmetric stable process of index α and T be the stable subordinator of index α. Then

$$x \text{ is } X\text{-regular for } K \cap [x,\infty)$$

if and only if

$$x \text{ is } T\text{-regular for } K \cap [x,\infty).$$

(d) Recurrent sets. A set B is said to be recurrent for X_t if and only if $\{t: X_t \in B\}$ is unbounded. There are criteria similar to (16) and (17) which provide a necessary and sufficient condition for a set to be recurrent for a stable process. We illustrate the general method by considering the special case of recurrence of a set for an asymmetric Cauchy process on the line.

Here the exponent takes the form

$$\psi(z) = |z|\{1 + i2\beta \operatorname{sgn}(z)\pi^{-1}\log|z|\}$$

with $0 < |\beta| \leq 1$. The transience criterion ensures that the process is indeed transient. When $\beta = 1$ (or $\beta = -1$) the process takes only positive (negative) jumps and the paths drift to $+\infty$ $(-\infty)$. In this case the process is said to be completely asymmetric.

THEOREM 8. Let X_t be an asymmetric, but not completely asymmetric, Cauchy process on the line. Then a closed set B is recurrent for X_t if and only if

$$\sum_{n\geq 1} \operatorname{cap}(B_n)/n = \infty$$

where $B_n = \{z \in B: 2^n \leq |z| < 2^{n+1}\}$.

Proof. First one needs the following well-known extension of the Borel-Cantelli lemma.

LEMMA 2. Let $\{A_i\}$ be a sequence of events such that:

(i) $P(A_i \text{ i.o.}) = 0$ or 1; and

(ii) there is an absolute constant M such that $P(A_i \cap A_j) \leq MP(A_i)P(A_j)$ whenever $i \neq j$.

Then

$$\sum P(A_i) = \infty \quad \text{implies} \quad P(A_i \text{ i.o.}) = 1.$$

Now let $A_i = \{F_{B_i} < \infty\}$ so that

$$\{B \text{ is recurrent}\} \equiv \{A_i \text{ i.o.}\}.$$

Equation (8) and the facts that

$$\lim_{x\to-\infty} \log|x|u(x) = c^- \quad \text{and} \quad \lim_{x\to\infty} \log x\, u(x) = c^+$$

where

$$c^- = \frac{\pi(1-\beta)}{4\beta^2} \quad \text{and} \quad c^+ = \frac{\pi(1+\beta)}{4\beta^2},$$

see Port and Stone ([42], Proposition 2), show that $P(A_i)$ is bounded above and below by multiples of $\operatorname{cap}(B_i)/i$.

When $\Sigma P(A_i) = \infty$ so is $\Sigma_i P(A_{ki+\ell})$ for some ℓ, $1 \leq \ell \leq k$. By an appropriate choice of k (dependent on β) one can ensure that the conditions of the last lemma are satisfied. The theorem then follows immediately.

§10. WHEN ARE SEMIPOLAR SETS POLAR?

No discussion of the subject matter of this paper would be complete without mention of one of the major outstanding problems. We drop now the restriction that the sets under consideration are compact. The thin and semipolar sets are defined by

$$\mathcal{T} = \{A: A^{(r)} = \emptyset\}$$

and

$$\mathcal{SP} = \{B: B \subseteq \bigcup_{i=1}^{\infty} T_i, \quad T_i \in \mathcal{T}\}.$$

We always have $\mathcal{P} \subseteq \mathcal{SP}$ and the inclusion can be strict (for example the Poisson process with drift). If the process is symmetric we have

$$\mathcal{P} = \mathcal{SP} \ . \tag{18}$$

One is interested in conditions on ψ under which there is this equality. The principal reason for this interest is that Hunt ([25], p.193) found it convenient to introduce this as a regularity condition (Hypothesis H) which ensured the validity of a number of desirable potential theoretic properties. See Hunt and Blumenthal and Getoor ([3], Chapter 6, especially section 4) for further details.

In [4] Blumenthal and Getoor established the equivalence of (18) and a certain maximum principle for a class of Markov processes. The first progress for a general Lévy process was due to Kanda ([31]) who showed that (18) holds if ψ satisfies the so-called sector condition

$$|\mathrm{Im}\psi(z)| \le M|1 + \mathrm{Re}\ \psi(z)|$$

and a few other restrictions. An alternative proof was given by Silverstein ([52]), using the theory of Dirichlet spaces and another exceptionally short proof was given by Rao ([49]).

Kanda ([32]) subsequently showed that, for a general class of Markov processes, (18) is equivalent to the statement that

$$\lambda \to \mathrm{cap}^{\lambda}(K)$$

is unbounded whenever it is non-zero. Note that it follows from Hunt ([25], eq. 19.7) that the latter function is monotone increasing.

Glover ([15]), again for a general class of Markov processes, showed that (18) is equivalent to the statement

$$E^{\lambda}(|\mu|) < \infty \quad \text{implies} \quad E^{\lambda}(\mu) \geq 0 \tag{19}$$

for all signed measures μ. He then gives specific (but very technical) conditions which ensure the validity of (19) for certain Lévy processes. Note that (19) is trivially implied by the sector condition.

It seems fair to say that there is, as yet, no satisfactory solution to the problem.

In this context one should also mention the note by Zabczyck ([61]).

REFERENCES

[1] BERG, C. and G. FORST. Potential Theory on Locally Compact Abelian Groups. Springer-Verlag, Berlin, 1975.

[2] BLUMENTHAL, R.M. Some relationships involving subordination. Proc. Amer. Math. Soc. 10 (1959) 502-510.

[3] BLUMENTHAL, R.M. and R.K. GETOOR. Markov Processes and Potential Theory. Academic Press, New York, 1968.

[4] BLUMENTHAL, R.M. and R.K. GETOOR. Dual processes and potential theory. Proc. 12th Biennial Seminar, Canad. Math. Congress (1970) 137-156.

[5] BRELOT, M. Les étapes et les aspects multiples de la théorie du potentiel. L'Enseignement mathém. XVIII (1972) 1-36.

[6] BRETAGNOLLE, J. Résultats de Kesten sur les processus à accroissements indépendants. Séminaire de Probabilités V, Lecture Notes in Mathematics 191 21-36, Springer-Verlag, Berlin, 1971.

[7] CHUNG, K.L. Probabilistic approach to the equilibrium problem in potential theory. Ann. Inst. Fourier 23 (1973) 313-322.

[8] CHUNG, K.L. Remarks on equilibrium potential and energy. Ann. Inst. Fourier 25 (1975) 131-138.

[9] CHUNG, K.L. Lectures from Markov Processes to Brownian Motion. Springer-Verlag, New York, 1982.

[10] CHUNG, K.L. and M.RAO. Equilibrium and energy. Probab. Math. Statist. 1 (1980) 99-108.

[11] COURANT, R., K. FRIEDRICHS and H. LEWY. Über die partiellen Differenzengleichungen der mathematischen Physik. Math. Annalen 100 (1928) 32-74.

[12] FUKUSHIMA, M. Potential theory of symmetric Markov processes and its applications. Lecture Notes in Mathematics 550 119-133, Springer-Verlag, Berlin, 1976.

[13] FUKUSHIMA, M. Dirichlet Forms and Markov Processes. North-Holland, Amsterdam, 1980.

[14] GETOOR, R.K. Some Asymptotic Formulas Involving Capacity. Z. Wahrscheinlichkeitstheorie 4 (1965) 248-252.

[15] GLOVER, J. Energy and the maximum principle for non symmetric Hunt processes. Theory Probab. and its Applications XXVI (1981) 745-757.

[16] HAWKES, J. Polar sets, regular points and recurrent sets for the symmetric and increasing stable processes. Bull. London Math. Soc. 2 (1970) 53-59.

[17] HAWKES, J. On the Hausdorff Dimension of the Intersection of the Range of a Stable Process with a Borel Set. Z. Wahrscheinlichkeitstheorie 19 (1971) 90-102.

[18] HAWKES, J. Potential theory of Lévy processes. Proc. London Math. Soc. (3) 38 (1979) 335-352.

[19] HAWKES, J. Transition and resolvent densities for Lévy processes. J. London Math. Soc. To appear.

[20] HAWKES, J. Fourier methods in the geometry of small sets. In preparation.

[21] HAWKES, J. Energy, capacity and polar sets for some Markov processes. In preparation.

[22] HAWKES, J. Harmonic analysis of Lévy sets. In preparation.

[23] HELMS, L.L. Introduction to Potential Theory. Wiley, New York, 1969.

[24] HUNT, G.A. Markov processes and potentials I and II. Illinois J. Math. 1 (1957) 44-93 and 316-369.

[25] HUNT, G.A. Markov processes and potentials III. Illinois J. Math. 2 (1958) 151-213.

[26] ITÔ, K. and H.P. McKEAN. Diffusion processes and their sample paths. Springer-Verlag, Berlin, 1965.

[27] KAC, M. Aspects Probabilistes de la théorie du potentiel. Publications du séminaire de Mathématiques Supérieures, Montreal, 1970.

[28] KAHANE, J.-P. Ensembles parfaits et processus de Lévy. Periodica Math. Hungar. 2 (1972) 49-59.

[29] KAKUTANI, S. Two-dimensional Brownian motion and harmonic functions. Proc. Acad. Tokyo 20 (1944) 706-714.

[30] KAKUTANI, S. Markoff processes and the Dirichlet problem. Proc. Acad. Tokyo 21 (1945) 227-233.

[31] KANDA, M. Two Theorems on Capacity for Markov Processes with Stationary Independent Increments. Z. Wahrscheinlichkeitstheorie 35 (1976) 159-165.

[32] KANDA, M. Characterization of Semipolar Sets for Processes with Stationary Independent Increments. Z. Wahrscheinlichkeitstheorie 42 (1978) 141-154.

[33] KANDA, M. On the class of polar sets for a certain class of Lévy processes on the line. J. Math. Soc. Japan 35 (1983) 221-242.

[34] KELLOGG, O.D. Foundations of Potential Theory. Springer-Verlag, Berlin, 1929; reprinted by Dover, New York, 1955.

[35] KESTEN, H. Hitting probabilities of single points for processes with stationary independent increments. Mem. Amer. Math. Soc. 93 (1969).

[36] KINGMAN, J.F.C. Recurrence properties of processes with stationary independent increments. J. Austral. Math. Soc. 4 (1964) 223-228.

[37] KINGMAN, J.F.C. Subadditive ergodic theory. Ann. Probab. 1 (1973) 883-909.

[38] LAMPERTI, J. Wiener's test and Markov chains. J. Math. Anal. Appl. 6 (1963) 58-66.

[39] LANDKOF, N.S. Foundations of Modern Potential Theory. Springer-Verlag, Berlin, 1972.

[40] McKEAN, H.P. A probabilistic interpretation of equilibrium charge distribution. J. Math. Kyoto Univ. 4 (1965) 617-625.

[41] OREY, S. Polar sets for processes with independent increments. In Markov processes and potential theory, ed. J. Chover, Wiley, New York, 1967.

[42] PORT, S.C. and C.J. STONE. The asymmetric Cauchy process on the line. Ann. Math. Statist. 40 (1969) 137-143.

[43] PORT, S.C. and C.J. STONE. Infinitely divisible processes and their potential theory I. Ann. Inst. Fourier (Grenoble) 21 (2) (1971) 157-275.

[44] PORT S.C. and C.J. STONE. Infinitely divisible processes and their potential theory II. Ann. Inst. Fourier (Grenoble) 21 (4) (1971) 179-265.

[45] PORT, S.C. and C.J. STONE. Brownian Motion and Classical Potential Theory. Academic Press, New York, 1978.

[46] PHILIPS, H.B. and N. WIENER. Nets and the Dirichlet problem. J. Math. and Phys. 2 (1923) 105-124.

[47] PRUITT, W.E. The Hausdorff dimension of the range of a process with stationary independent increments. J. Math. Mech. 19 (1969) 371-378.

[48] PRUITT, W.E. Some Dimension Results for Processes with Independent Increments. In Stochastic Processes and Related Topics, ed. M. Puri. Academic Press, New York, 1975.

[49] RAO, M. On a result of M. Kanda. Z. Wahrscheinlichkeitstheorie 41 (1977) 35-37.

[50] SILVERSTEIN, M.L. Symmetric Markov Processes. Lecture Notes in Mathematics 426, Springer-Verlag, Berlin, 1974.

[51] SILVERSTEIN, M.L. Boundary Theory for Symmetric Markov Processes. Lecture Notes in Mathematics 516, Springer-Verlag, Berlin, 1976.

[52] SILVERSTEIN, M.L. The sector condition implies that semipolar sets are quasi-polar. Z. Wahrscheinlichkeitstheorie 41 (1977) 13-33.

[53] SPITZER, F. Electrostatic capacity, heat flow and Brownian motion. Z. Wahrscheinlichkeitstheorie 3 (1964) 110-121.

[54] STEIN, E.M. and G. WEISS. Introduction to Fourier Analysis on Euclidean Spaces. Princeton University Press, Princeton, 1971.

[55] STRATTON, H.H. On dimension of support for stochastic processes with independent increments. Trans. Amer. Math. Soc. 132 (1968) 1-29.

[56] TAYLOR, S.J. Sample path properties of a transient stable process. J. Math. Mech. 16 (1967) 1229-1246.

[57] WERMER, J. Potential Theory. Lecture Notes in Mathematics 408, Springer-Verlag, Berlin, 1974.

[58] WHITMAN, W. Some strong laws for random walks and Brownian motion. Ph. D. Thesis, Cornell, 1964.

[59] WIENER, N. The Dirichlet problem. J. Math. Phys. 3 (1924) 127-146.

[60] ZABCZYCK, J. Sur la théorie semi-classique du potentiel pour les processus à accroissements indépendants. Studia Math. 35 (1970) 227-247.

[61] ZABCZYCK, J. A note on semipolar sets for processes with independent increments. Lecture Notes in Mathematics 472 277-283, Springer-Verlag, Berlin, 1975.

$BM(\mathbb{R}^3)$ and its area integral $\int \beta \times d\beta$

by

Gareth C. Price, L.C.G. Rogers, and David Williams

1. Let β be a $BM(\mathbb{R}^3)$, that is, a Brownian motion on $\mathbb{R}^3$. For the moment, regard β_0 as some fixed (deterministic) point of $\mathbb{R}^3$.

Let α denote the 'area integral' of β, defined by

(1.1) $$\alpha_t = \alpha_0 + \int_{(0,t]} \beta_s \times d\beta_s ,$$

where α_0 is some fixed point of $\mathbb{R}^3$, the $\times$ symbol signifies the vector product, and d signifies the Itô differential.

Since

$$d\langle \alpha^i, \alpha^j \rangle = -\beta^i \beta^j dt \qquad (i \neq j),$$

the path of α determines the path of β modulo a global (in t) sign change $\beta \mapsto -\beta$. For some remarkable examples of this kind of explicit construction of one process in terms of another, see Stroock and Yor [1].

We wish to investigate how much information the process $|\alpha|$ carries about β, but with a different interpretation of how this might be measured. In a sense we want to know how much freedom we have to 'perturb' β without changing $|\alpha|$. Now, let us be more precise.

(1.2) THEOREM. *Let $\tilde{\beta}$ be a Brownian motion relative to the augmented filtration determined by β. Let $\tilde{\alpha}_0$ be a fixed point of $\mathbb{R}^3$, and let*

$$\tilde{\alpha}_t = \tilde{\alpha}_0 + \int_{(0,t]} \tilde{\beta}_s \times d\tilde{\beta}_s .$$

Suppose that $|\tilde{\alpha}_t| = |\alpha_t|$, $\forall t$. _Then, on each component interval of the open set_ $\{t : \alpha_t . \beta_t \neq 0\}$, _the function_ $\tilde{\beta}$ _is a constant orthogonal transformation of_ β.

A much more complete description of the relation between β and $\tilde{\beta}$ will be given later.

Two of the results used in the proof of the theorem, Lemmas 1.3 and 1.4, have independent interest.

(1.3) LEMMA. _We have the following skew-product representation_:

$$|\alpha_t \times \beta_t| = r\left(\int_{(0,t]} \{|\alpha_s|^2 + |\beta_s|^4\} ds\right),$$

where r _is a BES(2) process_. _Thus_, $\alpha_t \times \beta_t$ _can never be zero at a positive time_.

Recall that a BES(2) process is a process identical in law to the radial part of 2-dimensional Brownian motion.

For the next lemma, we need some notation:

$O(3)$ denotes the group of orthogonal 3×3 matrices,

$o(3)$ denotes the Lie algebra of skew-symmetric 3×3 matrices,

a superscript T signifies transpose,

for a vector $\beta = (\beta^1, \beta^2, \beta^3)$ in $\mathbb{R}^3$, $V(\beta)$ denotes the element of $o(3)$ defined by

$$V(\beta) = \begin{pmatrix} 0 & -\beta^3 & \beta^2 \\ \beta^3 & 0 & -\beta^1 \\ -\beta^2 & \beta^1 & 0 \end{pmatrix},$$

so that $V(\beta)\gamma = \beta \times \gamma$, $\gamma \in \mathbb{R}^3$.

We let ∂ denote the Stratonovich differential.

(1.4) LEMMA. Let β and $\tilde{\beta}$ be two $BM(\mathbb{R}^3)$ processes. Suppose that

1.4 (i) $\tilde{\beta}$ is a Brownian motion relative to the augmented filtration generated by β,

1.4 (ii) $|\tilde{\beta}_t| = |\beta_t|$, $\forall t$.

Then there exists a previsible $O(3)$ valued process H such that

(1.5) $d\tilde{\beta} = Hd\beta$,

(1.6) $\tilde{\beta} = H\beta$.

Now make the extra assumption that H is a continuous semimartingale. Define a 3×3 matrix valued process A by

(1.7) $A_0 = 0, \quad \partial A = H^{-1}\partial H.$

Then A is $o(3)$ valued, and

(1.8) $\partial H = H\partial A.$

Moreover, A solves an Itô equation

(1.9) $dA = V(\beta)dx + V(\lambda)dt$,

where x is a 1-dimensional semimartingale with canonical decomposition

(1.10) $dx = \lambda.d\beta + df$,

where λ is a previsible $\mathbb{R}^3$ valued process, and f is a continuous (adapted) process of finite variation.

The switching between Itô and Stratonovich is a little annoying. However, (1.5) and (1.10) must be Itô equations, while the Stratonovich form of (1.7) and (1.8) best brings out their meaning. In Stratonovich form, equation (1.9) reads:

$$\partial A = V(\beta)\partial x + \frac{1}{2}V(\lambda)\partial t .$$

We emphasize that the 'converse' to Lemma 1.4 holds. Thus, take an arbitrary previsible $\mathbb{R}^3$ valued process λ, and an arbitrary continuous adapted process f of finite variation. Define x via (1.10), and A (with $A_0 = 0$) via (1.9). Next define H via (1.8) with H_0 an arbitrary element of $O(3)$. Finally, define $\tilde{\beta}$ via (1.6). Then (1.5) holds, so that $\tilde{\beta}$ is a $BM(\mathbb{R}^3)$ satisfying 1.4(i); and, of course, 1.4(ii) follows from (1.6).

Notation. We continue to use :
Greek letters for processes with values in $\mathbb{R}^3$;
capital Roman letters for 3×3 matrix valued processes;
small Roman letters for real valued processes.

For continuous semimartingales x and y, we write Itô's formula for the derivative of a product as

$$d(xy) = xdy + (dx)y + dxdy.$$

so that $dxdy = d\langle x,y\rangle$. This extends to 3×3 matrix valued continuous semimartingales as

$$d(XY) = XdY + (dX)Y + dXdY,$$

where, with X^i_j denoting the (i,j) th component of X,

$$(dXdY)^i_k = \sum_j d\langle X^i_j, X^j_k\rangle .$$

We make much use of the standard formulae:

$$(\alpha\times\beta)\times\gamma = (\alpha.\gamma)\beta - (\beta.\gamma)\alpha, \qquad (\alpha\times\beta).\gamma = \alpha.(\beta\times\gamma),$$

$$(\alpha\times\beta).(\gamma\times\delta) = (\alpha.\gamma)(\beta.\delta) - (\alpha.\delta)(\beta.\gamma),$$

etc..

2. <u>Proof of Lemma 1.3.</u> Let β be a BM($\mathbb{R}^3$), and let α be its area integral. Define

$$a = |\beta|^2, \qquad b = (\alpha.\beta), \qquad c = |\alpha|^2.$$

It is intuitively clear that the triple (a,b,c) is Markovian, and this is easily confirmed from the following calculations:

(2.1) $da = 2\beta.d\beta + d\beta.d\beta = 2\beta.d\beta + 3dt,$

(2.2) $db = \alpha.d\beta + (d\alpha).\beta + (d\alpha).(d\beta) = \alpha.d\beta,$

(2.3) $dc = 2\alpha.d\alpha + d\alpha.d\alpha = 2\alpha.(\beta\times d\beta) + (\beta\times d\beta).(\beta\times d\beta)$

$\qquad = 2(\alpha x\beta).d\beta + 2adt.$

What clinches the Markov property is of course that

(2.4) $u \equiv |\alpha\times\beta|^2 = |\alpha|^2|\beta|^2 - (\alpha.\beta)^2 = ac - b^2.$

Thus the diffusion process (a,b,c) has drift $(3,0,2a)$, and diffusion matrix

$$\begin{pmatrix} 4a & 2b & 0 \\ 2b & c & 0 \\ 0 & 0 & 4u \end{pmatrix}.$$

We do not actually use here the Markovian nature of (a,b,c), but it did suggest the skew-product formula.

From (2.4),

(2.5) $du = adc + cda + dadc - 2bdb - dbdb$

$\qquad = 2\{a(\alpha\times\beta) + c\beta - b\alpha\}.d\beta + (2a^2 + 3c - c)dt$

$\qquad = 2\{|\beta|^2(\alpha\times\beta) + (\alpha\times\beta)\times\alpha\}.d\beta + 2(|\beta|^4 + |\alpha|^2)dt.$

Thus

$$du - 2(|\beta|^4 + |\alpha|^2)dt = d(\text{local martingale}),$$

$$dudu = 4u(|\beta|^4 + |\alpha|^2)dt.$$

It is well known that these properties imply Lemma 1.3.

3. Proof of Lemma 1.4. Let β and $\tilde{\beta}$ be two $BM(\mathbb{R}^3)$ processes, with $\tilde{\beta}$ a Brownian motion relative to the augmented filtration determined by β. Then the martingale representation theorem guarantees that there exists a previsible $O(3)$ valued process H such that

(3.1) $d\tilde{\beta} = Hd\beta$.

Suppose further that $|\tilde{\beta}_t| = |\beta_t|$, $\forall t$. Then

$$d(\tilde{\beta}.\tilde{\beta}) = 2\tilde{\beta}.d\tilde{\beta} + 3dt = d(\beta.\beta) = 2\beta.d\beta + 3dt.$$

Hence

$$\tilde{\beta}.d\tilde{\beta} = (H^T\tilde{\beta}).d\beta = \beta.d\beta$$

and so $H^T\tilde{\beta} = \beta$, equivalently, $\tilde{\beta} = H\beta$, for almost all t. If we modify H on a set of measure zero, we do not affect (3.1). Hence, we can assume that

(3.2) $\tilde{\beta} = H\beta$ (for all t).

Now, we assume that H is a continuous semimartingale. Taking the Itô derivative of (3.2), and comparing with (3.1), we see that

(3.3) $(dH)\beta + dHd\beta = 0$.

It will be convenient for a moment to work with Stratonovich derivatives. From

$$HH^T = I,$$

it follows that

$$(\partial H)H^T + H\partial H^T = 0, \quad \text{so} \quad H^{-1}\partial H = -(\partial H^T)(H^T)^{-1}.$$

Let

$$A_0 = 0, \quad \partial A = H^{-1}\partial H = -\partial A^T.$$

Then, obviously, A is $o(3)$ valued, and

$$\partial H = H\partial A. \tag{3.4}$$

The Itô form of (3.4) reads

$$dH = HdA + \frac{1}{2}dHdA.$$

Thus (3.3) now yields

$$(3.5) \qquad (HdA)\beta + \frac{1}{2}(dHdA)\beta + dHd\beta = 0.$$

Let M be the martingale part of the 3×3 matrix valued process A, and let F be the continuous finite-variation part: $A = M + F$. On looking at the martingale part of (3.5), we see that

$$(HdM)\beta = 0, \quad \text{so} \quad (dM)\beta = 0.$$

It is easy to deduce, using the fact that M is skew-symmetric, that

$$dM = dmV(\beta),$$

where m is a 1-dimensional martingale. Necessarily, we have

$$dm = \lambda.d\beta$$

for some previsible $\mathbb{R}^3$ valued process λ.

We now have

$$dH = HV(\beta)dm + d(\text{finite variation}),$$

so

$$dHdA = HdAdA = HdMdM = H|\lambda|^2 V(\beta)^2 dt,$$

and

$$(dHdA)\beta = 0.$$

Moreover,

$$dHd\beta = HV(\beta)dmd\beta = HV(\beta)\lambda dt.$$

Substitution in (3.5) now gives

$$(HdF)\beta + HV(\beta)\lambda dt = 0,$$

so that

$$(dF)\beta + V(\beta)\lambda dt = (dF)\beta - V(\lambda)\beta dt = 0.$$

Since F is skew-symmetric, we must have

$$dF = V(\beta)df + V(\lambda)dt,$$

where f is a 1-dimensional continuous finite-variation process.

Lemma 1.4 is proved.

4. <u>Proof of Theorem 1.2</u>. Let β be a $BM(\mathbb{R}^3)$. Let $\tilde{\beta}$ be another $BM(\mathbb{R}^3)$ relative to the augmented filtration generated by β. We assume equality of the moduli of the area integrals:

$$|\tilde{\alpha}_t|^2 = |\alpha_t|^2, \quad \forall t.$$

By equation (2.3),

$$2(\tilde{\alpha}\times\tilde{\beta}).d\tilde{\beta} + 2|\tilde{\beta}|^2 dt = 2(\alpha\times\beta).d\beta + 2|\beta|^2 dt. \tag{4.1}$$

Equating the finite-variation parts gives:

$$|\tilde{\beta}_t| = |\beta_t|, \quad \forall t. \tag{4.2}$$

We can now apply the trivial first part of Lemma 1.4 to show that, for some previsible $O(3)$ valued process H,

$$d\tilde{\beta} = H d\beta, \tag{4.3}$$

$$\tilde{\beta} = H\beta . \tag{4.4}$$

On equating martingale parts at (4.1), we obtain

$$(\tilde{\alpha}\times\tilde{\beta}).d\tilde{\beta} = (\alpha\times\beta).d\beta,$$

whence (compare the argument leading to (3.2))

$$\tilde{\alpha}\times\tilde{\beta} = H(\alpha\times\beta), \tag{4.5}$$

for almost all t, and it can be assumed that (4.5) holds for all t.

It is obvious from (4.5) that

$$|\tilde{\alpha}\times\tilde{\beta}|^2 = |\alpha\times\beta|^2.$$

Take Itô derivatives using (2.5) to see that (again via the argument leading to (3.2))

$$|\tilde{\beta}|^2(\tilde{\alpha}\times\tilde{\beta}) + (\tilde{\alpha}\times\tilde{\beta})\times\tilde{\alpha} = |\beta|^2 H(\alpha\times\beta) + H\{(\alpha\times\beta)\times\alpha\} ,$$

so that, from (4.2) and (4.5),

$$|\tilde{\alpha}|^2\tilde{\beta} - (\tilde{\alpha}.\tilde{\beta})\tilde{\alpha} = |\alpha|^2 H\beta - (\alpha.\beta)H\alpha.$$

Thus, because of (4.4) and the given fact that $|\tilde{\alpha}| = |\alpha|$, we have

$$(\tilde{\alpha}.\tilde{\beta})\tilde{\alpha} = (\alpha.\beta)H\alpha. \tag{4.6}$$

Take the scalar product of (4.6) with $\tilde{\beta} = H\beta$, and recall that H preserves scalar products, to find that

$$(\tilde{\alpha}.\tilde{\beta})^2 = (\alpha.\beta)^2.$$

For $\alpha.\beta \neq 0$, define $e_t = (\tilde{\alpha}.\tilde{\beta})_t/(\alpha.\beta)_t = \pm 1$. Then (4.6) implies that

$$\tilde{\alpha} = eH\alpha. \tag{4.7}$$

But

$$(\tilde{\alpha}\times\tilde{\beta}) = H(\alpha\times\beta) = (\det H)(H\alpha)\times(H\beta) = (\det H)e(\tilde{\alpha}\times\tilde{\beta}),$$

and, for $\alpha.\beta \neq 0$,

$$e_t = \det H_t .$$

It is obvious from the definition of e that e is continuous, and therefore constant either at 1 or at -1, on component intervals of the set $\{t : \alpha_t.\beta_t = 0\}$.

The reader will be able to see that, to finish the proof, we need only show that *if* $\alpha_0 \times \beta_0 \neq 0$, *and* e *is globally constant* $(e_t = e_0, \forall t)$, *then* $H_t = H_0$, $\forall t$.

So *assume that* $\alpha_0 \times \beta_0 \neq 0$, *and* $e_t = e_0$, $\forall t$. Recall from Lemma 1.3 that then $\alpha_t \times \beta_t \neq 0$, $\forall t$. Then H_t is uniquely determined by the fact that it maps the orthogonal triple

$$(\beta_t, \alpha_t \times \beta_t, \beta_t \times (\alpha_t \times \beta_t))$$

into the triple

$$(\tilde{\beta}_t, \tilde{\alpha}_t \times \tilde{\beta}_t, e_0\tilde{\beta}_t \times (\tilde{\alpha}_t \times \tilde{\beta}_t)).$$

Hence H *is a continuous semimartingale*, and all the results of Lemma 1.4 apply. We use the notation of that Lemma.

From (4.7),

$$\tilde{\alpha} = e_0 H\alpha,$$

so that

$$d\tilde{\alpha} = e_0 H d\alpha + e_0 (dH)\alpha + e_0 dHd\alpha.$$

But

$$d\tilde{\alpha} = \tilde{\beta} \times d\tilde{\beta} = (H\beta)\times(Hd\beta) = (\det H)Hd\alpha = e_0 Hd\alpha ,$$

so that

$$(dH)\alpha + dHd\alpha = 0.$$

Thus,

(4.8) $$(HdA + \frac{1}{2} dHdA)\alpha + dHd\alpha = 0.$$

Looking at the martingale-differential part of (4.8), we see that

$$HV(\beta)\alpha\, dm = H(\beta\times\alpha)dm = 0, \quad \text{where} \quad dm = \lambda.d\beta.$$

Since $\beta\times\alpha$ is never zero, it follows that $dm = 0$. Thus, (4.8) reduces to the statement

$$HV(\beta)\alpha df = 0 = H(\beta\times\alpha)df,$$

and, again <u>because</u> $\beta\times\alpha$ <u>is never zero</u>, we have $df = 0$. Thus, $dA = 0$, and $H_t = H_0$, $\forall t$.

5. Example. The proof of Theorem 1.2 shows clearly how to construct an example to show what can 'go wrong' when $\alpha.\beta = 0$.

Let β be a $BM(\mathbb{R}^3)$ with $\beta_0 = 0$, and let $\alpha = \int \beta x d\beta$. Let

$$\tau = \inf\{t > 1 : \alpha_t.\beta_t = 0\}.$$

Let

$$H_t = \begin{cases} I, & t < \tau, \\ J, & t \geq \tau, \end{cases}$$

where J is specified by

$$J(\beta_\tau) = \beta_\tau, \quad J(\alpha_\tau \times \beta_\tau) = \alpha_\tau \times \beta_\tau, \quad J(\gamma_\tau) = -\gamma_\tau,$$

where

$$\gamma_\tau = \beta_\tau \times (\alpha_\tau \times \beta_\tau) = |\beta_\tau|^2 \alpha_\tau ,$$

since $\alpha_\tau . \beta_\tau = 0$. Note that

$$J(\alpha_\tau) = -\alpha_\tau .$$

Set $\tilde{\beta}_0 = 0$,

$$\tilde{\beta}_t = \int H_s d\beta_s .$$

Then

$$\tilde{\beta}_t = \begin{cases} \beta_t, & t < \tau, \\ \beta_\tau + J(\beta_t - \beta_\tau) = J\beta_t, & t \geq \tau . \end{cases}$$

Define $\tilde{\alpha} = \int \tilde{\beta} \times d\tilde{\beta}$. Then, since $\det I = 1$ and $\det J = -1$,

$$d\tilde{\alpha}_t = \begin{cases} d\alpha_t, & t < \tau , \\ -Jd\alpha_t, & t \geq \tau . \end{cases}$$

Thus

$$\tilde{\alpha}_t = \begin{cases} \alpha_t, & t < \tau, \\ \alpha_\tau - J(\alpha_t - \alpha_\tau) = -J\alpha_t, & t \geq \tau. \end{cases}$$

Finally, $$|\tilde{\alpha}_t| = |\alpha_t|, \ \forall t.$$

REFERENCE

[1] D.W. Stroock and M. Yor, Some remarkable martingales, Séminaire de Probabilités, XV Springer Lecture Notes in Math., 850, 1981.

University College
Singleton Park
SWANSEA SA2 8PP
Great Britain

THE UNIQUE FACTORISATION OF BROWNIAN PRODUCTS

by

Gareth C. Price

The object of this work is to investigate the connection between an n-dimensional real or complex Brownian Motion and the product of its components, and to discuss the extent to which the 'factorisation' of the product is unique.

We first make the following definition.

Let $\underset{\sim}{x}$ and $\underset{\sim}{y}$ be elements of $\mathbb{R}^n$ or $\mathbb{C}^n$. We say that $\underset{\sim}{x}$ and $\underset{\sim}{y}$ are <u>equivalent</u>, and write $\underset{\sim}{x} \sim \underset{\sim}{y}$, if $\prod_{j=1}^{n} x_j = \prod_{j=1}^{n} y_j$ and there is a permutation σ of $\{1,2,...,n\}$ such that $|y_j| = |x_{\sigma(j)}|$ $\forall$ i $\in \{1,2,...,n\}$. The equivalence class $\bar{\underset{\sim}{x}}$ containing $\underset{\sim}{x}$ under this equivalence relation may be termed a <u>factorisation</u> of $p \equiv \prod_{j=1}^{n} x_j$ by analogy with the factorisation of integers, where such an equivalence class is <u>unique</u>: every integer z has a unique factorisation into signed primes $(q_1,...,q_n) \in \mathbb{Z}^n$ in the sense that if $(r_1,...,r_n)$ is another such factorisation, $\prod_{i=1}^{n} q_j = \prod_{i=1}^{n} r_j$ and there is a permutation (σ) of $\{1,2,...,n\}$ such that $|q_j| = |r_{\sigma(j)}|$ for each $j \in \{1,2,...,n\}$.

The following result is intended to show that Brownian products behave like integers in the sense of their factorisation properties.

<u>Theorem 1</u>

Let $\underset{\sim}{\beta}$ and $\underset{\sim}{\gamma}$ be two $\mathbb{R}^n$- or $\mathbb{C}^n$-valued Brownian Motions on a complete probability space $(\Omega, \mathcal{F}, \mathbb{P})$, and suppose that

$\prod_{j=1}^{n}\beta_j = \prod_{j=1}^{n}\gamma_j$ for each $t \in \mathbb{R}^+$, $w \in \Omega$. Then the conclusion is that (almost surely) $\underset{\sim}{\beta} \sim \underset{\sim}{\gamma}$ for each $t \in \mathbb{R}^+$; in other words $\bar{\underset{\sim}{\beta}}$ is uniquely defined by $p \equiv \prod_{j=1}^{n}\beta_j$.

Proof

As the proof for the real case is simpler, but involving essentially the same argument, we give the proof only for the complex case. If $W_1 = U_1 + iV_1$ and $W_2 = U_2 + iV_2$ are continuous $\mathbb{C}$-valued semimartingales, we write

$$\langle W_1, W_2\rangle = \langle U_1, U_2\rangle - \langle V_1, V_2\rangle + i(\langle U_1, V_2\rangle + \langle V_1, U_2\rangle).$$

Suppose then that we are given an n-dimensional complex Brownian Motion $\underset{\sim}{\beta} \equiv (\beta(1), \beta(2), \ldots, \beta(n))$. Fix $k \in \{1,2,\ldots,n\}$ and let $\{i_1, i_2, \ldots, i_k\}$ be integers drawn from $\{1,2,\ldots,n\}$ such that $i_1 < i_2 < \ldots < i_k$. Define

$$a_k \equiv \sum_{i_1<i_2<\ldots<i_k} |\beta(i_1)|^2|\beta(i_2)|^2\ldots|\beta(i_k)|^2$$

and for an $r \in \{1,2,\ldots,k\}$ consider the terms of the martingale differential of $d(a_k)$ with the factor $|\beta(i_r)|^2$ differentiated. These are

$$2\sum_{i_1<i_2<\ldots<i_k} |\beta(i_1)|^2|\beta(i_2)|^2\ldots\widehat{|\beta(i_r)|^2}\ldots|\beta(i_k|^2$$
$$\times\ [x(i_r)dx(i_r) + y(i_r)dy(i_r)]$$

where $\beta(i_r) \equiv x(i_r) + iy(i_r)$ $[i = \sqrt{(-1)}]$ and the circumflex denotes the fact that the factor directly beneath it is omitted.

Similarly, for the martingale $p \equiv \prod_{j=1}^{n} \beta(j)$, the term of dp with the factor $\beta(i_r)$ differentiated is simply

$$\beta(1)\beta(2)\ldots\widehat{\beta(i_r)}\ldots\beta(n)\,[dx(i_r) + i\,dy(i_r)].$$

Hence since

$$\frac{d}{dt}\left\langle \int_0^t x(i_r)dx(i_r) + y(i_r)dy(i_r),\ \ \beta(i_r)\right\rangle = \beta(i_r),$$

it follows that $\frac{d}{dt}\langle a_k, p\rangle = 2(n-k+1)pa_{k-1}$, there being $(n-k+1)$ ways of choosing i_r with the other i_j's fixed. Hence by continuity of a_{k-1} together with the fact that $p = 0$ only on a set of Lebesgue measure zero, we see that if a_k is uniquely defined by p (and adapted to the filtration generated by p) then so is a_{k-1}. Since $a_n \equiv |p|^2$ it follows that all the a_k's are both uniquely defined by p and adapted to p.

Now form the polynomial equation

$$z^n - a_1 z^{n-1} + a_2 z^{n-2} - \ldots + (-1)^n a_n = 0.$$

From the elementary relationship existing between the roots of a polynomial and its coefficients, we see that the roots are precisely the $\{|\beta(j)|^2\}$ for $j \in \{1,2,\ldots,n\}$. Now replace $\underset{\sim}{\beta}$ by $\underset{\sim}{\gamma}$ throughout the above argument to see that the same set of roots will arise for $\underset{\sim}{\gamma}$ since the a_i's are exactly the same for $\underset{\sim}{\gamma}$, as $\underset{\sim}{\gamma}$ has the same product p as $\underset{\sim}{\beta}$. This proves the theorem.

University College of Swansea

SOME INTEGRAL EQUALITIES IN WIENER-HOPF THEORY

by

Neil Baker

1. Introduction.

In 1982, London, McKean, Rogers and Williams ([1,2]) published two papers on Wiener-Hopf problems for 1-dimensional diffusions. Rogers ([3,4,5]) has given a most illuminating extension of the theory presented in [1,2] in work on Wiener-Hopf factorization for Lévy processes.

The paper [2] by London et al. established the following solution to a concrete problem. Let B be a Brownian motion, and let

$$\phi_t = \int_0^t I_{(1,\infty)}(B_s)ds - \int_0^t I_{[0,1]}(B_s)ds ,$$

$$\tau_t^+ = \inf\{s : \phi_s > t\}, \qquad Y_t^+ = B(\tau_t^+).$$

Then, for $0 < x < 1$ and $0 < y < \infty$,

$$\mathbb{P}[Y_0^+ \in 1 + dy | B_0 = x] = \pi(x,y)dy, \tag{1.1}$$

where

$$\pi(x,y) = \frac{\cosh \frac{1}{2}\pi y(\cos \frac{1}{2}\pi x \sinh \frac{1}{2}\pi y)^{\frac{1}{2}}}{2^{\frac{1}{2}}(\sinh^2 \frac{1}{2}\pi y + \cos^2 \frac{1}{2}\pi x)} . \tag{1.2}$$

The problem just described is the "even half" of the problem discussed in this paper. Consider now the functional

$$\phi_t = \int_0^t I_{(1,\infty)}(B_s)ds + \int_0^t I_{(-\infty,-1)}(B_s)ds - K^2\int_0^t I_{[-1,1]}(B_s)ds. \tag{1.3}$$

Define

$$\tau_t^+ = \inf\{s : \phi_s > t\}, \qquad Y_t^+ = B(\tau_t^+).$$

Then $\mathbb{P}[Y_0^+ \in (-\infty,-1)\cup(1,\infty)] = 1$, and it is convenient to write

$$V_0 = \begin{cases} Y_0^+ - 1 & \text{if } Y_0^+ > 1, \\ Y_0^+ + 1 & \text{if } Y_0^+ < -1. \end{cases}$$

We prove that for $|x| < 1$, $|V| > 0$,

$$\text{(1.4)} \qquad \mathbb{P}[V_0 \in dV | B_0 = x] = \frac{(\sinh \frac{\pi |V|}{2K})^{\alpha} (\cos \frac{\pi x}{2})^{1-\alpha}}{2K(K^2+1)^{\frac{1}{2}}(\cosh \frac{\pi V}{2K} - \operatorname{sgn}(V) \sin \frac{\pi x}{2})}$$

where $\alpha = \frac{2\mu}{\pi}$ and $\mu = \cos^{-1}(K^2+1)^{-\frac{1}{2}}$, $0 < \mu < \frac{\pi}{2}$, $K > 0$.

An interesting feature of this problem is that it does not reduce to a classical Wiener-Hopf problem in terms of local time at a single boundary point. Thus the probabilistic structure is richer. It is hoped to publish soon a treatment of more interesting problems which do much to explain the strange combinations of exponential and trigonometric functions which appear in this theory.

2. The probabilistic approach.

Suppose that B is a Brownian motion on the real line. Let us define the following quantities:

$$N_+^+ = \max(B_t - 1, 0);$$

$$N_-^+ = \min(B_t + 1, 0); \tag{2.1}$$

$$B^- = \begin{cases} \min(B_t, 1) & \text{if } B_t > 0 \\ \max(B_t, -1) & \text{if } B_t < 0. \end{cases}$$

Also, let

$$I_+^+(t) = I_{(1,\infty)} \circ B_t; \quad I_-^+(t) = I_{(-\infty,-1)} \circ B_t; \quad I^-(t) = I_{[-1,1]} \circ B_t.$$

If we let L_1 and L_{-1} represent the local time of B, with the Itô-McKean normalization, at 1 and -1 respectively, then Tanaka's formula gives:

$$d(N_+^+ - L_1) = I_+^+ dB;$$

$$d(N_-^+ + L_{-1}) = I_-^+ dB; \tag{2.2}$$

$$d(B^- + L_1 - L_{-1}) = I^- dB.$$

Let us define a fluctuating functional ϕ by

$$\phi(t) = \phi_+^+(t) + \phi_-^+(t) - \phi^-(t),$$

where

$$\phi_+^+(t) = \int_0^t I_+^+(s)ds; \quad \phi_-^+(t) = \int_0^t I_-^+(s)ds; \quad \phi^-(t) = K^2\int_0^t I^-(s)ds$$

and let

$$\tau^+(t) = \inf\{s : \phi_s > t\}, \quad \tau^-(t) = \inf\{u : \phi_s < -t\}.$$

We may then define the time-changed processes

$$Y^+(t) = B \circ \tau^+(t) \in (-\infty,-1] \cup [1,\infty)$$

and

$$V^+(0) = \begin{cases} Y^+(0)-1, & \text{if } Y^+(0) > 1 \\ Y^+(0)+1, & \text{if } Y^+(0) < -1\,. \end{cases}$$

Hence $V^+(0) \in (-\infty,\infty)$.

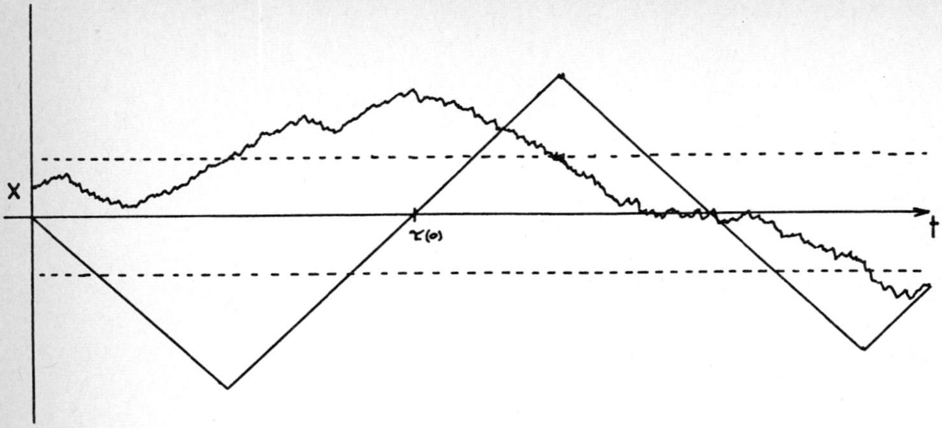

We wish to find a $\pi(x,v)$ such that

$$\mathbb{P}[V^+(0) \in dv \mid B(0) = +x] = \pi(x,v)dv$$

where $-1 < x < 1$ and $-\infty < v < \infty$.

Let $\gamma > 0$. Then we may write down the following exponential martingales corresponding to (2.2):

$$\exp[+i\gamma(N_+^+ - L_1) + \tfrac{1}{2}\gamma^2\phi_+^+];$$

$$\exp[+i\gamma(N_-^+ + L_{-1}) + \tfrac{1}{2}\gamma^2\phi_-^+];$$

$$\exp[K\gamma(B^- + L_1 - L_{-1}) - \tfrac{1}{2}\gamma^2\phi^-].$$

Since the square brackets processes $[N_+^+ - L_1, N_-^+ + L_{-1}]$, $[N_+^+ - L_1, B^- + L_1 - L_{-1}]$ and $[N^+ + L_{-1}, B^- + L_1 - L_{-1}]$ are all zero, it is natural to consider the process

(2.3) $$W = \exp[+i\gamma(N_+^+ + N_-^+ - iKB^-) + \tfrac{1}{2}\gamma^2\phi].$$

Applying Itô's formula, we get

$$dW = (i - k)\, W(dL_1 - dL_{-1}) + d(\text{local martingale}).$$

Now, L_1 grows only when $B = 1$, and, similarly, L_{-1} grows only when $B = -1$. Therefore we have

$$dW = (i - k)\gamma[\exp\{\gamma + \tfrac{1}{2}\gamma^2\phi\}dL_1 - \exp\{-\gamma + \tfrac{1}{2}\gamma^2\phi\}dL_{-1}]$$

$$+ \text{d(local martingale)},$$

or equivalently,

$$(1 - Ki)W_t = i(K^2 + 1)\gamma\int_0^t e^{\frac{1}{2}\gamma^2\phi}\{e^{\gamma}dL_1 - e^{-\gamma}dL_{-1}\}$$

$$+ \text{local martingale}.$$

Hence

$$M_t = \mathrm{Re}[(1 - Ki)W_t]$$

$$= \mathrm{Im}[(K + i)W_t] \text{ is a local martingale.}$$

Since M is bounded on $[0,\tau^+(0)]$ we have, for $-1 < x < 1$,

$$\mathbb{E}[M \circ \tau^+(0) | B(0) = +x] = \mathbb{E}[M(0) | B(0) = +x]$$

so that (cf (2.3))

$$\text{(2.4)} \qquad \mathrm{Im}[(K + i)\int_{y>0} e^{i\gamma y}e^{K\gamma}\pi(x,y)dy$$

$$+(K + i)\int_{y<0} e^{i\gamma y}e^{-K\gamma}\pi(x,y)dy] = e^{K\gamma x}$$

Now comes the tricky part!

If we write

$$\pi(x,y) = \pi^e(x,y) + \pi^0(x,y)$$

where

$$\pi^e(x,y) = \pi^e(x,-y) = \pi^e(-x,y)$$

and $$\pi^0(x,y) = -\pi^0(x,-y) = -\pi^0(-x,y)$$

then (2.4) may be written

$$(2.5)\qquad \mathrm{Im}\left[(K+i)\left[\int_{y>0} e^{i\gamma y}e^{K\gamma}(\pi^e+\pi^0)dy + \int_{y<0} e^{i\gamma y}e^{-K\gamma}(\pi^e+\pi^0)dy\right]\right] = e^{K\gamma x}$$

If we use the change of variable $v = -y$ above we get

$$(2.6)\qquad \mathrm{Im}\ (K+i)\left[\int_{v<0} e^{-i\gamma v}e^{K\gamma}(\pi^e-\pi^0)dv + \int_{v>0} e^{-i\gamma v}e^{-K\gamma}(\pi^e-\pi^0)dv\right] = e^{K\gamma x}\ .$$

Adding (2.5) and (2.6) gives

$$\begin{aligned}(2.7)\qquad \mathrm{Im}\ (K+i)\Bigg[\int_{y>0} &([\cos\gamma y\cosh K\gamma + i\sin\gamma y\sinh K\gamma]\pi^e \\ &+ [\cos\gamma y\sinh K\gamma + i\sin\gamma y\cosh K\gamma]\pi^0)dy \\ + \int_{y<0} &([\cos\gamma y\cosh K\gamma - i\sin\gamma y\sinh K\gamma]\pi^e \\ &- [\cos\gamma y\sinh K\gamma - i\sin\gamma y\cosh K\gamma]\pi^0)dy\Bigg] \\ &= e^{K\gamma x}\ .\end{aligned}$$

Let us now replace x by $-x$ in (2.7) to get

$$\begin{aligned}(2.8)\qquad \mathrm{Im}\ (K+i)\Bigg[\int_{y>0} &([\cos\gamma y\cosh K\gamma + i\sin\gamma y\sinh K\gamma]\pi^e \\ &- [\cos\gamma y\sinh K\gamma + i\sin\gamma y\cosh K\gamma]\pi^0)dy \\ + \int_{y<0} &([\cos\gamma y\cosh K\gamma - i\sin\gamma y\sinh K\gamma]\pi^e \\ &+ [\cos\gamma y\sinh K\gamma - i\sin\gamma y\cosh K\gamma]\pi^0)dy\Bigg] \\ &= e^{-K\gamma x}\ .\end{aligned}$$

Lo and behold! if we now substract (2.8) from (2.7) we get:

$$\sinh K\gamma x = \mathrm{Im}\ (K+1)\Big[\int_{y>0} [\cos\gamma y \sinh K\gamma + i \sin\gamma y \cosh K\gamma]\ \pi^0 dy$$

$$+ \int_{y<0} [-\cos\gamma y \sinh K\gamma + i \sin\gamma y \cosh K\gamma]\pi^0 dy\Big] .$$

Therefore

$$\sinh K\gamma x = \int_{y>0} (\cos\gamma y \sinh K\gamma + K \sin\gamma y \cosh K\gamma)\ \pi^0 dy \tag{2.9}$$

$$+ \int_{y<0} (-\cos\gamma y \sinh K\gamma + K \sin\gamma y \cosh K\gamma)\pi^0 dy .$$

Alternatively, adding (2.8) and (2.7) gives the "even" equation, namely,

$$\cosh K\gamma x = \int_{y>0} (\cos\gamma y \cosh K\gamma + K \sin\gamma y \sinh K\gamma)\ \pi^e dy \tag{2.10}$$

$$+ \int_{y<0} (\cos\gamma y \cosh K\gamma - K \sin\gamma y \sinh K\gamma)\ \pi^e dy .$$

Let us now, instead, derive the dual equations. We consider the following exponential martingales:

$$\exp[-\gamma\ (N_+^+ - L_1) - \tfrac{1}{2}\gamma^2\phi_+^+] ,$$

$$\exp[\gamma(N_-^+ + L_{-1}) - \tfrac{1}{2}\gamma^2\ \phi_-^+] ,$$

$$\exp[iK\gamma(B^- + L_1 - L_{-1}) + \tfrac{1}{2}\gamma^2\phi^-] ,$$

$$W_t = \exp[-\gamma(N_+^+ - N_-^+ - iKB^-) - \tfrac{1}{2}\gamma^2\phi\ ]. \tag{2.11}$$

Applying Itô's formula we get

$$dW_t = -\gamma W\,(dL_1 + dL_{-1}) + iK\gamma W(-\,dL_1 + dL_{-1})$$
$$+\ d(\text{local martingale}).$$

Since L_1 increases only when $B = 1$ and L_{-1} increases only when $B = -1$ we have, on integrating,

$$W_t = -\gamma \int_0^t e^{-\frac{1}{2}\gamma^2\phi}\Big[(e^{iK\gamma}dL_1 + e^{-iK\gamma}dL_{-1})$$
$$+\ iK(e^{iK\gamma}dL_1 - e^{-iK\gamma}dL_{-1})\Big]$$

(2.12)
$$+\ \text{local martingale}$$

$$= -\gamma \int_0^t e^{-\frac{1}{2}\gamma^2\phi}\Big[dL_1(\cos K\gamma - \sin K\gamma + i(\cos K\gamma + \sin K\gamma))$$
$$+\ dL_{-1}(\cos K\gamma - \sin K\gamma - i(\cos K\gamma + \sin K\gamma))\Big]$$
$$+\ \text{local martingale}\ .$$

Let us now choose $\cos K\gamma = \sin K\gamma$. Then $\mathrm{Re}\{W_t\}$ is clearly bounded on $[0,\tau^-(0)]$, and so we have

$$\mathbb{E}[W_{\tau^-(0)}|B(0) = y] = \mathbb{E}[W_0|B(0) = y],$$

or equivalently (here we appeal to the duality principle of [2] - see, for example (4.2) there),

$$\text{(2.13)}\qquad \mathrm{Re}\left[K^2 \int_x e^{iK\gamma x}\pi(x,y)dx\right] = \begin{cases} \mathrm{Re}[e^{-\gamma y}e^{iK\gamma}] & ,\ \text{if}\quad y > 0 \\ \mathrm{Re}[e^{\gamma y}e^{-iK\gamma}] & ,\ \text{if}\quad y < 0 \end{cases}$$

$$= \begin{cases} e^{-\gamma y}\cos K\gamma & ,\quad y > 0 \\ e^{+\gamma y}\cos K\gamma & ,\quad y < 0\ . \end{cases}$$

Let us now write $\pi(x,y) = \pi^e(x,y) + \pi^0(x,y)$. Then (2.13) becomes:

$$\text{(2.14.1)} \qquad \mathrm{Re}\left[K^2 \int_x e^{iK\gamma x}(\pi^e + \pi^0)dx\right] = \begin{cases} e^{-\gamma y}\cos K\gamma, & y > 0 \\ e^{\gamma y}\cos K\gamma, & y < 0 . \end{cases}$$

If we now replace x by $-x$ we get

$$\text{(2.14.2)} \qquad \mathrm{Re}\left[K^2 \int_x e^{-iK\gamma x}(\pi^e - \pi^0)dx\right] = \begin{cases} e^{-\gamma y}\cos K\gamma, & y > 0 \\ e^{\gamma y}\cos K\gamma, & y < 0 , \end{cases}$$

and, adding this to (2.14.1), then gives

$$\text{(2.15)} \qquad K^2 \int_x \cos K\gamma x \pi^e(x,y)dx = \begin{cases} e^{-\gamma y} \cos K\gamma, & y > 0 \\ e^{\gamma y} \cos K\gamma, & y < 0 . \end{cases}$$

If, instead, we choose $\cos K\gamma = -\sin K\gamma$ and take imaginary parts in (2.10) we obtain, in the same manner as above,

$$\text{(2.16)} \qquad K^2 \int_x \sin K\gamma x\, \pi^0(x,y)dx = \begin{cases} e^{-\gamma y}\sin K\gamma , & y > 0 \\ -e^{\gamma y}\sin K\gamma , & y < 0 . \end{cases}$$

Equations (2.15) and (2.16), with the respective conditions on γ, represent the required dual equations.

Theorem 46 of [1] shows that the solution π^e of (2.10) is <u>unique</u>. We leave aside the proof of uniqueness of π^0 until another occasion.

It will be appreciated that our verification of (2.15) and (2.16) - which is done here only for the case when $K = 1$ - is done to confirm the validity of the 'duality principle' of [2] in the current situation.

3. The Solution.

In order to give the flavour of the proofs we shall prove only the 'odd' case. The 'even' case may be proven similarly.

The odd case.

First note that (2.9) may be written as

$$(3.1)\qquad \int_{y>0} (\cos\mu \cos\theta y \sinh K\theta + \sin\mu \sin\theta y \cosh K\theta)\pi^{0}(x,y)dy$$

$$+ \int_{y<0} (-\cos\mu \cos\theta y \sinh K\theta + \sin\mu \sin\theta y \cosh K\theta)\,\pi^{0}(x,y)dy$$

$$= (K^2+1)^{-\frac{1}{2}} \sinh K\theta x$$

where $\cos\mu = (K^2+1)^{-\frac{1}{2}}$ and $\sin\mu = K(K^2+1)^{-\frac{1}{2}}$. Suppose that $x,y > 0$. Consider the contour integral

$$\oint \frac{(\sinh \frac{\pi z}{2K})^{\alpha}\, e^{i\theta z}}{(\sinh^2 \frac{\pi z}{2K} + \cos^2 \frac{\pi x}{2})}\, dz = \oint h(z)e^{i\theta z} dz ,$$

where $0 < \alpha < 1$, taken around

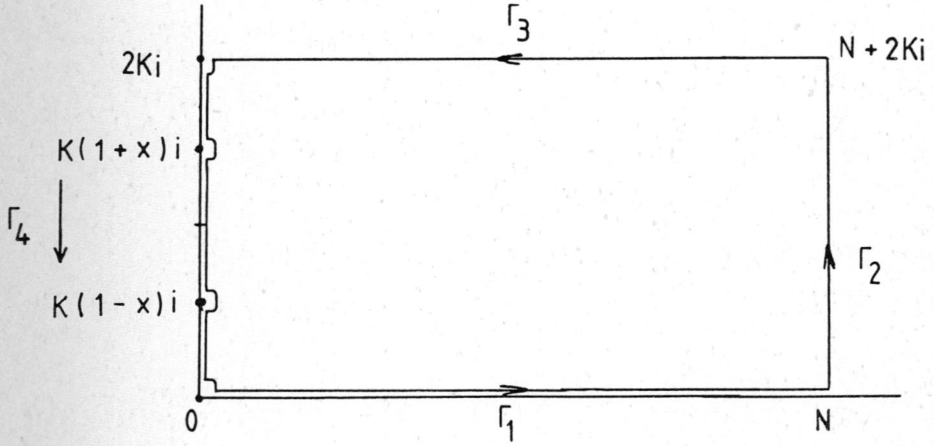

- this contour consisting of straight lines, half circles centred at $K(1-x)i$ and $K(1+x)i$, and quarter circles centred at 0 and $2i$. Since $\mathrm{Im}(\sinh\frac{\pi z}{2K}) \geq 0$ for all z in, and on, this contour, $(\sinh\frac{\pi z}{2K})^{\alpha}$ has a regular branch in this region. The branch that we shall use is obtained by taking the argument of $\sinh\frac{\pi z}{2K}$ in $[-\frac{\pi}{2}, \frac{3\pi}{2})$. Note also that there is no contribution from the quarter circles around the branch points since the integrand remains bounded as their radii tend to zero.

The points $K(1 \pm x)i$ are the poles of the integrand. The residues corresponding to these poles can be calculated to give

$$\operatorname{res} K(1+x)i = \frac{Ke^{-K\theta}e^{-K\theta x}}{-\pi(i\cos\frac{\pi x}{2})^{1-\alpha}\sin\frac{\pi x}{2}}$$

and

$$\operatorname{res} K(1-x)i = \frac{Ke^{-K\theta}e^{K\theta x}}{\pi(i\cos\frac{\pi x}{2})^{1-\alpha}\sin\frac{\pi x}{2}} .$$

Therefore, the contribution from the poles is given by

$$\pi i\Sigma \mathrm{res} = \frac{2i^{\alpha}Ke^{-K\theta}\sinh K\theta x}{(\cos\frac{\pi x}{2})^{1-\alpha}\sin\frac{\pi x}{2}} .$$

Let us now consider the contributions of the various straight lines. If we bear in mind that

$$\left|\sinh\frac{\pi z}{2K}\right| = (\sinh^2\frac{\pi x}{2K} + \sin^2\frac{\pi y}{2K})^{\frac{1}{2}} ,$$

it is clear that

$$\lim_{N\to\infty} |\Gamma_2| \le \lim_{N\to\infty} \int_0^{2K} |h(N+iy)| e^{-\theta y} dy = 0$$

Now,
$$\Gamma_1 = \int_0^N h(y) e^{i\theta y} dy$$

$$\Gamma_3 = - \int_0^N e^{i\theta y} e^{-2K\theta} h(y+2iK) dy$$

$$\Gamma_4 = - \pi i \Sigma \text{res} - i \int_0^{2K} e^{-\theta y} h(iy) dy .$$

But :

$$h(y+2iK) = i^{2\alpha} h(y)$$

and
$$h(iy) = \frac{i^{\alpha}(\sin \frac{y}{2K})^{\alpha}}{(\cos^2 \frac{\pi x}{2} - \sin^2 \frac{\pi y}{2K})} = i^{\alpha} g(y), \quad \text{say},$$

where g is real-valued. Therefore, Γ_3 and Γ_4 become

$$\Gamma_3 = - i^{2\alpha} e^{-2K\theta} \Gamma_1$$

and
$$\Gamma_4 = - \pi i \Sigma \text{res} - i^{\alpha+1} \int_0^{2K} e^{-\theta y} g(y) dy .$$

Hence, applying Cauchy's Theorem and letting $N \to \infty$, we get:

$$(1 - i^{2\alpha} e^{-2K\theta}) \int_0^{\infty} e^{i\theta y} h(y) dy = \pi i \Sigma \text{res} + i^{\alpha+1} \int_0^{2K} e^{-\theta y} g(y) dy.$$

If we now multiply both sides by $i^{-\alpha} e^{K\theta}$ and take real parts we obtain:

(3.2)
$$\text{Re}\left[(i^{-\alpha} e^{K\theta} - i^{\alpha} e^{-K\theta}) \int_0^{\infty} e^{i\theta y} h(y) dy\right]$$
$$= \frac{2K \sinh K\theta x}{(\cos \frac{\pi x}{2})^{1-\alpha} \sin \frac{\pi x}{2}} .$$

If we bear in mind that $i^{\alpha} = e^{i\frac{\pi\alpha}{2}} = \cos\frac{\pi\alpha}{2} + i\sin\frac{\pi\alpha}{2}$, then we may write (3.2) in the form

$$(3.3) \qquad \int_0^{\infty}(\cos\frac{\pi\alpha}{2}\cos\theta y \sinh K\theta + \sin\frac{\pi\alpha}{2}\sin\theta y\cosh K\theta)h(y)dy$$

$$= \frac{K\sinh K\theta x}{(\cos\frac{\pi x}{2})^{1-\alpha}\sin\frac{\pi x}{2}} .$$

We can also write (3.3) in the form

$$(3.4) \qquad -\int_{-\infty}^{0}(\cos\frac{\pi\alpha}{2}\cos\theta y\sinh K\theta - \sin\frac{\pi\alpha}{2}\sin\theta y\cosh K\theta)(-h(-y))dy$$

$$= \frac{K\sinh K\theta x}{(\cos\frac{\pi x}{2})^{1-\alpha}\sin\frac{\pi x}{2}} .$$

Hence, to obtain (3.1) we put $\alpha = \frac{2\mu}{\pi}$, where μ is defined as above, and add (3.3) and (3.4). This then gives $\pi^{o}(x,y)$, namely,

$$\pi^{o}(x,y) = \frac{\operatorname{sgn}(y)\sin\frac{\pi x}{2}(\sinh\frac{\pi|y|}{2k})^{\alpha}(\cos\frac{\pi x}{2})^{1-\alpha}}{2K(K^2+1)^{\frac{1}{2}}(\sinh^2\frac{\pi y}{2K} + \cos^2\frac{\pi x}{2})}$$

where $0 < |x| < 1$ and $0 < |y| < \infty$.

We can similarly show that

$$\pi^{e}(x,y) = \frac{\cosh\frac{\pi y}{2K}(\sinh\frac{\pi y}{2K})^{\alpha}(\cos\frac{\pi x}{2})^{1-\alpha}}{2K(K^2+1)^{\frac{1}{2}}(\sinh^2\frac{\pi y}{2K} + \cos^2\frac{\pi x}{2})} ,$$

where $0 < |x| < 1$ and $0 < |y| < \infty$.

If we now combine the results for both 'even' and 'odd' cases we obtain

$$\pi(x,y) = \pi^e(x,y) + \pi^\circ(x,y)$$

$$= \frac{(\cosh\frac{\pi y}{2K} + \operatorname{sgn}(y)\sin\frac{\pi x}{2})(\sinh\frac{\pi|y|}{2K})^\alpha(\cos\frac{\pi x}{2})^{1-\alpha}}{2K(K^2+1)^{\frac{1}{2}}(\sinh^2\frac{\pi y}{2K} + \cos^2\frac{\pi x}{2})}$$

$$= \frac{(\sinh\frac{\pi|y|}{2K})^\alpha(\cos\frac{\pi x}{2})^{1-\alpha}}{2K(K^2+1)^{\frac{1}{2}}(\cosh\frac{\pi y}{2K} - \operatorname{sgn}(y)\sin\frac{\pi x}{2})}$$

- a remarkable simplification. We can show that $\pi(x,y)$ integrates to 1 over $(-\infty,\infty)$ with respect to y by elementary integration.

4. The dual equations.

The case "K = 1" has certain symmetries not possessed by other cases, and so we present the solution of the dual equations only for this case. Again we shall only consider the 'odd' case, namely, (c.f. (2.16)),

$$(4.1) \qquad \int_x \sin\theta_n x\pi^\circ(x,y)dx = \begin{cases} e^{-\theta_n y}\sin\theta_n, & y > 0 \\ -e^{\theta_n y}\sin\theta_n, & y < 0 \end{cases}$$

where $\theta_n = (n-\frac{1}{4})\pi$.

We shall only consider the case $y > 0$ since the case $y < 0$ can be readily obtained from this. Also, for convenience we shall write 'θ' for 'θ_n'.

The odd case (K = 1)

Let us consider the contour integral

$$\oint_\Gamma \frac{\sin\frac{\pi z}{2}(\cos\frac{\pi z}{2})^{\frac{1}{2}}e^{iK\theta z}dz}{(\sinh^2\frac{\pi y}{2} + \cos^2\frac{\pi z}{2})} = \oint_\Gamma h(z)e^{iK\theta z}dz$$

taken around

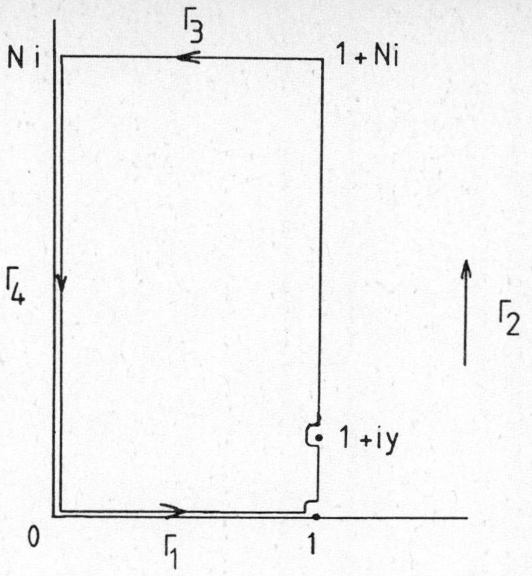

Now, $\mathrm{Im}(\cos\frac{\pi z}{2}) \le 0$ for all z in, and on, the contour Γ and so $(\cos\frac{\pi z}{2})^{\frac{1}{2}}$ has a regular branch in this region. When working on Γ we choose the branch so that $\frac{\pi}{2} \le \arg(\cos\frac{\pi z}{2}) < \frac{5\pi}{2}$.

The point $z = 1$ is a branch point for $(\cos\frac{\pi z}{2})^{\frac{1}{2}}$ but, as the integrand remains bounded as the radius of the quarter-circle tends to zero, there is no contribution from this point. The point $z = 1 + iy$ is a pole of the integrand. If we now use the fact that $\cos\theta = -\sin\theta$ and note that $\sin\frac{\pi}{2}(1+iy) = \cosh\frac{\pi y}{2}$ and $\cos\frac{\pi}{2}(1+iy) = -i\sinh\frac{\pi y}{2}$, we may calculate the contribution of the residue to be:

$$\pi i \,\mathrm{res} = \frac{\pi i\, e^{i\theta} e^{-\theta y} \cosh\frac{\pi y}{2}}{-\pi(-i\sinh\frac{\pi y}{2})^{\frac{1}{2}} \cosh\frac{\pi y}{2}}$$

$$= \frac{-i\sqrt{2}e^{-\theta y}\sin\theta}{(\sinh\frac{\pi y}{2})^{\frac{1}{2}}} \,.$$

Let us now consider the contributions from the various straight lines. If we bear in mind that

$$\left|\cos\frac{\pi z}{2}\right| = \left(\cos^2\frac{\pi u}{2} + \sinh^2\frac{\pi v}{2}\right)^{\frac{1}{2}},$$

it is clear that

$$\lim_{N\to\infty} |\Gamma_3| \le \lim_{N\to\infty} \int_0^1 |h(x+iN)|\ e^{-\theta N} dx = 0.$$

Before continuing we note the following:

$$h(1+ix) = \frac{\cosh\frac{\pi x}{2}\left(-i\sinh\frac{\pi x}{2}\right)^{\frac{1}{2}}}{\left(\sinh^2\frac{\pi y}{2} - \sinh^2\frac{\pi x}{2}\right)} = \frac{\cosh\frac{\pi x}{2}\left(\sinh\frac{\pi x}{2}\right)^{\frac{1}{2}}}{\left(\sinh^2\frac{\pi y}{2} - \sinh^2\frac{\pi x}{2}\right)}\, i^{3/2}$$

$$= i^{3/2} f(x), \quad \text{say},$$

and

$$h(ix) = \frac{i\left[-\left(\cosh\frac{\pi x}{2}\right)^{\frac{1}{2}}\right]\sinh\frac{\pi x}{2}}{\left(\sinh^2\frac{\pi y}{2} + \cosh^2\frac{\pi x}{2}\right)} = -ig(x), \quad \text{say},$$

where f and g are both real-valued.

We have

$$\Gamma_1 = \int_0^1 (-h(x))e^{i\theta x} dx$$

$$\Gamma_2 = -\pi i\,\text{res} + i\int_0^N h(1+ix)e^{i\theta} e^{-\theta x} dx$$

$$= -\pi i\,\text{res} + \int_0^N \sqrt{2}\,\sin\theta\, f(x)e^{-\theta x} dx$$

$$\Gamma_4 = -i \int_0^N (-ig(x))e^{-\theta x}dx$$

$$= - \int_0^N g(x)e^{-\theta x}dx.$$

If we now apply Cauchy's Theorem, let $N \to \infty$ and take imaginary parts we obtain

$$\int_0^1 h(x) \sin \theta x dx = \frac{\sqrt{2}\, e^{-\theta y} \sin \theta}{(\sinh \frac{\pi y}{2})^{\frac{1}{2}}}$$

or, equivalently,

$$\int_{-1}^1 h(x) \sin \theta x dx = \frac{2\sqrt{2}\, e^{-\theta y} \sin \theta}{(\sinh \frac{\pi y}{2})^{\frac{1}{2}}}.$$

If we now extend this to the case $y < 0$ and rearrange we see that

$$\pi^{\circ}(x,y) = \frac{\operatorname{sgn}(y) \sin \frac{\pi x}{2} (\sinh \frac{\pi|y|}{2} \cos \frac{\pi x}{2})^{\frac{1}{2}}}{2\sqrt{2}\, (\sinh^2 \frac{\pi y}{2} + \cos^2 \frac{\pi x}{2})}$$

as expected.

Acknowledgements.

I would like to thank Dr R.R. London for several discussions related to this work, and for checking the results obtained, and also David Williams, without whose encouragement this paper would not have been written.

This work was funded by the Science Research Council.

REFERENCES

1. R.R. London, H.P. McKean, L.C.G. Rogers and David Williams, 'A martingale approach to some Wiener-Hopf problems, I'. Séminaire de Probabilités XVI, Springer Lecture Notes in Math., 920, 1982, 41-67.

2. ______________, 'A martingale approach to some Wiener-Hopf problems, II', i.b.i.d., 68-90.

3. L.C.G. Rogers and David Williams, 'Time-substitution based on fluctuating additive functionals (Wiener-Hopf factorization for infinitesimal generators)', Séminaire de Probabilités XIV, Springer Lecture Notes in Math. 784, 332-342.

4. L.C.G. Rogers, 'Wiener-Hopf factorization for diffusions and Lévy processes', Proc. London Math. Soc. (3), 67 (1983), 177-191.

5. ______________, 'A new identity for real Lévy processes', Ann. Inst. H. Poincaré, Sec. B, Vol. 20.

Department of Mathematics and Computer Science
University College
Swansea
SA2 8PP
Great Britain

A DIFFERENTIAL EQUATION IN WIENER-HOPF THEORY

by

L.C.G. Rogers and David Williams

This is a heuristic introduction to some progress with certain calculations in Wiener-Hopf theory. Further details will be presented later.

PART 1. THE CASE WHEN THERE IS ONLY ONE BOUNDARY POINT

1. Let $B = \{B(t) : t \geq 0\}$ (also written $\{B_t : t \geq 0\}$) be a Brownian motion on $\mathbb{R}$. The symbol $\mathbb{P}_r$ denotes $\mathbb{P}[\cdot | B_0 = r]$, and $\mathbb{E}_r$ denotes $\mathbb{P}_r$ expectation. Let $V : \mathbb{R} \to \mathbb{R}$, with $V > 0$ on $(0,\infty)$ and $V < 0$ on $(-\infty,0)$. For $t \geq 0$, define:

$$\phi_t = \int_0^t V(B_s)ds, \quad \tau_t^+ = \inf\{u : \phi_u > t\}, \quad \tau_t^- = \inf\{u : -\phi_u > t\},$$

$$Y_t^+ = B(\tau_t^+) \in [0,\infty), \quad Y_t^- = B(\tau_t^-) \in (-\infty,0].$$

If $B_0 = x < 0$, then τ_0^+ is the half-winding time about the origin for the joint process $\{(\phi_t, B_t)\}$. Our primary concern is to calculate half-winding hitting probabilities:

$$\Pi^+(x,y) = \mathbb{P}_x[Y_0^+ \in dy]/dy, \quad \Pi^-(x,y) = \mathbb{P}_y[Y_0^- \in dx]/dx,$$

where $x < 0$, $y > 0$.

Back in 1963, McKean ([3]) solved this problem for the case when $V(r) = r$, $\forall r \in \mathbb{R}$. The joint process (ϕ_t, B_t), the phase picture for McKean's resonator, is then Gaussian (as well as Markov); and McKean exploits this fact via some sparkling analysis to show that in this case,

$$\Pi^{+}(x,y) = \frac{3|x|^{1/2}y^{3/2}}{2\pi(|x|^{3}+y^{3})} . \tag{1.1}$$

McKean's paper contains much more of interest.

2. Our method leads to a generalization of (1) which further highlights the bizarre nature of the distributions which arise in this subject.

Example 2. Let $\alpha > -1$, and let $\delta = \pi/(2+\alpha)$. Let $K > 0$. Let

$$V(r) = \begin{cases} r^{\alpha} & , r > 0 , \\ -K^{2+\alpha}|r|^{\alpha} & , r < 0 . \end{cases}$$

Then, for every starting point, it is almost surely true that τ_t^{+} and τ_t^{-} are finite for every t. Moreover, for $x < 0$, $y > 0$,

$$\Pi^{+}(x,y) = \rho(x,y)y^{\alpha} , \quad \Pi^{-}(x,y) = K^{2+\alpha}|x|^{\alpha}\rho(x,y) ,$$

where

$$\rho(x,y) = \frac{Cx^{\beta}y^{1-\beta}}{K^{2+\alpha}|x|^{2+\alpha} + y^{2+\alpha}} ,$$

where β is the unique solution in $(0,1)$ of the equation

$$\sin((1-\beta)\delta) = K\sin(\beta\delta) ,$$

and

$$C = \pi^{-1}(2+\alpha)K^{\beta}\sin(\beta\delta) .$$

Notes. a) For McKean's case, $\alpha = 1$, $\delta = \pi/3$, $K = 1$, $\beta = \frac{1}{2}$

b) For the case when $\alpha = 0$, N. Baker had obtained the solution by a different (complex-variable) method.

3. The processes Y^{+} and Y^{-} are Markovian. Let L^{+}(resp.,L^{-}) be the local time at 0 for $Y^{+}(Y^{-})$ in some arbitrary, but fixed, normalization. [Note. In certain extremely pathological cases, 0 may be visited by Y^{+} (or by Y^{-}) only finitely often in finite time intervals. The 'local time'

for Y^+ (or Y^-) then just counts the number of visits.] Let $J^+_y dy$ (resp., $J^-_x dx$) be the Lévy measure describing jumps made from 0 by $Y^+(Y^-)$. Define

(3.1) $$m^{\pm}(t) = \frac{d}{dt}\mathbb{E}_0[L^{\pm}(t)], \quad b^{\pm}_r(t) = \mathbb{P}_0[Y^{\pm}(t) \in dr]/dr .$$

Then we have the Fokker-Planck equation:

$$\frac{\partial b^{\pm}_r(t)}{\partial t} = \frac{1}{2}\frac{\partial^2}{\partial r^2}[|V_r|^{-1} b^{\pm}_r(t)] + m^{\pm}(t) J^{\pm}_r ,$$

where $V_r = V(r)$. For the better formulation of the Fokker-Planck equation, introduce the Radon-Nikodym derivatives:

$$\beta^{\pm}_r(t) = b^{\pm}_r(t)/|V_r| , \quad \Lambda^{\pm}_r = J^{\pm}_r/|V_r| .$$

Then

(3.2) $$\frac{\partial \beta^{\pm}_r(t)}{\partial t} = \mathcal{G}^{\pm}_r \beta^{\pm}_r(t) + m^{\pm}(t)\Lambda^{\pm}_r , \quad \mathcal{G}^{\pm}_r = \frac{1}{2}|V_r|^{-1}\frac{\partial^2}{\partial r^2} .$$

If $H = \inf\{s : B_s = 0\}$, and $h_r(t)dt = \mathbb{P}_r[|\phi(H)| \in dt]$, then it is clear that for $x < 0$, $y > 0$,

$$\Pi^+(x,y) = \int_0^\infty h_x(t) b^+_y(t)dt ,$$

so that

(3.3) $$\Pi^+(x,y) = \rho(x,y)V(y) ,$$

where

(3.4) $$\rho(x,y) = \int_0^\infty h_x(t)\beta^+_y(t)dt .$$

The symmetry properties discovered in [2] (see Note below for a correction to [2]) make it clear that

(3.5) $$\Pi^-(x,y) = V(x)\rho(x,y) ,$$

and that we have the following dual expression for ρ :

(3.6) $$\rho(x,y) = \int_0^\infty \beta^-_x(t)h_y(t)dt .$$

For $x < 0$,

(3.7) $$\frac{\partial}{\partial t} h_x(t) = \mathcal{G}_x^- h_x(t) .$$

Using (3.7) and the '+' verson of (3.2), we have

$$(\mathcal{G}_x^- + \mathcal{G}_y^+)\rho(x,y) = \int_0^\infty \left(\beta_y^+ \frac{\partial h_x}{\partial t} + h_x \frac{\partial \beta_y^+}{\partial t} \right) dt - \Lambda^+(y) \int_0^\infty h_x(t) m^+(t) dt$$

$$= -\Lambda^+(y) \int_0^\infty h_x(t) m^+(t) dt ,$$

since $\beta_y^+(t) h_x(t)$ vanishes when $t = 0$ and when $t = \infty$. By symmetry, we must have

$$-\Lambda^+(y) \int_0^\infty h_x(t) m^+(t) dt = -\Lambda^-(x) \int_0^\infty h_y(t) m^-(t) dt ,$$

so that, for some constant R,

(3.8) $$\int h_x(t) m^+(t) dt = R\Lambda^-(x) , \qquad \int h_y(t) m^-(t) dt = R\Lambda^+(y) .$$

Thus, we obtain the following differential equation for ρ :

(3.9) $$\boxed{(\mathcal{G}_x^- + \mathcal{G}_y^+)\rho(x,y) = -R\Lambda^-(x)\Lambda^+(y) .}$$

This corresponds to the Ricatti equation (2,4) of [2].

In §5 we extend (3.9) to the case when there are several boundary points, and prove that ρ is the minimal (Green's function integral) solution.

Note: Correction to [2]. Of course, in §3 of [2], adjoints should be taken relative to the measure $|V|m$ rather than m. (An early draft had a preliminary transformation to reduce $|V|$ to 1 for determination of Π .)

4. Study of Example 2A. We have $\alpha > -1$, and

$$V(r) = r^\alpha , \quad r > 0 ; \qquad V(r) = -A|r|^\alpha , \quad r < 0 ,$$

where $A > 0$. Recall that

$$\phi_t = \int_0^t V(B_s)ds\,, \quad \tau_0^\pm = \inf\{u : \pm\phi_u > t\}\,, \quad Y_0^\pm = B(\tau_0^\pm)\,.$$

With the Brownian scaling in mind, let $c > 0$, and let

$$\tilde{B}_t = c^{-1}B(c^2t)\,,$$

$$\tilde{\phi}_t = \int_0^t V(\tilde{B}_s)ds\,, \quad \tilde{\tau}_0^\pm = \inf\{u : \pm\tilde{\phi}_u > t\}\,, \quad \tilde{Y}_0^\pm = \tilde{B}(\tilde{\tau}_0^\pm)\,.$$

Then

$$\tilde{\phi}_t = c^{-2-\alpha}\phi(c^2t)\,, \quad \tilde{\tau}_t^\pm = c^{-2}\tau^\pm(c^{2+\alpha}t)\,, \quad \tilde{Y}_t^\pm = c^{-1}Y^\pm(c^{2+\alpha}t)\,.$$

Suppose for a moment that $B_0 = 0$, so that B and $\tilde{B}$ are identical in law. To avoid too heavy a notation, let us write J^+ for the Lévy measure of Y^+ at 0 (as a measure), as well as writing $J^+(\cdot)$ for the density of J^+ relative to Lebesgue measure. In short, $J^+(dx) = J^+(x)dx$. Let $y > 0$, and let

$$T_y = T(y) = \inf\{t : Y_{t-}^+ = 0;\ Y_t^+ > y\}\,.$$

Then, for $z > y$,

$$J^+(z,\infty)/J^+(y,\infty) = \mathbb{P}_0[Y^+(T_y) > z]\,.$$

But, with the obvious notation, $\tilde{T}(y) = c^{-2-\alpha}T(cy)$, and $\tilde{Y}^+(\tilde{T}_y) = c^{-1}Y^+(T_{cy})$. Thus,

$$J^+(z,\infty)/J^+(y,\infty) = \mathbb{P}_0[Y^+(T_{cy}) > cz] = J^+(cz,\infty)/J^+(cy,\infty)\,,$$

so that $J^+(y) \propto y^\eta$ for some η. Thus, for some constants ε and θ,

$$\Lambda^+(y) \propto |y|^\theta\,, \quad \Lambda^-(x) \propto |x|^\varepsilon\,.$$

The fundamental equation (3.9) therefore takes the form:

$$(4.1)\qquad \frac{1}{2A|x|^\alpha}\frac{\partial^2\rho}{\partial x^2} + \frac{1}{2|y|^\alpha}\frac{\partial^2\rho}{\partial y^2} = -R|x|^\varepsilon|y|^\theta\,.$$

The Brownian scaling gives us further information: for $x < 0$, $y > 0$,

$$(4.2)\qquad \mathbb{P}[Y_0^+ \le y \,|\, B_0 = x] = \mathbb{P}[\tilde{Y}_0^+ \le y \,|\, \tilde{B}_0 = x] = \mathbb{P}[Y_0^+ \le cy \,|\, B_0 = cx]\,.$$

Hence

$$\Pi(x,y) = c\Pi(cx,cy)\ , \quad \text{and} \quad \rho(x,y) = c^{1+\alpha}\rho(cx,cy)\ .$$

On taking $c = 1/|x|$, we see that

$$\rho(x,y) = |x|^{-1-\alpha}\rho(-1,|y/x|) = |x|^{-1-\alpha}k(u)\ , \tag{4.3}$$

where $u = |y/x|$, and $k(u) = \rho(-1,u)$.

Substitution of (4.3) into (4.1) yields:

$$\frac{d^2}{du^2}[(A + u^{2+\alpha})k(u)] = -2AR|x|^{3+\alpha+\varepsilon}y^{\theta+\alpha}\ .$$

Since the left-hand side is a function of u alone, we must have

$$\frac{d^2}{du^2}[(A + u^{2+\alpha})k(u)] = -2ARu^{\theta+\alpha}\ ,$$

and $\theta + \alpha = -3 - \alpha - \varepsilon$. Thus, for some constants a and b ,

$$(A + u^{2+\alpha})k(u) = -\frac{2AR}{(\theta+\alpha+2)(\theta+\alpha+1)}u^{\theta+\alpha+2} + a + bu\ .$$

The fact that

$$\int_{y=0}^{\infty}\rho(x,y)y^{\alpha}dy < \infty \quad \text{and} \quad \int_{x=-\infty}^{0}\rho(x,y)|x|^{\alpha}dx < \infty$$

implies that $a = b = 0$ and $-2 < \theta + \alpha < -1$. We have shown that

$$\rho(x,y) = \frac{C|x|^{\beta}|y|^{1-\beta}}{K^{2+\alpha}|x|^{2+\alpha}+|y|^{2+\alpha}}$$

where $1 - \beta = \theta + \alpha + 2$, so that $0 < \beta < 1$, and we have written $A = K^{2+\alpha}$.

The scaling (4.2) shows that the expression $\mathbb{P}_x[\tau_0^+ < \infty]$ is independent of x . It is therefore intuitively obvious, and an amusing exercise for the reader to prove that for $x < 0$,

$$1 = \mathbb{P}_x[\tau_0^+ < \infty] = \int_0^{\infty}\Pi^+(x,y)dy = \int_0^{\infty}\rho(x,y)y^{\alpha}dy$$

$$= C\int_0^{\infty}\frac{y^{1-\beta}}{K^{2+\alpha}+y^{2+\alpha}}y^{\alpha}dy = \delta CK^{-\beta}\operatorname{cosec}(\beta\delta)\ ,$$

where $\delta = \Pi/(2+\alpha)$. See §3.123 of Titchmarsh [4]. Similarly, for $y > 0$.

$$1 = \int_0^\infty K^{2+\alpha} |x|^\alpha \rho(x,y)\,dx = \delta C K^{1-\beta} \operatorname{cosec}((1-\beta)\delta) .$$

Hence β is the unique solution in $(0,1)$ of the equation.

$$\operatorname{cosec}(\beta\delta) = K \operatorname{cosec}((1-\beta)\delta) ,$$

and then

$$C = \pi^{-1}(2+\alpha)K^\beta \sin(\beta\delta) .$$

PART II. THE CASE OF FINITELY MANY BOUNDARY POINTS.

5. In this section, we prove in particular that ρ is the minimal positive solution of (3.9). However, we work in a more general situation in which there are finitely many boundary points.

We repeat that this is a preliminary, and heuristic, report on this topic.

Suppose that E^+ is a closed subset of $\mathbb{R}$ of the form:

$$E^+ = \mathbb{R} \cap ([c_1, d_1] \cap [c_2, d_2] \cap \ldots . \cap [c_n, d_n]) ,$$

where $n \in \mathbb{N}$, and $-\infty \le c_1 < d_1 < c_2 < d_2 < \ldots < c_n < d_n \le \infty$. Let E^- be the closure of $\mathbb{R} \backslash E^+$, so that

$$E^- = (-\infty, c_1] \cup [d_1, c_2] \cup \ldots \cup [d_{n-1}, c_n] \cup [d_n, \infty) .$$

Let

$$\Gamma = E^+ \cap E^- = \mathbb{R} \cap \{c_1, d_1, c_2, d_2, \ldots, c_n, d_n\}$$

be the set of 'boundary' points. Henceforth, we shall use $i, j, \ldots,$ to represent typical points of Γ.

We write x for a typical point of E^-, y for a typical point of E^+.

Let V be a function on $\mathbb{R}$ which is strictly positive on the interior of E^+, and strictly negative on the interior of E^-.

For $t \ge 0$, define:

$$\phi_t = \int_0^t V(B_s)ds , \quad \tau_r^+ = \inf\{u : \phi_u > t\} , \quad \tau_t^- = \inf\{u : -\phi_u > t\} ,$$

$$Y_t^+ = B(\tau_t^+) , \quad Y_t^- = B(\tau_t^-) .$$

The process Y^+ is a Markov process on E^+. We introduce a number of entities associated with the process Y^+.

For $j \in \Gamma$, let L_j^+ be the local time (see Note at start of §3) at j for Y^+, in some arbitrary but fixed normalization. For $i, j \in \Gamma$, and $t > 0$, define

$$M^+_{ij}(t) = \frac{d}{dt}\mathbb{E}_i L^+_j(t) ,$$

and let $M^+(t)$ be the $\Gamma\times\Gamma$ matrix with (i,j)th component $M^+_{ij}(t)$.

Let

$$T^+_\Gamma = \inf\{t \ge 0 : Y^+_t \in \Gamma\} .$$

For $y \in \mathrm{Int}(E^+)$, and $i \in \Gamma$, let

$$h^+_{yi}(t) = \frac{d}{dt}\mathbb{P}_y[T^+_\Gamma \le t;\ Y^+(T^+_\Gamma -) = i] .$$

Let $h^+_{y\cdot}(t)$ be the row vector $\{h^+_{yi}(t) : i \in \Gamma\}$. Let $J^+_{iy}dy$ be the Lévy measure describing jumps made by Y^+ from i, and let $J^+_{\cdot y}$ be the column vector $\{J^+_{iy} : i \in \Gamma\}$. Introduce the Radon-Nikodym derivative

$$\Lambda^+_{iy} = |V(y)|^{-1}J^+_{iy} .$$

Define b^+ and β^+ via

$$b^+_{iy}(t)dy = \beta^+_{iy}(t)|V(y)|dy = \mathbb{P}_i[Y^+(t) \in dy] .$$

Introduce the column vectors $b^+_{\cdot y}(t)$, $\beta^+_{\cdot y}(y)$, $\Lambda^+_{\cdot y}$ in the obvious way.

The Fokker-Planck equation

$$\frac{\partial}{\partial t} b^+_{\cdot y}(t) = \frac{1}{2}\frac{\partial^2}{\partial y^2}[|V(y)|^{-1}b^+_{\cdot y}(t)] + M^+(t)J^+_{\cdot y}(t)$$

holds, and, as at (3.2) transforms to

$$\text{(5.1)}\qquad \frac{\partial}{\partial t}\beta^+_{\cdot y}(t) = \mathcal{G}^+_y\beta^+_{\cdot y}(t) + M^+(t)\Lambda^+_{\cdot y}(t) ,$$

where

$$\mathcal{G}^+_y = \frac{1}{2}|V(y)|^{-1}\frac{\partial^2}{\partial y^2} , \quad \text{for } y \in \mathrm{Int}(E^+) .$$

Let ${}_\Gamma p^+$ be the taboo transition function on $E^+\times E^+$:

$${}_\Gamma p^+(t,y_1,y_2)dy_2 = \mathbb{P}_{y_1}[Y^+(t) \in dy_2;\ T^+_\Gamma > t] .$$

Then, the symmetry property:

$$|V(y_1)|\,{}_\Gamma p^+(t,y_1,y_2) = |V(y_2)|\,{}_\Gamma p^+(t,y_2,y_1)$$

is well known, and it is obvious that, for $y_1 \in E^+, j \in \Gamma$ and $s,t > 0$,

(5.2) $$\int_{E^+} {}_\Gamma p^+(t,y_1,y_2) h^+_{y_2 j}(s)\,dy_2 = h^+_{y_1 j}(t+s) .$$

We can intoduce analogous concepts for Y^-.

Now, for $x \in E^-$, $y \in E^+$,

$$\Pi(x,y) = \int_0^\infty h^-_{x\cdot}(t) b^+_{\cdot y}(t)\,dt ,$$

so that

$$\rho(x,y) = \int h^-_{x\cdot}(t) \beta^+_{\cdot y}(t)\,dt .$$

By standard theory of excursions, last-exit decompositions, etc.,

$$b^+_{\cdot y}(t) = \int_0^t M^+(s) g^+_{\cdot y}(t-s)\,ds ,$$

where

$$g^+_{iy}(r) = \int_{y_1} \Lambda^+_{iy_1} |V(y_1)|\, {}_\Gamma p^+(r,y_1,y)\,dy_1$$

$$= |V(y)| \int_{y_1} {}_\Gamma p^+(r,y,y_1) \Lambda^+_{iy_1}\,dy_1 .$$

Hence

(5.3) $$\rho(x,y) = \int_{t=0}^\infty h^-_{x\cdot}(t) \beta^+_{\cdot y}(t)\,dt ,$$

where

(5.4) $$\beta^+_{\cdot y}(t) = \int_0^t M^+(s) \gamma^+_{\cdot y}(t-s) ,$$

where

(5.5) $$\gamma^+_{\cdot y}(r) = \int_{y_1} {}_\Gamma p^+(r,y,y_1) \Lambda^+_{\cdot y_1}\,dy_1 .$$

As at (3.7), we have

$$\frac{\partial}{\partial t} h^-_{x\cdot}(t) = \mathcal{g}^-_x h^-_{x\cdot}(t) .$$

Exactly as in the argument following (3.7), we can deduce from (5.3) and (5.1) that

$$\left(\mathcal{g}_x^- + \mathcal{g}_y^+\right)\rho = \left(\int h_{x\cdot}^-(t)M^+(t)dt\right)\Lambda_{\cdot y}^+ .$$

And we can again appeal to the symmetry result in [2] to obtain

$$(5.6) \qquad \left(\int h_{x\cdot}^-(t)M^+(t)dt\right)\Lambda_{\cdot y}^+ = \left(\int h_{y\cdot}^+(t)M^-(t)dt\right)\Lambda_{\cdot x}^- .$$

We claim that for some constants $a_i (i \in \Gamma)$,

$$(5.7) \qquad \int h_{x\cdot}^-(t)M_{\cdot i}^+(t)dt = a_i\Lambda_{ix}^- ,$$

$$(5.8) \qquad \int h_{y\cdot}^+(t)M_{\cdot i}^-(t)dt = a_i\Lambda_{iy}^+ .$$

This is one of several claims in this paper for which full justification will have to wait to a later paper. The reader should believe our results because the analogues for symmetric Markov chains are true, and we have tested out that one can force through weak-convergence results.

Let us explain briefly a direct method of deducing (5.7) and (5.8) from (5.6) in the case when i is a regular boundary point both for Y^+ and Y^- (so that each of these processes has a true continuous local time at i). As mentioned in a Note at the start of §3, this will be the situation in all but extremely pathological cases. The point is that as $y \to i$,

$$\Lambda_{jy}^+ = o(\Lambda_{iy}^+), \quad j \neq i ,$$

and

$$\lim \int h_{yj_1}^+(t)M_{j_1j_2}^-(t)dt \quad \begin{matrix} = & \infty \text{ if } j_1 = j_2 = i , \\ < & \infty \text{ otherwise} . \end{matrix}$$

These results allow us to infer (5.7) from (5.6).

It will simplify the algebra to assume, as we may plainly do, that the normalizations of L_i^+ and L_i^- are made compatible for each i, so that

$$a_i = 1 , \ \forall i .$$

Substitute (5.8) in (5.5), and use (5.2) to obtain

$$\{\gamma^{+}_{y}(r)\}^{*} = \int_{u=0}^{\infty} h^{+}_{y}(r+u)M^{-}(u)\,du ,$$

the * signifying transpose. Substitute in (5.4), and then substitute the resulting equation in (5.3) to obtain (with $\langle\cdot,\cdot\rangle$ the scalar product of vectors on Γ) :

$$\rho(x,y) = \int_{t=0}^{\infty}\int_{s=0}^{t}\int_{u=0}^{\infty}\left\langle h^{-}_{x\cdot}(y)M^{+}(s)\, ,\; h^{+}_{y\cdot}(t+u-s)M^{-}(u)\right\rangle du\,ds\,dt$$

$$= \int_{v=0}^{\infty}\left\langle\int_{s=0}^{\infty} h^{-}_{x\cdot}(s+v)M^{+}(s)ds,\; \int_{s=0}^{\infty} h^{+}_{y\cdot}(u+v)M^{-}(u)du\right\rangle dv$$

$$= \int_{v=0}^{\infty}\left\langle\gamma^{-}_{x\cdot}(v)\, ,\; \gamma^{+}_{y\cdot}(v)\right\rangle dv .$$

Finally, it now follows from (5.5) that

$$\rho(x,y) = \int_{x_1}\int_{y_1} G(x,y;x_1,y_1)\,\left(\sum_i \Lambda^{-}_{ix_1}\Lambda^{+}_{iy_1}\right)dx_1dy_1 ,$$

<u>where</u> G <u>is the Green's function</u>:

$$G(x,y;x_1y_1) = \int_{t=0}^{\infty} {}_{\Gamma}p^{-}(t,x,x_1)\;{}_{\Gamma}p^{+}(t,y,y_1)\,dt$$

<u>of the operator</u> $(\mathcal{G}^{-}_{x} + \mathcal{G}^{+}_{y})$ <u>on</u> $\mathrm{Int}(E^{-}\times E^{+})$. <u>Thus</u>, ρ <u>is the minimal positive solution of</u>

$$\boxed{(\mathcal{G}^{-}_{x} + \mathcal{G}^{+}_{y})\rho = \sum_i \Lambda^{-}_{ix}\Lambda^{+}_{iy} .}$$

REFERENCES

1. R.R. London, H.P. McKean, L.C.G. Rogers, D. Williams, 'A martingale approach to some Wiener-Hopf problems, I', Seminaire de Probabilites XVI, 41-67, Springer Lecture Notes in Math. 920, 1982.

2. R.R. London, H.P. McKean, L.C.G. Rogers, D. Williams, 'A martingale approach to some Wiener-Hopf problems, II', ibid. 68-90.

3. H.P. McKean, 'A winding problem for a resonator driven by a white noise', J. Math. Kyoto Univ. 2-2, 1963, 227-35.

4. E.C. Titchmarsh, The theory of functions (Oxford, 1939).

Vol. 952: Combinatorial Mathematics IX. Proceedings, 1981. Edited by E. Billington, S. Oates-Williams, and A.P. Street. XI, 443 pages. 1982.

Vol. 953: Iterative Solution of Nonlinear Systems of Equations. Proceedings, 1982. Edited by R. Ansorge, Th. Meis, and W. Törnig. VII, 202 pages. 1982.

Vol. 954: S.G. Pandit, S.G. Deo, Differential Systems Involving Impulses. VII, 102 pages. 1982.

Vol. 955: G. Gierz, Bundles of Topological Vector Spaces and Their Duality. IV, 296 pages. 1982.

Vol. 956: Group Actions and Vector Fields. Proceedings, 1981. Edited by J.B. Carrell. V, 144 pages. 1982.

Vol. 957: Differential Equations. Proceedings, 1981. Edited by D.G. de Figueiredo. VIII, 301 pages. 1982.

Vol. 958: F.R. Beyl, J. Tappe, Group Extensions, Representations, and the Schur Multiplicator. IV, 278 pages. 1982.

Vol. 959: Géométrie Algébrique Réelle et Formes Quadratiques, Proceedings, 1981. Edité par J.-L. Colliot-Thélène, M. Coste, L. Mahé, et M.-F. Roy. X, 458 pages. 1982.

Vol. 960: Multigrid Methods. Proceedings, 1981. Edited by W. Hackbusch and U. Trottenberg. VII, 652 pages. 1982.

Vol. 961: Algebraic Geometry. Proceedings, 1981. Edited by J.M. Aroca, R. Buchweitz, M. Giusti, and M. Merle. X, 500 pages. 1982.

Vol. 962: Category Theory. Proceedings, 1981. Edited by K.H. Kamps D. Pumplün, and W. Tholen, XV, 322 pages. 1982.

Vol. 963: R. Nottrot, Optimal Processes on Manifolds. VI, 124 pages. 1982.

Vol. 964: Ordinary and Partial Differential Equations. Proceedings, 1982. Edited by W.N. Everitt and B.D. Sleeman. XVIII, 726 pages. 1982.

Vol. 965: Topics in Numerical Analysis. Proceedings, 1981. Edited by P.R. Turner. IX, 202 pages. 1982.

Vol. 966: Algebraic K-Theory. Proceedings, 1980, Part I. Edited by R.K. Dennis. VIII, 407 pages. 1982.

Vol. 967: Algebraic K-Theory. Proceedings, 1980. Part II. VIII, 409 pages. 1982.

Vol. 968: Numerical Integration of Differential Equations and Large Linear Systems. Proceedings, 1980. Edited by J. Hinze. VI, 412 pages. 1982.

Vol. 969: Combinatorial Theory. Proceedings, 1982. Edited by D. Jungnickel and K. Vedder. V, 326 pages. 1982.

Vol. 970: Twistor Geometry and Non-Linear Systems. Proceedings, 1980. Edited by H.-D. Doebner and T.D. Palev. V, 216 pages. 1982.

Vol. 971: Kleinian Groups and Related Topics. Proceedings, 1981. Edited by D.M. Gallo and R.M. Porter. V, 117 pages. 1983.

Vol. 972: Nonlinear Filtering and Stochastic Control. Proceedings, 1981. Edited by S.K. Mitter and A. Moro. VIII, 297 pages. 1983.

Vol. 973: Matrix Pencils. Proceedings, 1982. Edited by B. Kågström and A. Ruhe. XI, 293 pages. 1983.

Vol. 974: A. Draux, Polynômes Orthogonaux Formels – Applications. VI, 625 pages. 1983.

Vol. 975: Radical Banach Algebras and Automatic Continuity. Proceedings, 1981. Edited by J.M. Bachar, W.G. Bade, P.C. Curtis Jr., H.G. Dales and M.P. Thomas. VIII, 470 pages. 1983.

Vol. 976: X. Fernique, P.W. Millar, D.W. Stroock, M. Weber, Ecole d'Eté de Probabilités de Saint-Flour XI – 1981. Edited by P.L. Hennequin. XI, 465 pages. 1983.

Vol. 977: T. Parthasarathy, On Global Univalence Theorems. VIII, 106 pages. 1983.

Vol. 978: J. Ławrynowicz, J. Krzyż, Quasiconformal Mappings in the Plane. VI, 177 pages. 1983.

Vol. 979: Mathematical Theories of Optimization. Proceedings, 1981. Edited by J.P. Cecconi and T. Zolezzi. V, 268 pages. 1983.

Vol. 980: L. Breen. Fonctions thêta et théorème du cube. XIII, 115 pages. 1983.

Vol. 981: Value Distribution Theory. Proceedings, 1981. Edited by I. Laine and S. Rickman. VIII, 245 pages. 1983.

Vol. 982: Stability Problems for Stochastic Models. Proceedings, 1982. Edited by V.V. Kalashnikov and V.M. Zolotarev. XVII, 295 pages. 1983.

Vol. 983: Nonstandard Analysis-Recent Developments. Edited by A.E. Hurd. V, 213 pages. 1983.

Vol. 984: A. Bove, J.E. Lewis, C. Parenti, Propagation of Singularities for Fuchsian Operators. IV, 161 pages. 1983.

Vol. 985: Asymptotic Analysis II. Edited by F. Verhulst. VI, 497 pages. 1983.

Vol. 986: Séminaire de Probabilités XVII 1981/82. Proceedings. Edited by J. Azéma and M. Yor. V, 512 pages. 1983.

Vol. 987: C.J. Bushnell, A. Fröhlich, Gauss Sums and p-adic Division Algebras. XI, 187 pages. 1983.

Vol. 988: J. Schwermer, Kohomologie arithmetisch definierter Gruppen und Eisensteinreihen. III, 170 pages. 1983.

Vol. 989: A.B. Mingarelli, Volterra-Stieltjes Integral Equations and Generalized Ordinary Differential Expressions. XIV, 318 pages. 1983.

Vol. 990: Probability in Banach Spaces IV. Proceedings, 1982. Edited by A. Beck and K. Jacobs. V, 234 pages. 1983.

Vol. 991: Banach Space Theory and its Applications. Proceedings, 1981. Edited by A. Pietsch, N. Popa and I. Singer. X, 302 pages. 1983.

Vol. 992: Harmonic Analysis, Proceedings, 1982. Edited by G. Mauceri, F. Ricci and G. Weiss. X, 449 pages. 1983.

Vol. 993: R.D. Bourgin, Geometric Aspects of Convex Sets with the Radon-Nikodým Property. XII, 474 pages. 1983.

Vol. 994: J.-L. Journé, Calderón-Zygmund Operators, Pseudo-Differential Operators and the Cauchy Integral of Calderón. VI, 129 pages. 1983.

Vol. 995: Banach Spaces, Harmonic Analysis, and Probability Theory. Proceedings, 1980–1981. Edited by R.C. Blei and S.J. Sidney. V, 173 pages. 1983.

Vol. 996: Invariant Theory. Proceedings, 1982. Edited by F. Gherardelli. V, 159 pages. 1983.

Vol. 997: Algebraic Geometry – Open Problems. Edited by C. Ciliberto, F. Ghione and F. Orecchia. VIII, 411 pages. 1983.

Vol. 998: Recent Developments in the Algebraic, Analytical, and Topological Theory of Semigroups. Proceedings, 1981. Edited by K.H. Hofmann, H. Jürgensen and H.J. Weinert. VI, 486 pages. 1983.

Vol. 999: C. Preston, Iterates of Maps on an Interval. VII, 205 pages. 1983.

Vol. 1000: H. Hopf, Differential Geometry in the Large, VII, 184 pages. 1983.

Vol. 1001: D.A. Hejhal, The Selberg Trace Formula for PSL(2, ℝ). Volume 2. VIII, 806 pages. 1983.

Vol. 1002: A. Edrei, E.B. Saff, R.S. Varga, Zeros of Sections of Power Series. VIII, 115 pages. 1983.

Vol. 1003: J. Schmets, Spaces of Vector-Valued Continuous Functions. VI, 117 pages. 1983.

Vol. 1004: Universal Algebra and Lattice Theory. Proceedings, 1982. Edited by R.S. Freese and O.C. Garcia. VI, 308 pages. 1983.

Vol. 1005: Numerical Methods. Proceedings, 1982. Edited by V. Pereyra and A. Reinoza. V, 296 pages. 1983.

Vol. 1006: Abelian Group Theory. Proceedings, 1982/83. Edited by R. Göbel, L. Lady and A. Mader. XVI, 771 pages. 1983.

Vol. 1007: Geometric Dynamics. Proceedings, 1981. Edited by J. Palis Jr. IX, 827 pages. 1983.

Kaolin Deposits of the World

B—Oversea Countries

GENERAL EDITOR

MIROSLAV MALKOVSKÝ

EDITED BY

JOSEF VACHTL

INTERNATIONAL GEOLOGICAL CONGRESS

Report of the Twenty-Third Session Czechoslovakia 1968

PROCEEDINGS OF SYMPOSIUM I

Kaolin Deposits of the World B–Oversea Countries

ACADEMIA

PRAGUE 1969

Printed in Czechoslovakia

Contents

Review of Kaolin Deposits of Asia, Africa, America and Australia

MILOŠ KUŽVART

This short introducing outline is a compilation of data contained in individual national reports published in this volume, or of data available to us, if no paper dedicated to regional part of the Symposium on kaolin deposits reached us in time. The main source of information about kaolin producers not participating in this Symposium was a report compiled by Geofond, Praha, according to all published and unpublished data accessible in Czechoslovakia, reports of Czechoslovak experts included.

Most important kaolin deposits outside Europe are in the USA, China and India.

Kaolin deposits of Asia

Hydrothermal kaolin deposits in Japan are mostly associated with metallic deposits in volcanic rocks (Itaya, Iki, Seta, Ebara), granites and quartz porphyries (Taishu, Kanpaku). Weathered granites are mined near Kakino and Motomiya. Quaternary volcanic ash was the parent rock of deposits Ina and Yame. China gave name to kaolin (china clay, china stone), which was first mined in Kaoling hills in the Kiangsi province and used for production of china-ware as far back as 210 B. C. Kaolins from the deposits Tying-to-chen, Fu-Chow, Minsa, Sinuzi and Su-chow (Kiangsi province) are weathered crystalline rocks. They contain 70—78 % of clay substance, 7—25 % of quartz and 8—20 % of muscovite. Other properties: 35.13—39.90 % Al_2O_3 (refractoriness 1760—1770 °C); 0.19—1.08 % Fe_2O_3, max. 0.06 % TiO_2. Other kaolin—producing provinces in China: Hopei, Shantung, Anhwei (deposit Kimen), Kiang-su, Fukien, Kwantung, Shansi, Szechwan, Honan, Shensi, Hunan, Liaoning, Kirin.

Philippines have only small kaolin deposits, mostly hydrothermally altered volcanic rocks (Batangas, Laguna, Benguet, Iloilo). The deposit Bukidnon is considered to have originated by weathering. "White clay" deposit was dis-

covered in Jurang near Singapore. Kaolinized granite aplites covered by lateritic soil are mined in Prachin Buri near Bangkok in Thailand.

More than 75 kaolin quarries are worked in India. Kaolinization of gneisses contemporary with laterization in Assam (Garo Hills) is dated as pre-Eocene. Gneisses and arkosic sandstones were the parent rocks of kaolin in Rajmahal Hills (Bhagalpur District, Bihar), hydrothermal alteration of granites is supposed to be the main agent of kaolinization in the southern part of Singhbhum District (deposits Hat Gamaria, Karamjia, Dumbria, Bhonda etc.). In Distict Ranchi (locality Bagru) kaolinized granite-gneisses, granites and pegmatites are the lower part of lateritic profile. In West Bengal, there are kaolin deposits in districts Bankura (localities Dalambhija, Manipur, Kharichanda, Kharigard and others), Birbhum (Kharia, Kamarpore and others) and Purulia. Kaolin deposits in Bombay Presidency (Karalgi, Kapoli, North Kanara, Ratnagiri, Panhala, Gudalkop, Bhudargah, Fort) form lower part of laterite profile on gneisses. Lateritic hardpan on some deposits was eroded. Kaolin deposits are also in Travancore (kaolinized gneisses: Changanacheri, Trivandrum), Mysore (kaolinized gneisses and granite-gneisses: Arjunabettahali, Gullahalli, Bageshapura, Malnad, Narasimharajpur, Koppa, Thirthahalli, Melkote and others), Madras (Trivellore, Sriperumbudur taluks, Tindvanam, Cuddalore), Andhra Pradesh (Nellore, Visakphapatnam, Salem, Cuddapah, Nilgiris), Kérala (Cannanore), Kashmir and other states.

Kaolin deposits in Boralesgamuva in Ceylon are part of lateritic profile on gneisses. Pegmatites near Metiyagoda weathered to a depth of 6—10 m. Deferrization occurred under the influence of moor waters (Fe_2O_3 in kaolin only 0.06 %, TiO_2 0 %).

Residual kaolins in Swat in Pakistan are suitable for the production of porcelain, sanitary ceramics and wall-tiles. Schists and sheared granites in Harara district (Ahl) are weathered to a depth of 3 m. The kaolin contains about 30 % of clay substance. Kaolin occurrences are also in Kohat (kaolin for paper industry), in Nagar Parker (3 m of kaolin on the base of laterite), in South Waziristan (kaolinized biotite granite near Kutmar Sar, Preghal) and in other places.

The most important kaolin deposit in Iran – Simerom southwest of Isfahan – is the product of deep weathering. Other deposits – Zonous and Isyso in the far northwest – are of hydrothermal origin.

With the exception of the deposit Arnavutköy (weathered porphyrite) the Turkish deposits are of hydrothermal origin. The parent rocks are volcanic tuffs (Bursa-Mustafakemalpasha), volcanic dikes (Sindirgi:Düvertepe; Ushak-Karatchayir), granite, rhyolite and porphyry (Eskishehir-Mihalichik).

Kaolin deposits of Africa

It is a striking feature that in the recent humid tropics of Africa there are few important kaolin deposits. It is caused probably by the fact that the recent position of equator and humid tropics is relatively very young (Upper Neogene-Quaternary). The deferrization by leaching of red tropical weathering crust (red earth) is now less thorough than in past ages (mainly Upper Carboniferous and Upper Mesozoic-Lower Cenozoic). During past ages only in Triassic the equator and with it the tropical belt was in the zone of recent tropics in Africa.

In North Africa, kaolin deposits and occurrences are known in Egypt, Morocco, Algeria and Tunis. Granite and crystalline schists are the parent rocks of deposits near Assuan in Egypt (Wadi el Heita, Sikket el Arud, Kor el Dabaa and Wadi Abu Aggag). They are covered by Nubian sandstone (Cretaceous). The deposit in Sinai (with outcrops near Wadi Budra, Abu Natash, Wadi Gonema, Abu Zenima: Sabba, Alfy), described sometimes as kaolin, is a claystone intercalation in Carboniferous sandstones.

Kaolin occurrences in Morocco are presumably of hydrothermal origin (Oulmes, Beni Snassen), sometimes combined with tectonic predisposition (Guenfouda, Sibara). Occurrence near Tiouanamane is described as the product of weathering of granite. Algerian kaolin deposits are situated in Djebel Debar (Guelma). In Tunis, only small occurrences of residual kaolin probably of hydrothermal origin are known (Ain Draham sud, Ain El Berr, Kef el Maad).

In West Africa, kaolin occurs in Mali (near the Guinean frontier), Niger, Ghana and Nigeria. Biotite, granite, aplite and pegmatite, weathered to a depth of 15 m and covered by sedimentary kaolin, ferruginous sandstone, iron ore, sandy clay and laterite, can be traced in Niger in steep slopes of rivers Niger, Gerubi, Sirba and Diamangou (localities Niamey, Tilabery, Gogore, Youri, Say). Kaolin contains 30—50 % of clay substance. Kaolin deposits near Abadzi, Saltpond and the occurrence near Kromantin in coastal Ghana originated through weathering of pegmatite. Kaolin from Saltpond deposit contains halloysite. Near Kibi in the tropical forest zone, there is a deposit of kaolinized phyllite situated on the top of an inselberg. The most important kaolin deposits in Nigeria are on Jos plateau. Their parent rock is granite of Jurassic age. A small deposit near Oshiele in Abeokuta province originated through weathering of gneisses. Kaolin near Nsukka is situated on the base of a laterite profile.

Kaolin deposits in Ethiopia are in Erithrea (Hamasien). All other occurrences (Decano in Kaffa province, Addis Ababa, Sodere and Ambo in Shoa province) have no industrial significance. Weathered granites are the parent rock of a deposit near Opete in Kenya. The kaolin contains 10—15 % of clay substance. The depth of weathering is 10—35 m (Kitandani hill). Similar occurrences are in the SE of Machakos, near Ndi and to the S of Mount Kenya (near Karatina). Pyroclastic rocks were the parent rock at Kerita soda spring near Nairobi. Fu-

maroles were the main agent of kaolinization of obsidian near Eburru in the Rift Valley and Orgaria near Naivasha Lake.

Tanzanian deposits of sedimentary kaolin are located in Pugu Hills near Dar es Salaam. Kaolin sandstone of Neogene age is approximately 180 m thick, contains 30 % of clay substance and its reserves are estimated at about 2 000 million tons. In Matamba area, kaolin 6—15 m thick was formed on granite and gabbro as a product of weathering. Similar deposit is near Malangali (Iringa District).

In Kongo kaolin from weathered granites and pegmatites is occasionally exploited in Maniema, Kivu and North Katanga.

Pegmatites are also the parent rocks in Mozambique (deposit Moriangane near Vila Manica).

Kaolin deposits in Botswana are in southern and southeastern parts of the country. The deposit near Kanya (Bangwaketse district) is probably formed by weathered granite of the Gaberones type. Granite is the parent rock of a small deposit near Mochudi (Bakhatla district). Kaolinized shales of the Karoo formation are known from Palapya (Bamangwato district).

In Swaziland, the kaolin deposits are located near Mbabane and Mankaiana. Basic dykes on the latter locality in Mahlangatsha Mountains are kaolinized to a depth of 20 to 75 m.

Kaolin deposits in the Republic of South Africa occur in Vanrhynsdorp District on Archaean granites, in south-western part of the Cape Province on Proterozoic Cape granite, in southern part of the Cape Province on Bokkeveld shales (Devonian to Carboniferous) and in south-eastern part of the Cape Province on Dwyka or Witteberg shales (Carboniferous). In the Republic of Malgasia there are kaolin deposits formed on leptynites and partially rewashed on short distance near Ampanihy (Andrakaraka, Andranofotsy, Masiadolo, Sokaginadra, Sihanamovo, Terahanaombitelo). The kaolin 1—3 m (rarely up to 23 m) thick is covered by thick sandy overburden. Kaolin is bound also on two pegmatite veins near Andilana.

Kaolin deposits of America

Canadian kaolin deposits are mostly small (e. g. kaolinized granulite near Rémi-d'Amherst, Quebec), rich in sand grains, mica or pyrite (e. g. southern part of Saskatchevan) or too remote from consumers (northern part of Saskatchevan).

The USA are the most important producer, exporter and consumer of kaolin in the world. Substantial part of the whole production is not residual, but "sedimentary" kaolin, i. e. kaolinitic clay. The production started as far back as 17th

century (North Carolina). Until the discovery of kaolin in Cornwall, American kaolin was exported to Great Britain. Most important deposits are nowadays in the south-eastern states—Georgia, South and North Carolina and Florida. To a lesser extent, kaolin is mined in Pennsylvania, California and other states. The most famous deposits of residual kaolin are in Spruce Pine District (N. Car.). Mylonitized granites weathered to a depth of 2—10 m during the Tertiary are covered by overburden 10 m thick in average. High content of alumina makes the kaolin a source of metal aluminium. Residual kaolins from Piedmont Province in Alabama derived by weathering of granite, gneiss and pegmatite are suitable as fillers and refractory materials. Hydrothermal kaolin is known from Medley in Texas. Near Spokane in Washington, there are kaolins on pegmatite and aplite veins. Granites and gneisses were the parent rocks of kaolin deposit near Troy (Idaho).

Kaolin deposits in Mexico are both of hydrothermal and climatic origin. The parent rock is in most cases rhyolite or rhyolitic tuff of Tertiary age. The deposits are situated in almost all states.

Cuban kaolin deposits are on Isla de Pinos and in provinces Pinar del Rio and Oriente. The parent porphyroid-like rock of Santa Barbara deposit on Isla de Pinos weathered to a minable depth of 40 m (in average 25 m). Kaolin contains locally up to 80 % of clay substance. Another deposit with average thickness of kaolin 8 m is called Km 13. Similar porphyroid-like very fine-grained rocks near Mantua in the Province Pinar del Rio gave rise to kaolin 5 m thick. The deposit Dumaňuegos in Oriente has a thickness of 10—15 m. Its parent rock was aphanitic rhyolite. The most striking feature of Cuban kaolins is a long interval between the point of sintering and deformation.

Jamaican kaolin deposits belong to the residual type (weathered adamellite near Above Rocks), hydrothermal type (altered granite near Jobs Hill [dickite]) and sedimentary type (kaolinitic sands along Black River). None is as yet exploited.

Kaolin in Costa Rica of inferior quality is suitable for production of pottery only. In Colombia, kaolin is mined in Palmar Creek, Santuario District (weathered pegmatite dyke) and near Chivatá, Tunja District (sedimentary kaolin). Another important deposit is near Mondoňedo, Mosquera District (1—4 m thick sedimentary kaolins). The parent rocks of other occurrences of residual kaolin are granite, pegmatite, rhyolitic porphyry, volcanic tuffs and porphyritic dykes.

Little is known about kaolin deposits in Venezuela. The biggest deposit seems to be Cerro La Bandera near Upata in Bolivar Province. Guayana's famous bauxite deposits (alumina-rich laterites) are accompanied near Hope Mine on Demerara River in the lying side by kaolinized crystalline rocks. Similar deposits were discovered during the exploitation of bauxite near Mackenzie and Kwakwani. Best kaolin is reported from Orealla on Courantyne River (4 m thick-

ness of kaolin, 12 m thickness of hanging sands). Sedimentary kaolin is a part of White Sand Formation (Plio-Pleistocene) between Courantyne and Demerara Rivers. Most important occurrences of kaolin in Surinam belong to the same type as the above mentioned deposits on Demerara River in Guayana. The thickness of bauxite layer on Surinamese kaolin occurrences Moengo, Paranam and Onverdacht is 3—6 m, the depth of kaolinization is 45 m. Kaolin accompanying quartz veins (some of them gold-bearing) is known from Gross, De Jong-Zuid (parent rock: sericite phyllite) and Kilometer 116.5 along the railroad. Occurrences along the rivers Arrawarra, Nickerie, Maratakka, Pankoekoe, Saramacca, Surinam and Lower Tempati are described in literature (mineralogy and technology included), but their industrial significance and reserves are not yet known. Most kaolin deposits in Brazil are kaolinized pegmatites in gneisses. The depth of weathering is in average 25 m, the content of clay substance is up to 60 %. Most deposits are in Minas Gerais (Juiz de Fora, Bicas, Mar de Espanha), Saõ Paulo, Parana and Rio Grande.

Some important kaolin deposits in Argentina are weathered feldspathic rocks: quartz porphyry in the mine Blaya Dougnac near Trelew and porphyry tuffs in the mine Don Emilio (Chubut). Hydrothermal kaolin accompanying ore veins is known from the deposit Mutquin near Siján (Catamarca). Similar origin can be ascribed to kaolin in the deposits Cerro Segundo and Maria Eugenia near Balcarce (kaolinized granite). Other occurrences of kaolin are known from the department Chacabuco (Virgen del Valle, Guebrada de los Montes).

The kaolin deposits and occurrences of central Chile belong to hydrothermal and subaerial weathering type. The hydrothermal zones in the volcanic rocks are several meters broad (deposits Mercedes and Mary), in tonalite — 25 m (La Chica). Known depth of alteration is 60 m at the maximum (Mercedes). Influence of weathering after hydrothermal alteration is quite probable. This view is corroborated by the occurrence of granites weathered to a depth of 5—18 m (kaolin deposits Blanquita, Andelina, San José, San Miguel, Lo Pequén, Landa) in the same area.

Kaolin deposits of Australia

Most of more than one hundred Australian kaolin deposits and occurrences are bound to the base of laterite profile on dykes, granites, metamorphosed and sedimentary rocks, only a few deposits can be linked with hydrothermal activity.

New Zealand deposits consist of hydrothermal halloysites of Northland, marginal hydrothermal kaolins in rhyolites in Coromandel Peninsula, and redeposited products of Mesozoic weathering in the South Island.

Statistical data

Production of kaolin is mentioned in most papers in this volume and is summarized in the volume dealing with the genesis of the kaolin deposits (R. A. Healing — P. C. Wright: International aspects of kaolin).

The reserves of kaolin can be, in most cases, estimated only very roughly. Reserves of hundreds of millions of tons have probably China and Tanzania (?), tens of millions of tons—Guyana, India, Japan and the USA, millions of tons—Ceylon, Chile, Malgasia, New Zealand, Niger, Nigeria, Philippines, South Africa and Venezuela, hundreds of thousands of tons—Argentina, Australia, Kenya, Mexico, Pakistan, Swaziland and Turkey.

ASIA

1969 XXIII INTERNATIONAL GEOLOGICAL CONGRESS (1968) — VOL. 16 PAGE 17—24

Kaolin Deposits of Ceylon

D. B. PATTIARATCHI
Ceylon

Summary—The occurrence of good quality kaolin in Ceylon has been known for over fifty years or more. It occurs disseminated in weathered rock or as distinct layers, lenses or pockets beneath surface soil or alluvium. The underlying bed-rock over 9/10 of the Island is of Pre-Cambrian age and all stages of kaolinisation of the felspar are seen in deep-weathering profiles of the Central Highlands and the south-western sector of the Island. Kaolin is also found as bands of pipe—clay in the Gondwana succession of the Tabbowa basin but the economic importance of these occurrences is not known. The best known deposit in the Island is the Boralesgamuwa kaolin field which is exploited at present. The other known deposits are at Metiyagoda and Iranaimadu. The refined Boralesgamuwa kaolin is used in the manufacture of ceramics, and as a filler in the paper, rubber and insecticide industries.

Location of deposits

The known deposits of kaolin in Ceylon are shown in Fig. 1. The largest known deposit is the Nugegoda-Boralesgamuwa kaolin field which occurs in the low-lying coastal plain of the Colombo District (Fig. 2). The best known part of this clay-field occurs around Boralesgamuwa where detailed investigations were carried out in 1952 and again in 1966 and are still in progress. The kaolin occurs in layers and lenses beneath 5 to 15 ft. of surface soil or peaty alluvium. The thickness of the kaolin layer varies from 4 to 19 ft. and with detailed investigations it has been possible to demarcate areas where the thickness of clay is uneconomic for exploitation or discoloured. The results of recent investigations of a part of the clay field are shown in Figs. 3 and 4 and indicate the variability of the thickness of the kaolin bed and overburden.

The Metiyagoda kaolin field is situated about 5 to 6 miles from Ambalangoda in the Southern Province. Sandy and iron-stained kaolin has been found to extend from 4 to 10 ft. below the surface to a depth of over 40 ft. The area is swampy and although extensive investigations have not been undertaken the deposit is classified as one of intermediate size (1—5 million tons).

The Iranaimadu deposit is situated in the Northern Province and preliminary

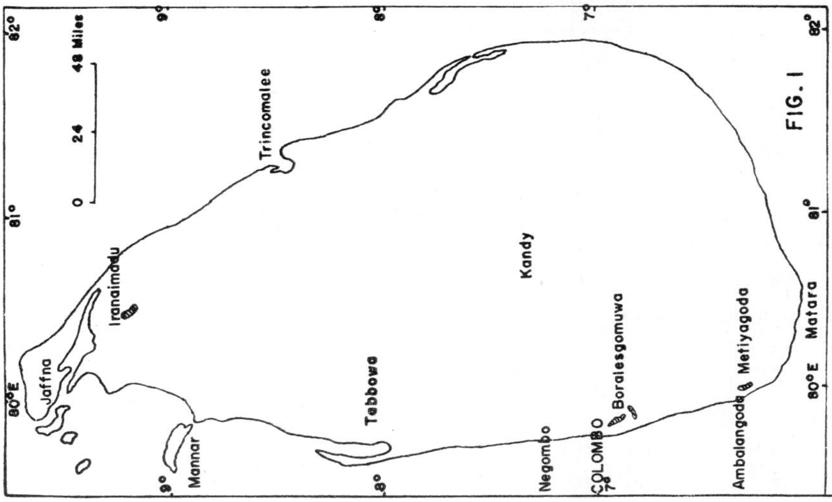

Fig. 1—Map of Ceylon showing kaolin deposits

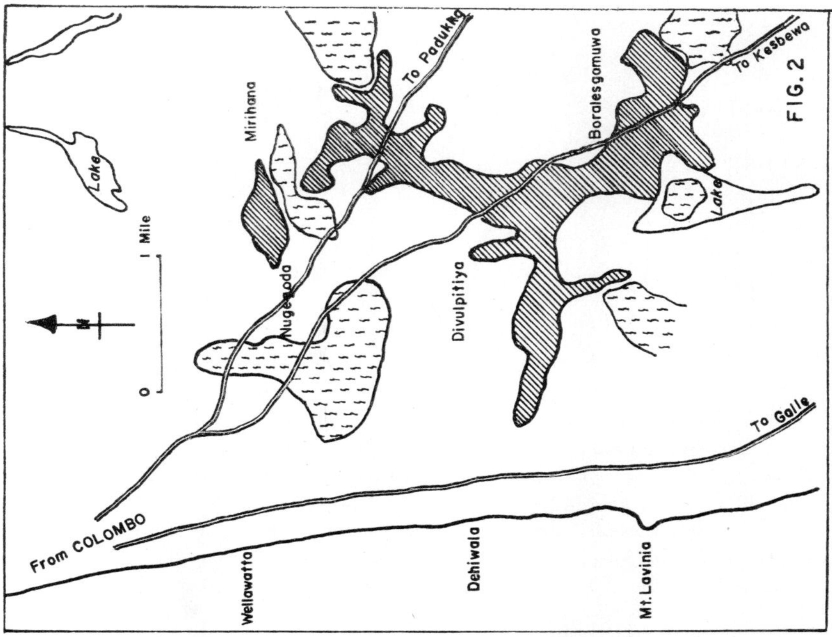

Fig. 2—Nugegoda-Boralesgamuwa kaolin deposits

LEGEND : Kaolin field; Archaean gneisses & granites; Main road

borings have indicated that the kaolin bed is covered by 10—15 feet of laterite; consequently the economic value of this clay field at present is doubtful.

Description of deposits

The known deposits are covered by overburden; the age of the latter is not known with certainty but is of probable Pliocene or Pleistocene age.

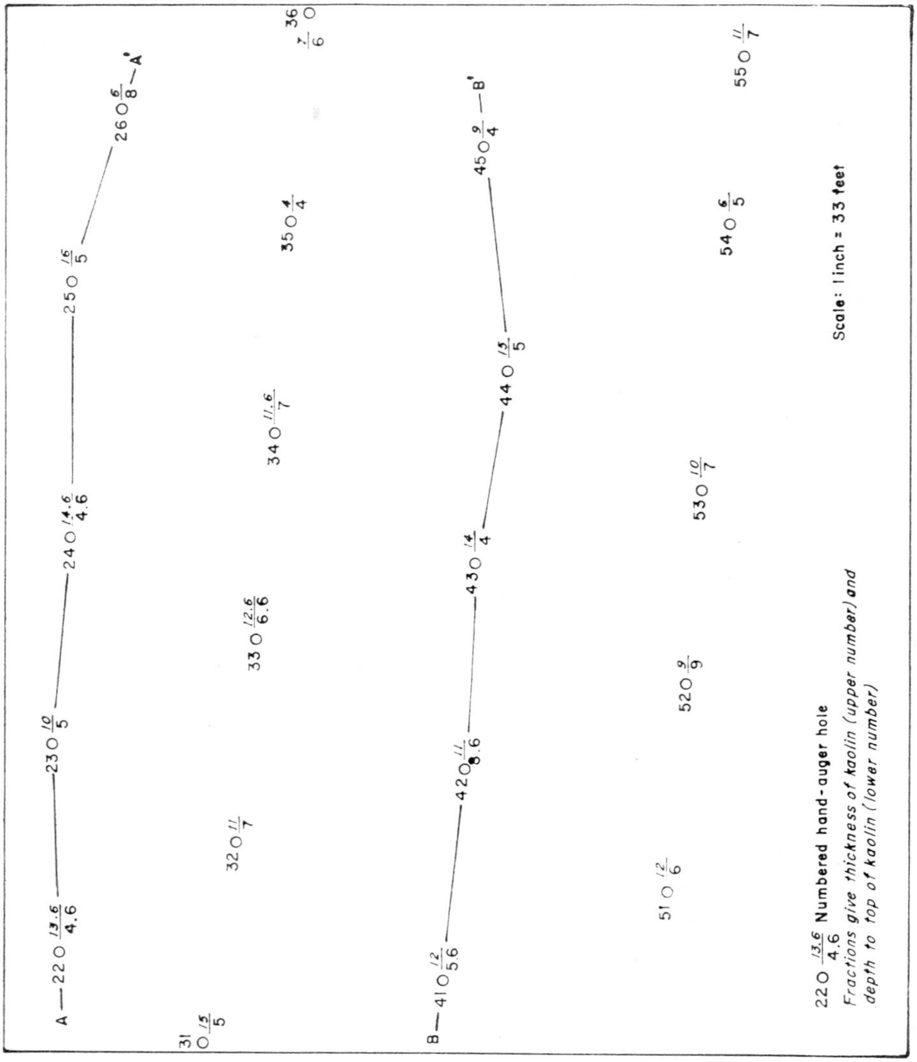

Fig. 3—Boralesgamuwa kaolin field—Map showing the depth to the top of the kaolin bed and the thickness of kaolin in selected bore-holes

Of the known deposits only the Metiyagoda kaolin deposit is associated with a pegmatite which was mined for moonstones in the past. This deposit is regarded as a residual deposit with perhaps some local transport, with sorting and re-deposition at the margins of the deposit.

In the case of the Boralesgamuwa kaolin field no deep borings have been carried out to examine the underlying bedrock, and exploitation has so far been only up to about 20—25 feet below ground level. Recent work has indicated that certain parts of this kaolin field may be of residual origin, but sorting and re-deposition in other parts of the clay field cannot be ruled out.

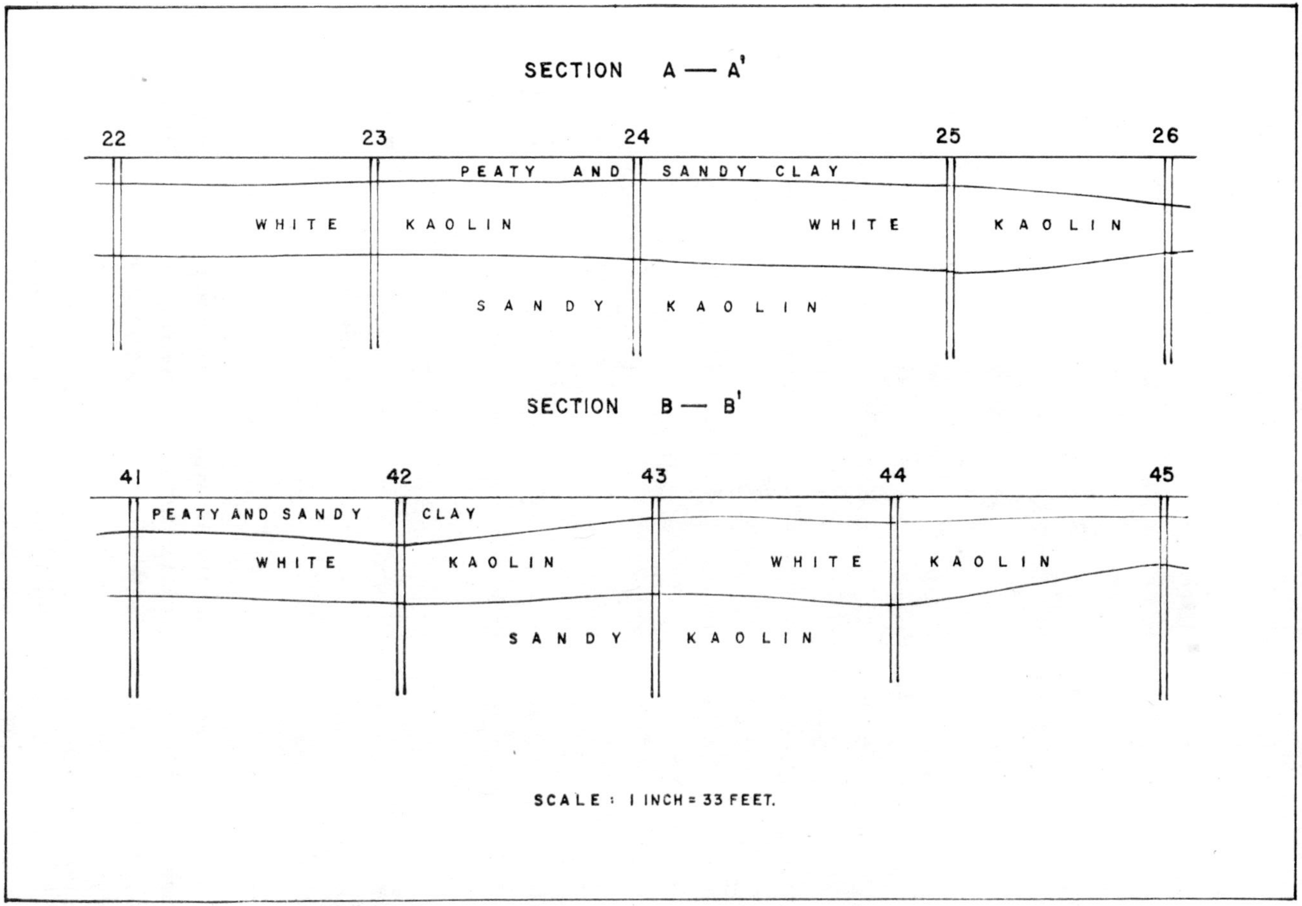

Fig. 4—Boralesgamuwa kaolin field—Sections along two lines of bore-holes shown in Fig. 3

Both these deposits occur in areas where the regional geology and geomorphology are similar. They occur in the coastal plain af the Island where the main features are silted marshes and swamps representing extinct ancient river valleys interrupted by knobs of high ground generally capped by laterite or hard rock, the latter of Precambrian age. Extensive investigations of the laterite profiles in the surrounding area indicate that laterite is absent in the low-lying areas; and the surface sandy clays associated with peaty earths pass down directly into soft kaolinitic material.

In the formation of kaolinite the environment has been one where constituents other than Al and Si have been almost entirely removed. This can be achieved in a stable land area with a humid climate where rainfall exceeds evaporation and the pH of the groundwater is acid. Under these conditions the felspar of the underlying bed-rock would decompose into kaolin and although this indicates an in situ origin for the Boralesgamuwa and Metiyagoda deposits it is very probable that subsequently some sorting and re-deposition by running water took place in some of the low-lying areas as is evident from layers and pockets of kaolin which are over 15 ft. thick and almost free of grit.

Composition, mineralogy and chemical properties of kaolin

The percentage of kaolin in the various horizons of the Boralesgamuwa kaolin field is variable from as much as 20 to 85 %, the average clay content in the exploited area being 40—50 % (Fig. 5). Similar conditions exist in the Metiyagoda deposit where the kaolin content in different horizons varies from 20 to 80 %.

In the washed samples of kaolin from Boralesgamuwa the clay substance is over 95 %.

X'ray and differential thermal analyses have shown that kaolinite is the chief constituent of the washed kaolin with 1 to 2 % quartz as the main impurity. The D.T.A. curves show the characteristic endothermic effect around 560 °C. The curve is typical of kaolinite with a little absorbed water (Herath 1963).

The accessory minerals in the raw clays of the Boralesgamuwa clay field include quartz, felspar, garnet, pyrite, magnetite and ilmenite. Recent work by the author has shown the presence of monazite in the heavy mineral concentrates obtained from the sand fraction of the Boralesgamuwa deposit, and extraction of the monazite from the sand fraction may be of economic value in the future. The impurities affecting the quality of the refined clay is iron, which occasionally discolours the product.

Three sets of typical chemical analyses of washed kaolin are given below:

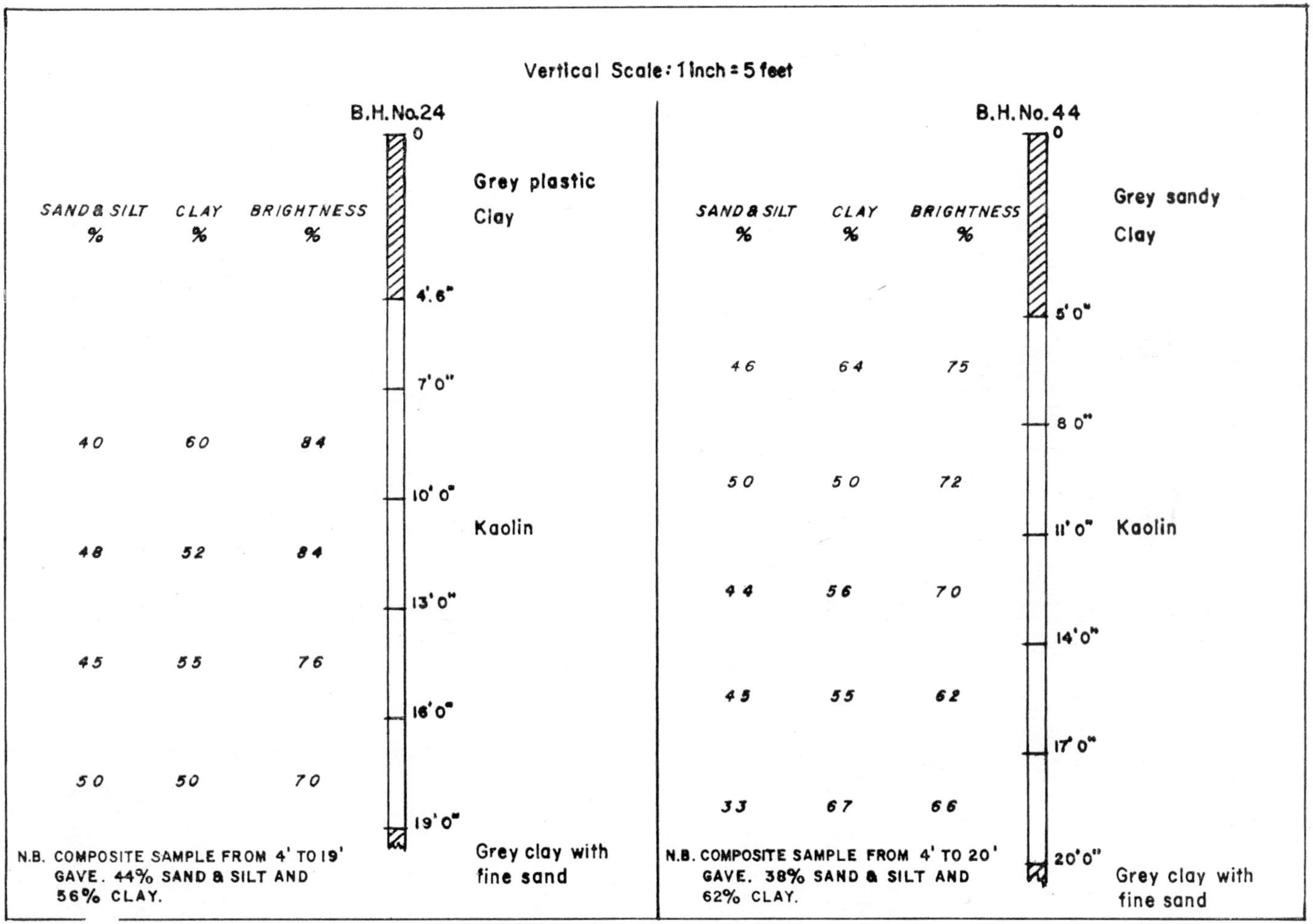

Fig. 5—Boralesgamuwa kaolin field—Two typical bore-hole logs and the test results of the respective samples

	Borales-gamuva[1]	Metiyagoda[2]	Iranaimadu[2]
SiO_2	46.63	48.35	45.95
Al_2O_3	38.4	37.35	37.18
Fe_2O_3	0.64	0.06	0.05
TiO_2	0.33	—	1.39
CaO	Traces	0.64	0.25
MgO	0.25	0.22	0.26
Na_2O	0.06	} 0.19	1.14
K_2O	0.17		0.36
H_2O	14.25	—	—
H_2O^-	1.15	—	—
Total	99.72	—	—
Loss on ignition	15.80	13.44	13.86

Analyst: [1] Mr. N. R. de Silva, [2] Geol. Surv. Dep.

Utilisation of kaolin

The deposit exploited on a commercial scale at present is the Boralesgamuwa deposit. The refined kaolin is used in the Island for manufacture of whiteware pottery, paper, chemical insecticides, rubber goods and paints.

The properties of the refined Boralesgamuwa kaolin are given below:

1. Residue on 0.076 mm. — 0.18 %
2. Particle size distribution of a typical sample of refined kaolin (Andreasen pipette method)
   N. B. The particle size finer than 2/μ varies from 60 to 70 %
   - < 25 μ — 100.00 %
   - < 10 μ — 97.5 %
   - < 5 μ — 96.3 %
   - < 2 μ — 88.2 %
   - < 1 μ — 76.0 %
3. Contraction: Plastic to dry (110 °C) 5.0 %
   Dry to fired (1260 °C) 12.6 %
4. Water of plasticity (Pfefferkorn) 38.9 %
5. Brightness — 75 % (at 457 mμ or Wratten light filter No. 49)
6. Acid soluble iron % — .02 approx.
7. Apparent porosity (at 1100 °C) 37 %, (at 1410 °C) 1 %
8. Fusion point—1750 °C.

Table 1—Production and Utilization of Kaolin in Ceylon — 1963—1967 (in long tons)

Year	Production	Utilization				
		Paper industry	Ceramic industry	Insecticides	Rubber industry	Other industries
1963	386	6	—	105	20	20
1964	1591	115	63	675	76	40
1965	542	212	127	220	80	30
1966	1335	422	425	770	97	60
1967	2570	935	664	730	181	245

Economic evaluation of deposits

The best known deposit is the Boralesgamuwa deposit. The present area demarcated for exploitation is six acres in which 145,000 tons of raw kaolin were proved. The total quantity of refined kaolin available in this clay field is estimated at 8 million tons.

The probable reserves of the Metiyagoda kaolin are estimated at 1 million tons.

Mining conditions

The Boralesgamuwa deposit occurs in swampy ground where the water-table is high and mining is by open cast methods. The pits are de-watered by pumping and much of the seepage water is confined to the overburden.

Statistical data

There are no records of the output of kaolin during the period 1900—1962. Small quantities of refined kaolin were probably produced during the period 1945—1962 but no records are available. In 1963 the Ceylon Ceramics Corporation commenced the commercial production of kaolin at Boralesgamuwa and Table I gives the production figures for the period 1963—1967.

Since the commencement of the Boralesgamuwa Refinery no appreciable tonnages of kaolin have been imported to Ceylon. The requirements of the local paper, ceramic, insecticide and rubber industries are at present met entirely by the Boralesgamuwa Refinery.

Acknowledgements

The author thanks Mr. L. J. D. Fernando, Director, Geological Survey for kind permission to publish this paper and his colleagues of the Geological Survey who have contributed to the study of Ceylon kaolin and on whose work he has drawn in writing this paper. The author also acknowledges with thanks the statistical data supplied by Mr. S. Pallewatte, Manager, Boralesgamuwa Kaolin Refinery, and the kind permission of Mr. S. N. B. Wijeyekoon, late Chairman, Ceylon Ceramics Corporation to publish the data.

References

Herath, J. W. and Sirimanne C. H. L. (1961): The Geology and some technological aspects of the Boralesgamuwa Kaolin deposit of Ceylon: Indian Ceramics, Vol. 8, pp. 177—180.

Herath, J. W. (1963): Kaolin in Ceylon: Economic Geology, Vol. 58, pp. 769—773.

Kaolin Resources of Ceylon—in ECAFE publication No. E/CN. 11/I & T/75 of 30. 10. 52.

Unpublished reports with Geological Survey Department, Ceylon.

Wadia, D. N. (1941): Geology of Colombo and environs: Spolia Zeylanica, V, XXIII, pt. 1.

1969 XXIII INTERNATIONAL GEOLOGICAL CONGRESS (1968) — VOL. 16 PAGE 25—28

Kaolin Deposits of Iran

N. KHADEM
Iran

There are many small kaolin deposits in Iran although the clay materials yet used for pottery are generally inferior. The Geological Survey was established only five years ago and there has not been an opportunity for systematic research.

Nevertheless, the most significant deposit in terms of size and quality was discovered by the Geological Survey in 1967 at Simiron southwest of Isfahan in central Iran. The reserves are of the order of 15,000,000 tons.

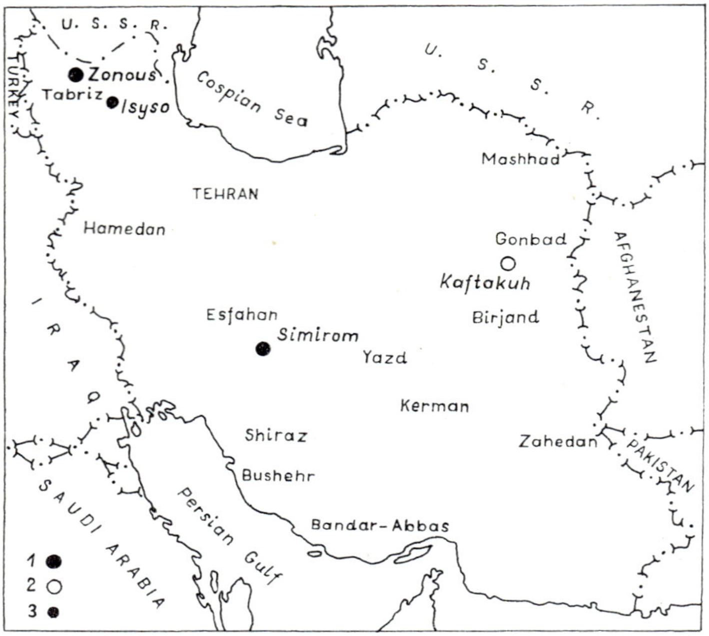

Fig. 1—Main known kaolin deposits in Iran
1 - large deposits; *2* - intermediate deposits; *3* - small deposits

Two other significant deposits are in the far northwest at Zonous (reserves—several million tons of low grade material) and Isyso (a relatively small deposit), the former having long been used as a source of material for the glazed tile industry and the latter as refractory material. Some impure deposits in eastern Iran in the area of Gonabad are used in Mashad for refractory purposes.

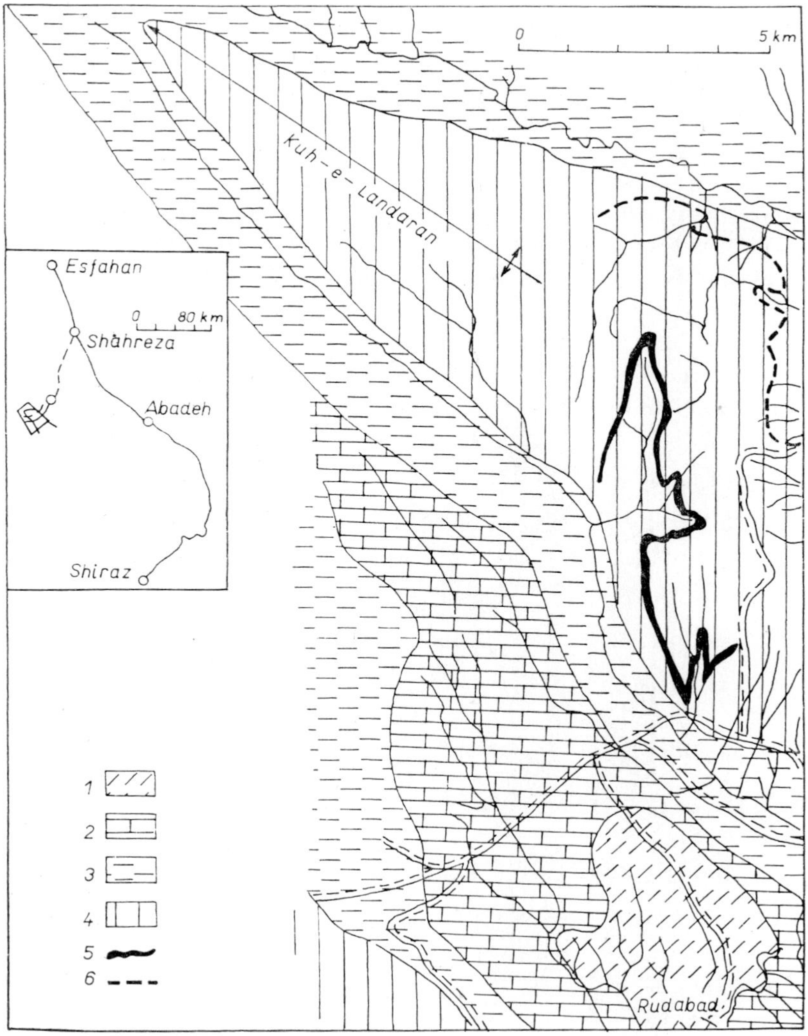

Fig. 2—Simirom clay deposits in Iran
1 - Eocene-Paleocene; *2* - Maestrichtian-Campanian limestone; *3* - Campanian-Santonian marls; *4* - Santonian-Albian limestone; *5* - kaolin horizon (proved); *6* - kaolin horizon (probable)

The Zonous and Isyso deposits are the product of hydrothermal action. The Simirom deposit is apparently the product of deep surface weathering at a diastem in the Cretaceous limestone sequence; in a 6 m zone, 30 % of the material consist of true kaolin lying in packets in a high alumina ferrous clay. This has been tested for fine ceramic and refractory purposes, but is essentially refractory material so far as is yet known—much further laboratory testing is called for before the full potentialities of this newly found deposit are known. The deposits in eastern Iran are also of two types, one at Kartakuh being of lenticular form at the Jurassic-Cretaceous unconformity and the other, at Gona-

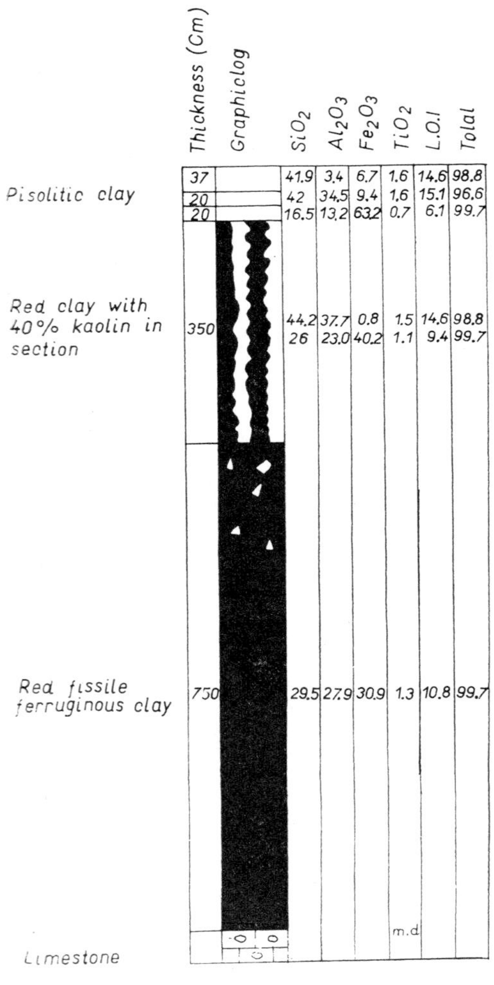

Fig. 3—Detailed section through Simirom kaolin deposit in Iran

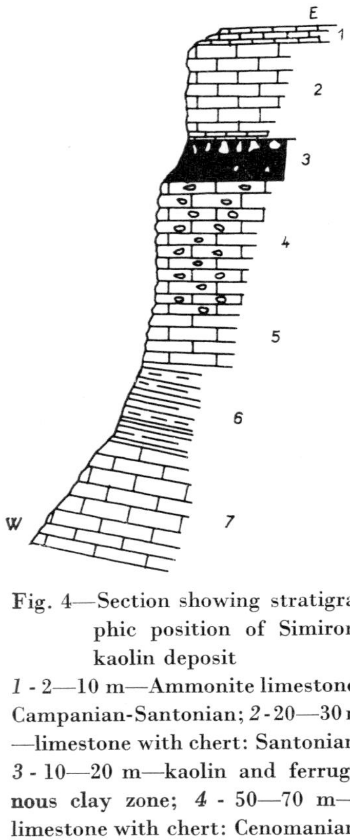

Fig. 4—Section showing stratigraphic position of Simirom kaolin deposit
1 - 2—10 m—Ammonite limestone: Campanian-Santonian; *2* - 20—30 m—limestone with chert: Santonian; *3* - 10—20 m—kaolin and ferruginous clay zone; *4* - 50—70 m—limestone with chert: Cenomanian; *5* - 10—30 m—bituminous limestone: Cenomanian; *6* - 20—50 m—black limestone and shale: Albian; *7* - limestone with rudist debris: Aptian

bad, is a hydrothermal alteration of rhyolite. Two chemical analyses are as follows:

	Simirom	Isyso
SiO_2	42.99	56.00
TiO_2	2.00	—
Al_2O_3	37.00	30.03
Fe_2O_3	1.04	1.32
FeO	0.06	—
MnO	n. d.	—
MgO	0.53	0.31
CaO	0.36	0.86
Na_2O	0.21	—
K_2O	0.27	—
H_2O^+	12.88	11.98
H_2O^-	1.64	—
CO_2	0.30	—
Organic matter	14.62	—

1969 XXIII INTERNATIONAL GEOLOGICAL CONGRESS (1968) — VOL. 16 PAGE 29—37

Kaolin Deposits of Japan

N. FUJII, T. OKANO and Y. SHIMAZAKI
Japan

Brief summary of the deposits

The kaolin deposits of Japan are classified by their genesis as follows.

Kaolin deposits	Hydrothermal deposits	
	Sedimentary deposits	"Kibushi-clay" "Gaerome-clay" Bedded white kaolin clay
	Residual deposits	

Kaolinic clays used mainly for pottery and white wares, and for paper clay are included in this report, and material used exclusively for fire clay are excluded.

"Kibushi and gaerome-clays" are usually classified as material for fire clay, but since more than 30 percent is used for pottery and white ware, they are included in this paper, also.

Hydrothermal deposits.—These deposits are vein and massive deposits occurring mostly in volcanic and granitic rocks. There are a large number of these deposits, but only a few are of large scale. Many of the material mined for use in paper industry come from the deposits of this type.

Sedimentary deposits.—"Kibushi-clay" is a general term designating soft and plastic underclay. "Gaerome-clay" is formed by secondary deposition of decomposed granitic rocks and it contains coarse quartz grains. The clay and quartz are separated by washing before use. These "Kibuschi-and Gaerome-clays" are the most important sources of kaolin in Japan, and they are used mainly for refractorics, pottery, and white ware. The bedded kaolin clay occurs in Tertiary terrigenous sediments. It is believed to have been formed by the kaolinization of tuffaceous material and is used mainly for pottery and white ware.

Residual deposits.—In Japan, there are residual deposits formed by in situ weathering of granitic rocks, pegmatites, and of Quaternary volcanic ashes. The

clays of granitic and pegmatitic origin are used for pottery and white ware, but the scale of the deposits is small. Those of the volcanic origin are abundant, but many of them contain impurities which makes utilization difficult.

The distribution of the major kaolin deposits of Japan is shown in the Fig. 1.

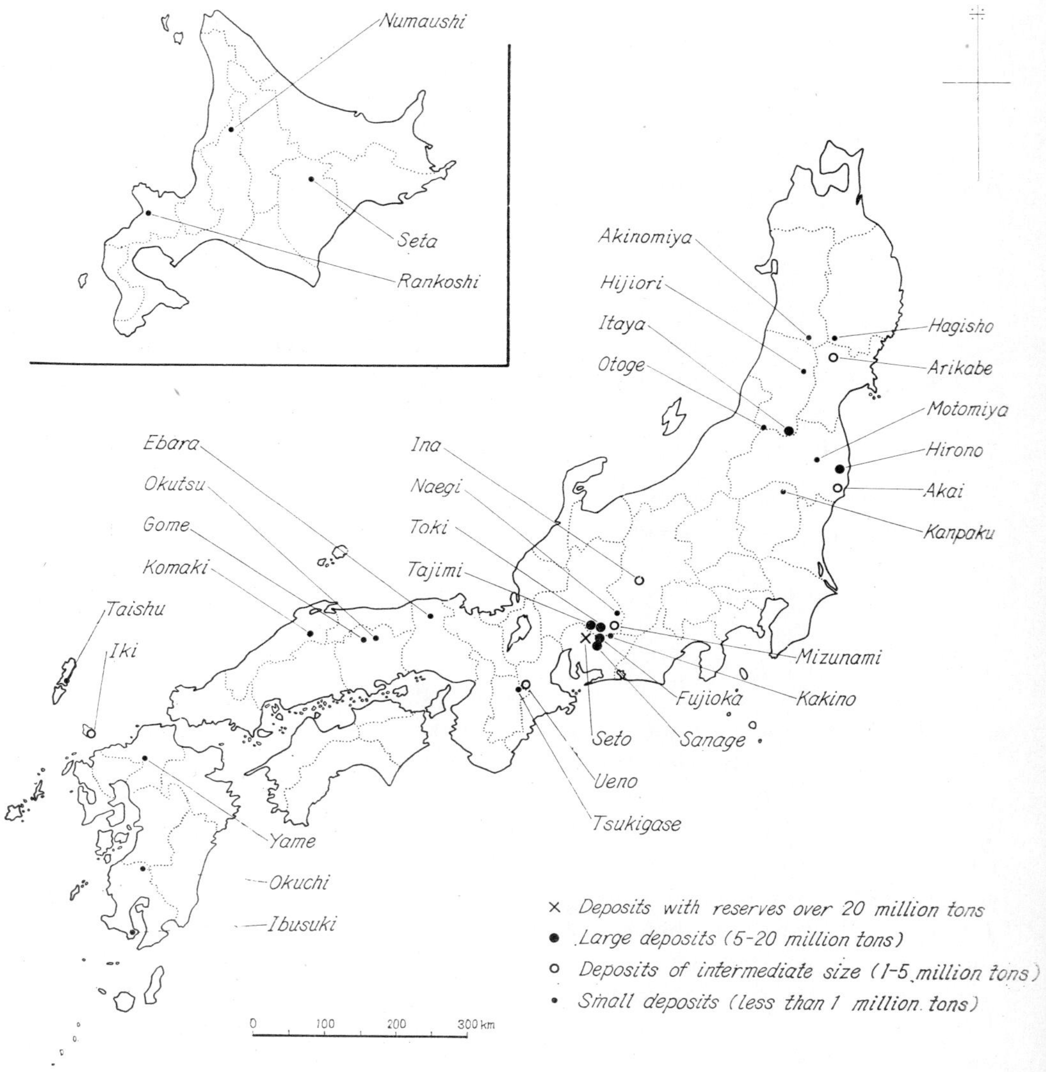

Fig. 1—Kaolin deposits in Japan

Geologic conditions

The number of kaolin deposits in Japan is very large, and the geologic conditions of the representative deposits of various types are presented below.

Hydrothermal deposits—I (in volcanic rocks)

These deposits were formed by the hydrothermal alteration of volcanic and pyroclastic rocks of Cretaceous to Quaternary period. Many of these are closely associated with metallic deposits of gold, mercury, iron sulphides, iron, and others.

Some of these deposits have vein forms such as those of the Seta mine, but most of them are irregular beds or massive deposits. The scale of these deposits is mostly small.

Itaya mine consists of exceptionally large deposits of this type and the monthly production is about 15,000 tons. There are three major deposits and the largest one has the confirmed dimensions of 350 m east-west, 300 m north-south, and the depth is over 100 m. The deposits are the hydrothermal alteration product of andesitic lava and pyroclastics of Quaternary eruption. The occurrence is complex as the silicified and kaolinized zones are intricately interwoven. Montmorillonitized zone is developed in the fringes of the deposits. Many parts of the kaolinized zone contain sericite.

Hydrothermal deposits—II (in granitic rocks)

There are hydrothermal kaolin deposits formed in the acidic intrusives such as granite and quartz porphyry of Cretaceous to early Tertiary age. Most of them are of massive or vein form.

The deposit of the Taishu mine is well known in this group. The deposit of this mine was formed by halloysitization of quartz porphyry which intruded almost concordantly into Paleogene strata. There are scores of deposits of this type with various dimensions and the largest ones are in the order of 100 m × 200 m. Similar deposits are Kanpaku, Okutsu, Komaki and others.

Sedimentary deposits ("Kibushi-clay", "Gaerome-clay", and bedded white kaolin deposits)

There are two types of deposits of this category, one is those consisting mainly of detrital origin minerals and the other is those composed mainly of authigenic

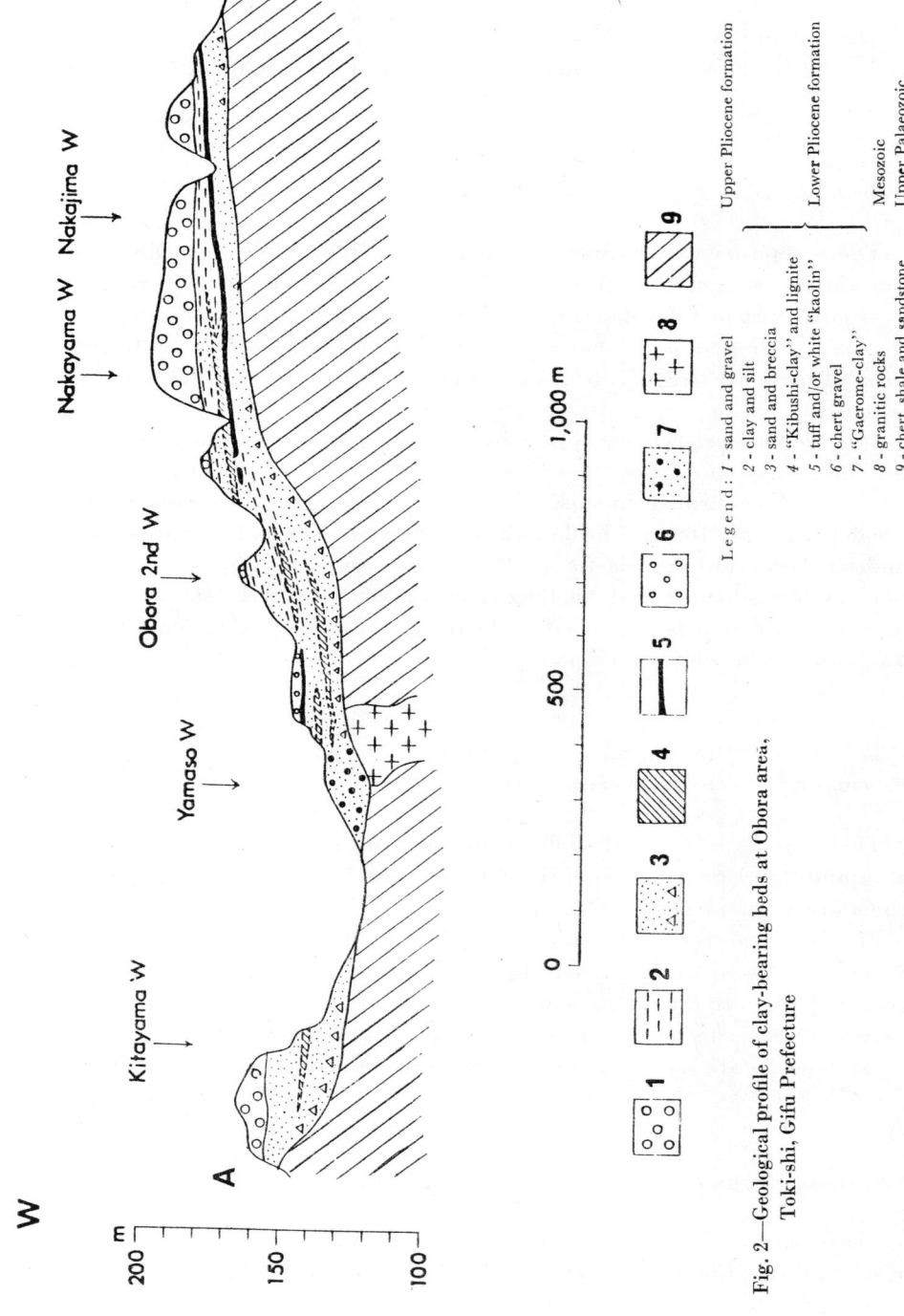

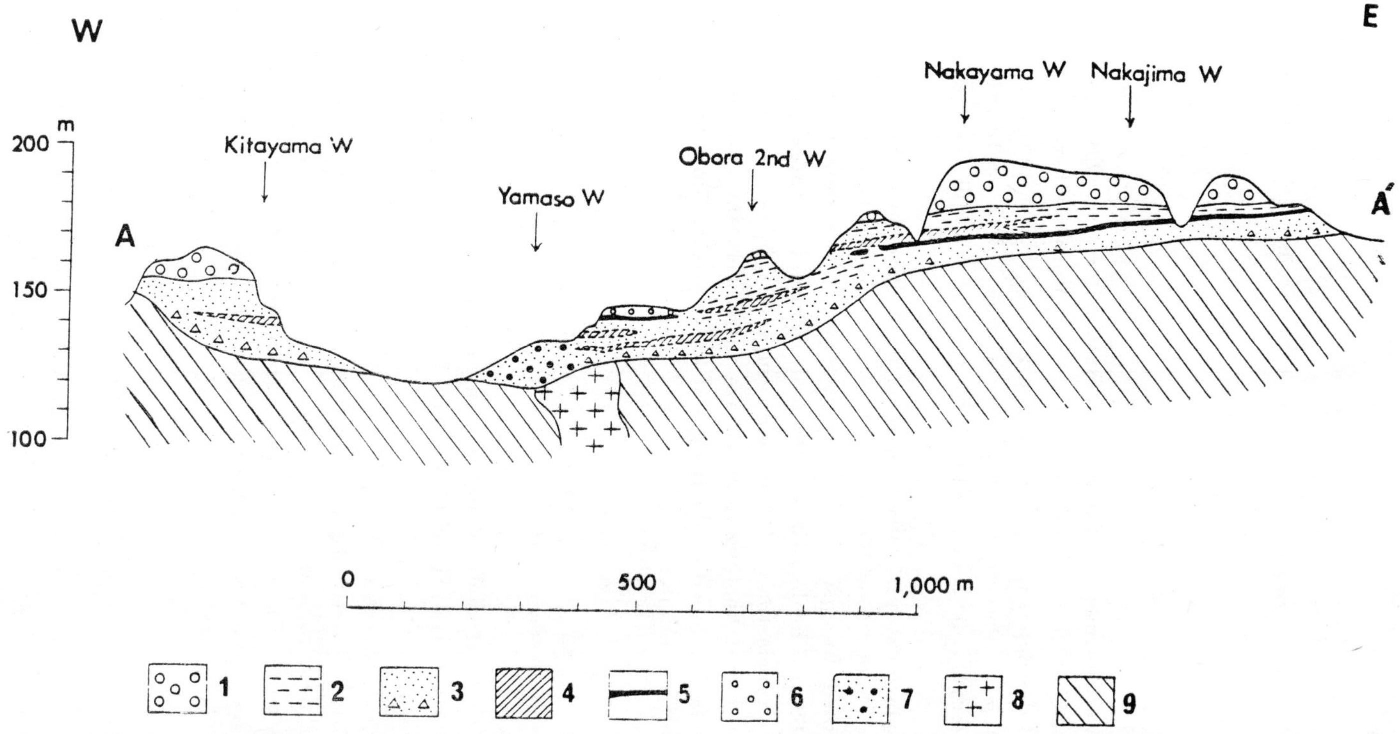

Fig. 2—Geological profile of clay-bearing beds at Obora area, Toki-shi, Gifu Prefecture

Legend:
- *1* - sand and gravel — Upper Pliocene formation
- *2* - clay and silt
- *3* - sand and breccia
- *4* - "Kibushi-clay" and lignite
- *5* - tuff and/or white "kaolin"
- *6* - chert gravel
- *7* - "Gaerome-clay"

(2–7: Lower Pliocene formation)

- *8* - granitic rocks — Mesozoic
- *9* - chert, shale and sandstone — Upper Palaeozoic

kaolin minerals. Majority of these deposits occur in post-Miocene terrigenous sediments. The important deposits of this type are intercalated in lower Pliocene strata and the major localities are Seto, Sanage, and Fujioka of Aichi Prefecture, Tajimi, Toki, and Mizunami of Gifu Prefecture.

The lower Pliocene series of these areas consist of lacustrine sedimentary rocks and the size of the individual sedimentary basins which are scattered in the region is in the order of 2—5 km. "Kibushi-clay" and "Gaerome-clay" are believed to have originated from detrital minerals while the bedded white kaolin clays were mostly formed authigenously.

"Kibushi-clay" occurs in the mother rocks of brown coal. The maximum thickness of individual beds is about 3 m. The clay consists mostly of disordered kaolinite and quartz grains are contained as impurity in most cases. In areas where Miocene series are distributed below the Pliocene beds, montmorillonite which is believed to be an alteration product of the Miocene strata is mixed in the kaolin clays.

"Gaerome-clay" occurs at the base of the lower Pliocene series and the maximum thickness is 20 m. It is believed to have been deposited from weathered and decomposed granitic rocks without sorting. It contains many coarse grains of quartz, unaltered feldspars, and mica minerals. Normally, the major constituent minerals are disordered kaolinite and partially dehydrated halloysite. These "Gaerome-clays" are developed only in sedimentary basins where the basement consists of granitic rocks such as Seto, Toki, and Fujioka.

The bedded white kaolin usually occurs in thin beds of 20—30 cm. They are distinguished from other clays in the lower Pliocene by the fresh white color. It is believed to be the product of alteration of tuffaceous material after deposition. The major clay minerals are halloysite and meta-halloysite with small amount of quartz. The occurrence of this clay is limited to the basins of Tajimi, Toki, and Naegi. Similar clay is distributed in the post-Miocene sediments of Hokkaido (Numaishi) and Northeast Japan (Hagisho, Arikabe, etc.).

Residual deposits

There are many kaolin deposits within the granitic basement below the lower Pliocene which contains "Kibushi-clay" and "Gaerome-clay". These deposits in the granitic rocks seem to have been formed by in situ weathering. Kakino and Mitsukuri kaolin deposits are the representative deposits of this type. Most of the deposits have thickness of several meters.

The deposits of the Motomiya mine is the only case of kaolin formed by the weathering of feldspars in pegmatites in Japan. On the other hand, Ina kaolin

is an example of the deposits formed by the weathering of Quaternary volcanic ashes. It is used for paper clay and the thickness of the deposits is in the order of 2 meters.

Mineralogical and geochemical conditions

The mineral compositions and the kinds of kaolin minerals are shown below according to the types of the deposits.

Mineral composition

Type of deposit	Name of deposit	Main	Accessories
Hydrothermal deposit			
In volcanic rocks	Itaya	K, (S), q	py
	Iki	H, c	K, q
	Seta	K, q	py
	Ebara	K, (D)	dias, q, B
In granitic rocks	Taisyu	H	q, f, S
	Kanpaku	K	H, dias, G, py
Sedimentary deposit			
"Kibushi-clay"	Tajimi	K, (q)	M, I
	Seto	K, (q)	I
	Ueno	K,	q, f, I
"Bedded white kaolin" clay	Tajimi	H	q
	Arikabe	H, (q)	
Residual deposit			
Weathered granite	Kakino	K, mH, q	I, f
	Motomiya	K, H	I, q, pl
Weathered volcanic ash	Ina	H, mH, q	pl
	Yame	H, mH, q	pl, h, limo

B—boehmite; c—crystoballite; D—dickite; dias—diaspore; f—feldspar; G—gibbsite; H—halloysite; mH—metahalloysite; h—hornblende; I—illite; K—kaolinite; limo—limonite; M—montmorillonite; S—sericite; q—quartz; py—pyrite

Chemical composition

	Hydrothermal deposits			Sedimentary deposits				Residual deposits
	Itaya*	Iki (washed)	Kanpaku	Tajimi (K) **	Seto (K)	Toki (G) (washed)	Tajimi (Shinmei) (W)	Motomiya (washed)
SiO_2	46.50	44.60	41.55	47.69	45.48	47.51	44.51	57.00
TiO_2	0.10	0.23	0.67	0.94	0.62	0.46	0.11	—
Al_2O_3	37.01	36.86	41.50	30.60	32.15	36.60	35.48	32.55
Fe_2O_3	0.69	3.11	0.46	1.28	0.74	1.24	0.11	0.54
FeO	—	tr	—	—	—	—	—	—
CaO	0.43	0.09	0.11	0.60	0.29	0.22	0.99	1.10
MgO	0.95	0.03	0.36	0.66	0.24	0.21	tr	tr
NaO_2	0.54	0.26	0.02	0.10	0.19	0.04	tr	1.94
K_2O	5.83	0.20	0.00	1.12	0.54	0.60	tr	tr
$H_2O(+)$	7.30	—	—	10.62	—	—	13.28	5.17
$H_2O(—)$	0.46	—	—	4.64	—	—	5.74	2.00
In. loss	—	14.50	15.30	—	19.91	13.44	—	—
Total	100.27	99.88	99.97	98.25	100.16	100.32	100.22	100.30

(K) "Kibushi-clay"
(G) "Gaerome-clay"
(W) "Bedded white kaolin" clay
* sericitic clay
** including organic materials

Utilization of kaolin

The major uses of kaolin clays are as follows.
Various properties of the clays are omitted from the table:

Kinds of clay		Use
Hydrothermal kaolin	In volcanic rocks	Paper clay
	In granitic rocks	Pottery, white ware, paper clay
Sedimentary kaolin	Kibushi-clay	Refractories, pottery, white ware
	Gaerome-clay	Pottery, white ware, refractories
Residual kaolin	Weathered granite	Pottery, white ware
	Weathered volcanic ash	Paper clay

Economic evaluation of deposits

The reserves of the kaolin deposits are evaluated as follows:

Gaerome-clay	approximately	31 million tons
Fire clay and kaolin	approximately	88 million tons

The demand and supply of kaolin is more or less balanced, but the kaolin for white wares is insufficient in quantity and the import from Korea, Hongkong and other countries is increasing every year. Also Georgia kaolin and others are being imported as paper clay.

Most of the hydrothermal and residual deposits are mined by open pit methods. However, in about a third of the mines, underground cutting is employed.

Statistical data

Output of kaolin clay (unit 1,000 metric tons)

Year	Hydrothermal kaolin clay	Sedimentary kaolin clay				Residual kaolin clay
		"Kibushi-clay"	"Gaerome-clay"		Bedded white kaolin	
			crude	washed		
1952	24	358	109		3.4	0.3
1953	27	316	110		6.6	0.1
1954	32	294	356		4.1	0.1
1955	35	330	397		4.9	0.7
1956	45	399	449		5.5	1.1
1957	54	469	479		4.8	2.2
1958	57	321	469		5.9	3.8
1959	76	405	485		5.7	1.4
1960	82	456	627		3.2	0.9
1961	96	534	715		5.3	0.7
1962	134	439	696	183	9.4	0.6
1963	169	369	774	219	13.0	8.4
1964	180	393	838	260	14.8	9.6
1965	169	427	846	262	16.5	9.6

As mentioned earlier, "Kibushi and Gaerome-clays" are usually classified as fire-clay, and thus the statistical figures of other sources such as the Minerals Yearbook may not agree completely.

Export and import of kaolin clay (unit 1,000 metric tons)

Year	1952	1953	1954	1955	1956	1957	1958
Export	1	2	21	539	422	208	1,185
Import	4,438	7,313	10,127	11,071	12,368	15,906	14,506
Year	1959	1960	1961	1962	1963	1964	1965
Export	199	1,407	2,428	5,225	3,046	2,780	2,835
Import	15,049	22,195	36,759	35,975	49,066	53,769	60,068

History of investigation and exploitation

It is believed that the use of kaolinic clay for earthen wares and pottery began more than a thousand years ago. The active use of the material for refractories and pottery and white ware, however, began about 1880. It is after the second world war that kaolin became used in large quantities for paper clays, pyrophyllite was mostly used before that.

The first systematic investigation of kaolin deposits was carried out by the Geological Survey of Japan during the period of 1920—1925, large scale investigations are being carried out actively after the war and various data on the distribution, mineral composition and others are being accumulated. Also geological and geochemical studies on the genesis of various types of deposits are in progress.

References

Clay Sci. Assoc. Jap. ed. (1967): Handbook of Clays.
Geological Survey of Japan (1950): Mineral Resources of Japan, Part B, III, 137p.
Muraoka, M., (1952): On the Fire Clay in Japan, Geol. Sur. Japan, Rept., No. 145, 81p.
Sueno T. and Iwao, S. (1958): Clay and its Uses, 470 p.

1969 XXIII INTERNATIONAL GEOLOGICAL CONGRESS (1968) — VOL. 16 PAGE 39—42

Kaolin Deposits of the Philippines

FRANCISCO A. COMSTI
Philippines

Summary—Philippine kaolin deposits are generally small and scattered. Excepting a few, most of the kaolin deposits were formed in either pyroclastic or volcanic flow rocks. They are mostly hydrothermal in origin and localized in Late Tertiary to Quaternary volcanic rocks, along or close to volcanic-tectonic zones. Genetic relations with the waning phase of volcanism are clear especially near existing hot springs and fumaroles where the formation of the kaolin is still an active process. Tropical weathering contributed mainly to the formation of the Bukidnon kaolin deposit. The excess alumina (Al_2O_3) content of this deposit may be due to formation of gibbsite (?) from the alteration of kaolin.

These deposits are mostly utilized by our ceramic industry particularly in the manufacture of refractories and whitewares.

Geological conditions

The Philippines is an archipelago of more than 7,000 islands. Volcanic rocks abound in the larger islands and many volcanic cones of Late Tertiary to Quaternary age which are oriented along tectonic zones are prominent. The kaolin of Batangas, Laguna, Benguet and other smaller deposits are hydrothermal in origin formed in situ along or close to these tectonic zones in pyroclastic and volcanic flow rocks. They are often close to acidic thermal springs and fumaroles where kaolin is presently forming. The Iloilo deposits are also hydrothermal in origin formed in volcanics and diorites. The kaolin deposits are close to the surface, irregular in shape and each deposit contains variable amounts of either silica, sulfur or pyrite or mixtures of the impurities. The Batangas kaolin contains patches of alunite. Some deposits contain cristobalite.

The Bukidnon kaolin deposits are situated in a tropical forest on the slopes of an extinct volcano. Although hydrothermal alteration may have contributed to the formation of the deposit, it is believed that the major factor is weathering in situ of dacite(?) and/or rhyolite(?) flows. The deposits are mixtures of hydrated halloysite (endellite), kaolinite and gibbsite(?). Gibbsite(?) is believed to have formed by prolonged leaching of the kaolin by an annually well distributed tropical rainfall. Under the same climatic conditions, extensive ferruginous laterites were formed from ultrabasic rocks in Surigao, 200 km northeast.

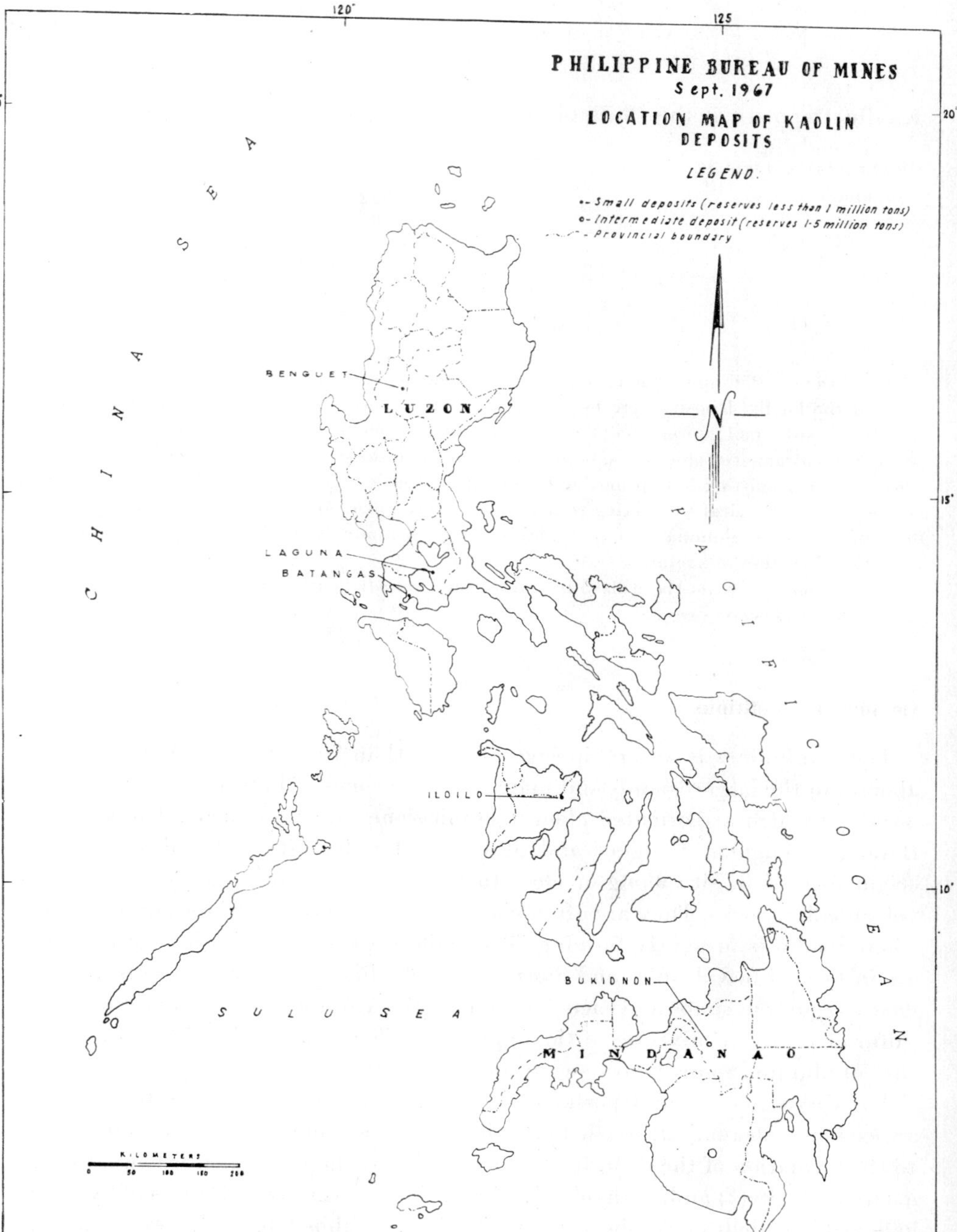

PHILIPPINE BUREAU OF MINES
Sept. 1967
LOCATION MAP OF KAOLIN DEPOSITS
LEGEND
Small deposits (reserves less than 1 million tons)
Intermediate deposit (reserves 1-5 million tons)
Provincial boundary
BENGUET
LUZON
LAGUNA
BATANGAS
ILOILO
BUKIDNON
MINDANAO
CHINA SEA
PACIFIC OCEAN
SULU SEA
CELEBES SEA
KILOMETERS
0 50 100 150 200
120°
125
20°
15°
10°
5°
N
JEC '67

The following are the chemical analyses (Table 1) of some of the kaolin deposits.

Table 1—Chemical analyses*

	Bukidnon	Iloilo	Laguna
SiO_2	38.75	44.07	46.84
TiO_2	—	—	1.11
Al_2O_3	43.66	39.92	34.43
Fe_2O_3	1.28	1.03	1.74
CaO	0.34	0.53	0.26
MgO	0.32	0.20	0.26
Na_2O	—	—	} 0.64
K_2O	—	—	
Loss on ignition	15.58	14.65	13.74

* dry basis
— dry not analyzed

Utilization of kaolin

The bulk of the kaolin deposits are utilized in the manufacture of refractory bricks, vitreous tiles, wall tiles and sanitary wares. Very little is used in the paper and rubber industry.

Since the kaolin is not beneficiated or dressed, selective mining is practiced to insure purity acceptable to the ceramic industry.

The Bukidnon kaolin has a short plasticity, high shrinkage and high refractoriness. It has a P.C.E. of Cone No. 35, reported in terms of Orton Standard Pyrometric Cones.

Economic evaluation of deposits

The following is a list of only the more important deposits (see Fig. 1) and their corresponding reserves:

	Positive reserves (A)	Probable reserves (C^1)
Bukidnon	2,500,000 M. T.	no estimate
Batangas	50,000 M. T.	50,000 M. T.
Iloilo and offshore islands	100,000 M. T.	100,000 M. T.
Laguna	100,000 M. T.	100,000 M. T.
Benguet	30,000 M. T.	no estimate

The hydrothermal kaolins are generally impure in their natural state. Their chemical analyses vary because of variable admixtures of quartz, pyrite and oxides of iron. The residual kaolin of Bukidnon has less variation in composition. Iron oxides give an "off color" when fired.

Small scale mining with pick and shovel is used. Many deposits are far and separated by sea from the ceramic plants. These factors coupled with poor transportation system make the price of the clay costly.

References

Abadilla, Quirico A. (1935): Geology of the white clay deposit in Siruma Peninsula, Camarines Sur, Luzon. Philippine Journal of Science, Vol. 57, 2, pp. 227—233.

Antonio, I. S. (1959): The Sipit clay deposits at Bitin, Bay, Laguna. Unpublished report, Philippine Bureau of Mines files, Manila.

Burgess, B. C. and Vera, E. C. (1956): Ceramic mineral deposits of the Philippines, reports revised and edited. Unpublished report, Philippine Bureau of Mines files, Manila.

Gorriceta, A. J. (1966): Geological study of white clay deposits in Puting Lupa, Calamba, Laguna. Unpublished report, Philippine Bureau of Mines files, Manila.

Llave, C. A. (1967): Geological investigation of clay materials in Barrio Tikalaan, Talakag, Bukidnon. Unpublished progress report, Philippine Bureau of Mines files, Manila.

1969 XXIII INTERNATIONAL GEOLOGICAL CONGRESS (1968) — VOL. 16 PAGE 43—51

Kaolin Deposits of Turkey

ISMAIL SEYHAN
Turkey

Summary—In this paper are discussed both the known primary kaolin deposits of Turkey and secondary kaolins, as well as kaolinitic fine ceramic clays that have white fired colour. A list of the deposits is given; locations and the reserves are shown on a map. Deposits of the same origin located in the same region are divided into various groups and dealt with as a kaolin province. The geological conditions of the provinces are discussed more than those of the deposits separately. Chemical and mineralogical compositions and technological qualities of these kaolins are also reported. In the closing chapter, the kaolin production of Turkey and the use of kaolin in various branches of industry are discussed. Clays used for producing coarse ceramic, illite and bentonite deposits of large reserves known as kaolin that have white fired colour are not mentioned here.

Geography

In Turkey, detailed investigations for kaolin deposits have been carried out so far only in the vicinity of ceramic factories. According to the results obtained so far, secondary kaolin and kaolinitic fine ceramic clay deposits of medium size with a visible and probable reserve of 1—5 million tons are restricted only to a few occurrences.

Primary kaolin deposits

ISTANBUL region:	Arnavutköy
	Beykoz
ÇANAKKALE region:	Çan—Dumanköy
	Çan—Bahadirli village
	Bayramiç-Söğüt gediği village
BİLECİK region:	Küreköy
	Deresakari village — Söğüt
	Aşaği village
BURSA region:	Mustafakemalpaşa-Mineviz village

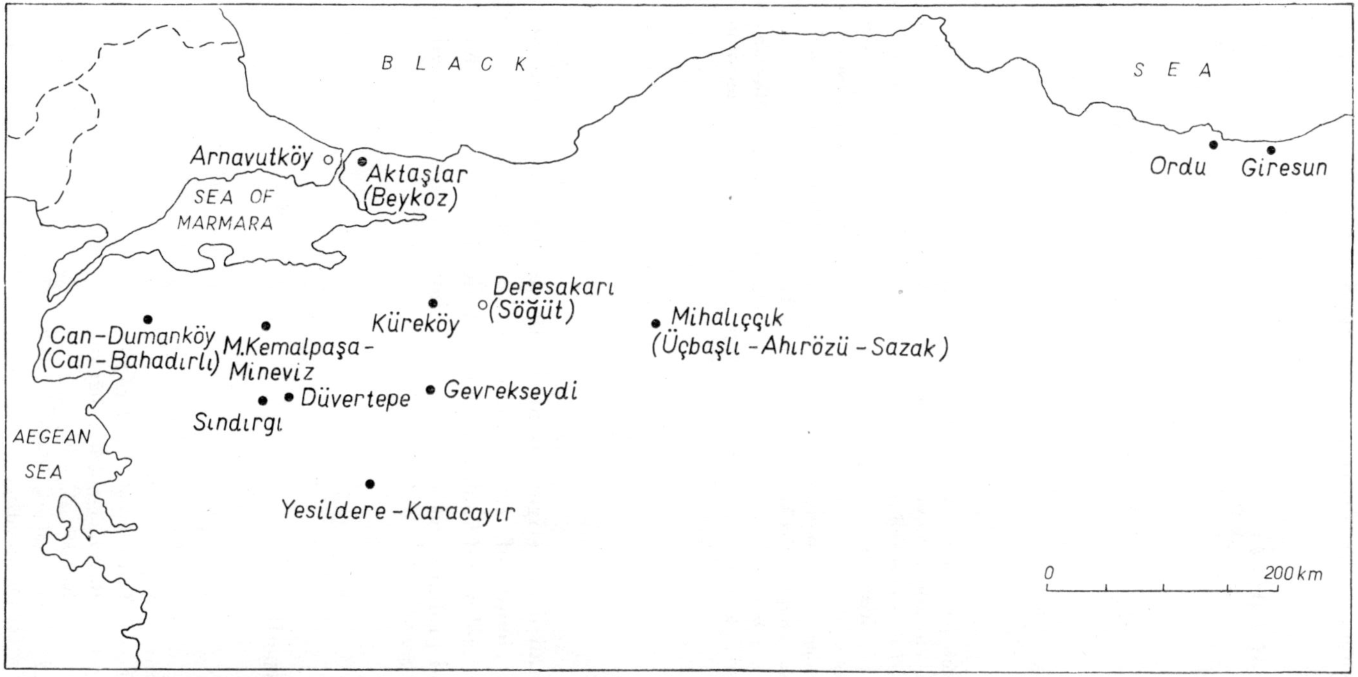

Fig. 1—Kaolin deposits of Turkey

● — Unexploited or small deposits with visible and probable reserves of less than 1 million tons,

○ — Medium deposits with reserves of 1—5 millions tons

UŞAK region:	Karaçayir village
KÜTAHYA region:	Gevrekseydi village
BALIKESİR region:	Sindirgi —Düvertepe
	İvrindi—Yenice
ESKIŞEHİR region	Mihaliççik-Ahirözü-Üç başli-Sazak

Geology

Istanbul-Arnavutköy kaolins have resulted from the alteration of host rocks which, in respect of petrography, can be determined as granodiorite-porphyrite or porphyritic granodiorite. This granodiorite-porphyrite which is a dike striking towards NNW-SSE and dipping almost vertically—has probably cut through Paleozoic schists and graywackes belonging to the Devonian. The petrographic studies of the host rock have shown that plagioclase has the composition of andesine and oligoclase, and calcite and sericite have been formed in spots through the decomposition of these. Kalifeldspars have been altered less and were transformed into sericite and kaolinite. The amount of quartz in the rock is very small and has not changed its character. Even in the poorest specimens the porphyritic structure can be recognized. Mafic minerals have been limonitized and under opaque separation have been chloritized. Kaolinization occurred during pre-Pliocene because there is kaolin detritus at the base, in the fresh-water sediments of Pliocene. The thickness of this Pliocene series covering all the kaolin deposits reaches up to 9 meters, while the thickness of the kaolin itself under this cover changes from 5—7.5 meters. In some places, kaolinization has penetrated up to 60 meters of depth. Although this point and the kaolinization, which is homogeneous and irregular in both horizontal and vertical directions, show that the above mentioned granodiorite-porphyrite has been kaolinized through ascending solution, it is most probable that a kaolinization occurred also through descending solution.

In particular, the presence of an old peneplain in the area in question and the fact that the alteration of feldspars diminishes towards depth can be pointed out as a good evidence of the above mentioned events. In the Arnavutköy region only, there are five separate kaolin occurrences of different character. This differentiation noticed in the other occurrences of Istanbul is caused through the alteration of the composition of the host rock and is also due to dragging. Kaolinization is closely connected with the tectonic and geomorphological development of the region. As the result of erosion which took place in Neogene and probably pre- or Early Pliocene, the deposit region had formed a valley and the valuable kaolin material has come out during this period. Later on, these valleys were filled again with a sedimentary cover. Then the whole region was covered with a Pliocene sedimentary cover. The dragging of primary kaolins in some parts

has affected their quality and as a result of the secondary alteration of kaolin, in some parts, for example in the Beykoz-Aktaşlar region, turning into bauxite was observed.

In Bursa-Mustafakemalpaşa region, kaolins consisted of altered tuffs of volcanic origin. In fault fissures, ferrous crusts are formed, and kaolins assume a green colour. Kaolinite is found together with quartz, sericite, and in spots with alunite.

In Balikesir region, at Sindirgi-Düvertepe, the occurrences are formed through the alteration of a volcanic dike which extends along N-S direction for a few km. At the contact of the occurrence, dark green amphibolite schist and extra-hard silicified hornfels are found. The Ivrindi-Yeniceköyü kaolin, which is also in the same region, cannot be used owing to antimonite needles which are intergrown with it. In almost all of the mines some antimony needles and spots that occur through the alteration of those needles can be seen. In a distance of about 200 m from the mines, an abandoned antimony mine is situated and kaolinization is probably due to the hydrothermal alteration during the formation of antimony.

The kaolin occurrence in Usak-Karaçayir village, is due to a dacite dike with an almost vertical dip, which is injected between metamorphic schists and outcrops on a steep slope. Those kaolins and kaolinitic clays that are known to exist in Çan, Dumanköy Çanakkale, Kütahya and Eskişehir regions, are formed through the alteration of tuffs of rhyolite, andesite and dacite character.

In Eskişehir-Mihaliççik region, kaolins occurred from the decomposition of acid rocks by hydrothermal effects, generally as veins of ENE-WNW strike. Some of these veins are themselves in acid rocks, but from the point of view of quality and reserves, they are of no importance. Though they include very little quartz, there is plenty of iron in them. Kaolin deposits which have larger reserve possibilities, are of the same character, like the acid rocks such as granite, rhyolite, porphyry, aplite-granite, and they are formed at the fault zones that are at their contact with the Neogene basin. Though the thickness appears as 20 m, it is probable that it may be much more than that. Mineralization extends for hundreds of metres, at Ahirözü and Uçbaşli kaolin mines, which are under the hard dolomitic red coloured conglomerates of Neogene. It is believed that the deposit is 100 m wide and 550 m long. Acid rocks are rich in quartz and they are of various character; this and their contact metamorphism with serpentines had some deleterious effects on the quality of kaolins.

Mineralogical and geochemical conditions

The mineralogical composition of kaolins of Turkey is examined by DTA-method. According to DTA-contours, the main clay mineral is generally kaolinite and the first anomaly is seen at about 100 °C (evaporation of hygroscopic water), molecule water is separated at 550 °C by an endotherm reaction, and the exothermic reaction which is suspected of having a relation with mullite formation, occurs at 950 °C. The exothermic reaction that occurs at 450 °C at the DTA-contours of those, is due to the burning temperature of the coal and organic materials which are found in the clay. The DTA analysis of the raw material which is called kaolin in Uşak-Karaçayir region and is used in fine ceramics, has revealed that this material is a bentonite containing calcite and kaolinite in it. It is found out that the clays of Balikesir-Bigadiç region, which are found together with colemanites, are bentonites, which have a tectonic mineral. In the occurrences in Eskişehir-Mihaliççik region, vermiculite and sepiolite were discovered. Üçbaşli and Ahirözü kaolins of this region are divided into two groups as hard and soft, and they contain 40 %—60 % quartz in them. Sazak kaolins are pure kaolins without any quartz, but sometimes they contain iron oxide, as much as 18 %, and for that reason they cannot be used. When we make a summary, we can say that, in Turkey, the main clay mineral of kaolinitic clays and primary kaolins which are used in fine ceramics, is kaolinite, and besides this, apart from mica, feldspar and quartz, according to the conditions of the region, there are small amounts of minerals as montmorillonite, hectorite, vermiculite, illite and sepiolite. As other accessory materials and colour material which effects the technology of kaolin, and as traces of elements, there are limonite, calcite, bauxite, antimonite, pyrite and organic materials and tuffs, gypsum, alunite and opal. In Table 1 average chemical composition of some kaolins of Turkey are shown. Quantity and quality of kaolin and content of clay differ locally. Total amount of earth alkali oxides is not more than 1 %, and total amount of alkali oxides is not more than 3 %, but the amount of iron oxide is usually far less than 1 %. In kaolin investigations following methods are applied: chemical analysis, differential-thermo-analysis, electron-microscopic and rational analysis, and samples are made subjects of further various technological tests in ceramic laboratory.

Use of kaolin

Kaolins of Instanbul-Arnavutköy region are divided into four groups according to their technological qualities.

1—Kaolins of the first group are plastic and have fluidity, and of white colour

Table 1—Chemical analysis and technology of some of the kaolins of Turkey

Name of the deposit	Shrinkage 110 °C %	Shrinkage 850 °C %	Shrinkage 1300°C %	Water absorption 850 °C %	Water absorption 1300°C %	Colour after burning 1300 °C	Required water % for normal mud	Loss on ignition %	SiO_2 %	Al_2O_3 %	Fe_2O_3 %	MgO %	CaO %	SO_3 %	Alkali accessories etc.
Bilecik–Sögüt dere sakari	1.0	1.0	0.0	26.90	33.70	white	32	5.57	77.74	16.25	0.15	0.26	0.33	—	—
Balikesir Ivrindi	2.0	3.0	7.0	28.4	21.8	white	22	11.90	54.76	32.95	0.21	—	0.10	0.61	antimonite,
	6.0	10.0	23.0		13.8		58	15.50	43.25	38.56	0.17	—	0.35	2.56	alunite
Bursa — Mustafa Kemal Paşa Minevis köyü	2.0	2.5	2.5	33.6	30.1		42	7.23	73.77	18.96	0.11	0.14	0.10	0.75	
	3.5	3.5	7.4	29.9	20.0		25	12.07	66.49	19.89	0.11	0.12	—	5.28	(alunite)
	3.5	5.0	6.5	27.7	23.9	white	19	10.82	59.62	29.29	0.13	—	0.07	0.90	
	3.0	5.5	13.0	30.0	16.4		13	12.59	52.05	35.65	0.12	0.24	0.14	0.23	
	2.5	3.5	18.5		8.9		20	16.60	49.46	32.41	0.15	—	0.14	5.43	
	3.0	3.0	9.5	33.3	19.0		10	12.88	57.26	28.50	0.10	—	—	2.98	
Eskişehir–Mihaliççik –Ahirözü–Yagaslan	4.0	6.0	13.5	26.30	9.95	yellowish white	33	12.69	49.38	36.22	0.48	0.49	0.37	—	TiO_2 = 0.71 %
Bilecik Küreköy—aplite								3.57	74.03	17.28	0.30	0.36	0.29	—	
Canakkale Duman								9.80	56.80	29.40	0.70	0.20	0.40	—	K_2O = 23 %
Bilecik Aşagiköy								6.22	73.67	17.91	1.27	0.36	0.09	—	K_2O = 0.18 Na_2O = 0.39 %
Kütahya Pinarli								17.72	41.61	36.91	0.41	2.01	0.73	—	K_2O = 0.46 %
Istanbul Arnavutköy								7.66	65.97	24.45	1.70	0.47	—	—	
Uşak								5.07	73.13	14.12	0.78	1.48	1.81	—	
Ordu								10.04	52.8	33.4	0.5	0.3	0.6	—	alkali 1.8
Giresun								9.8	59.3	24.2	0.8	0.6	2.7	—	alkali 2.1
Edremit								12.83	50.56	35.69	0.35	—	0.17	—	
Bayramiç								11.60	53.60	34.10	0.60	0.40	0.3	—	

when they are fired. They are used in manufacturing first class and second class porcelain, and of kitchen utensils.

2—Second group kaolins are sintering with a light grey colour when burned at SK 13. It seems possible that they could be used in manufacturing technical porcelain and kitchen utensils.

3—Third group kaolins turn to sinters with a grey colour at 1250 °C and are used in manufacturing kitchen utensils.

4—Kaolins of this group have usually a dark fired colour. They are sintering and dissolving very early. They can be taken into consideration for coarse ceramic but not for fire resistant products. Istanbul-Beykoz kaolins include bauxite and they reduce a great deal when they burn. For this reason, they are used in manufacturing fireclay.

Kaolinitic clays of Bilecik-Esriköy region can be used for the manufacture of glazed tiles and tiles; they are used in Yildiz and Bozüyük ceramic factories of Sümerbank. The kaolinitic clays of Bilecik-Inhisar region are suitable for manufacturing of earthware, porcelain, lavabo, etc., and are used in Yildiz, Tuzla and Eczacibaşi ceramic factories.

The alunitic kaolins and clays of Bursa-M. Kemalpaşa region, are used in Yildiz porcelain factory and in Balikesir Cement factory. Kütahya-Gevrekseydi region kaolins contain plenty of alunite and they are sufficient for the Izmit paper-mill.

Balikesir-Ivrindi region kaolins cannot be used, because they contain antimonite, but Balikesir-Sindirgi (Düvertepe) kaolins of Çanakkale-Çan region, are used in Çan ceramic factory for manufacturing tiles and faience.

Ahirözü kaolins of Eskişehir-Mihaliççik region, due to their content of excessive iron oxide, are used in Bozüyük factory in manufacturing tiles and glazed tiles and in Yarimca ceramic factory for fine ceramic products. Uçbaşli kaolins are used for manufacturing porcelain in Tuzla ceramic factory, but as they contain excessive quartz, they are left aside till a washing basin is built. Sazak kaolins are found unsuitable for manufacturing fine ceramic products, because they are impure.

The technological qualities of some of the kaolins which have been discussed so far are shown in Table 1. Kaolins are sometimes cleansed and separated from impurities such as sand in simple maturing basins close to the kaolin mine. Quartzose and sometimes hard kaolins of Mihaliççik-Ahirözü are ground, washed and separated from quartz, e. g. in modern plants of the Bozüyük ceramic factory. Also Uşak-Karaçayir kaolins are completely cleansed from quartz in situ, in the plants established there.

Economical value of the deposits

Detailed investigations of reserves of the kaolin deposits mentioned above have not been carried out. According to the records of the Mineral Research and Exploration Institute, a total amount of 102 kaolin occurrences have been recorded, and as a result of the investigations carried out in 8 of them, about 2.5 million tons of reserves have been established. Proven (*) or probable reserves of some occurrences are listed below:

(*)	Istanbul-Arnavutköy	:	1,000,000 t kaolin
(*)	Çömlekçi bayiri	:	30,000 t kaolin
(*)	Uskumruköy	:	30,000 t kaolin
	Beykoz	:	270,000 t kaolin containing bauxite
	Çanakkale-Çan (Dumanköy)	:	600,000—1,000,000 t kaolin
	Çan-Bahadirli köyü	:	15,000 t kaolin
	Bayramiç Sögütgediği K.	:	20,000 t kaolin
(*)	Eskişehir-Mihaliççik	:	300,000 t kaolin
(*)	Bilecik-Küreköyü	:	320,000 t kaolin-aplite

In Turkey, production of kaolin is by opencast and underground mining. Most of the mines are out of production in winter. Underground mining is applied at Çan- (Dumanköy) and Bilecik (Söğüt) and Mihaliççik occurrences which have large reserves and where the overburden is thick. At Söğut, there is a network of galleries connected by passages (sometimes so narrow that only one man can pass with difficulty) which are more or less parallel to each other and with a few degrees inclination. There, because of the weight of the conglomeratic mass which lies on them, the argillaceous formation has become a syncline.

In the Mihaliççik-Ahirözü kaolin exploitation, there are various galleries. Their length is nearly 300 metres and they are reinforced in places. There, in three months, by employing 20 workers, 1500 tons of kaolin were produced. This kaolin, 80 TL per ton, is sent to the Yarimca ceramic factory, the construction of which is nearly completed. At Sazak and Uçbaşli kaolin occurrences of the same region underground mining is applied. Extracted material is usually loaded to trucks, but sometimes, instead of trucks animal power is used because of bad roads. Sent to stations, materials are forwarded from there to the place where they are to be used.

Statistical information

There is not any exact statistical information concerning the yearly needs of various branches of industry or the figures showing the raw and washed kaolin

production of the mines. However, about 25,000 tons of kaolin are absorbed by local enterprises. (Import and export figures are not comprised in the above estimation). According to the estimations, in 1970, 92,000 tons of kaolin and 98,000 tons of kaolinitic clay will be needed.

History of kaolin investigations and bibliography

In Turkey, kaolin investigations had started in 1896, with the construction and working of Istanbul Yildiz porcelain factory, for the requirements of the Palace. Fine ceramic production is earlier than this, it goes back to the 12th century and reached its climax in the 16th century during the rise of the Ottoman Empire.

But with regard to the rapid progress of the ceramic industry during the last decade, real investigations of kaolin and clay were carried out by the specialists of these factories and by the members of the M. T. A. Institute.

The same applies at present. Now, investigations show a great progress, but no exact figures or information are available because the investigations are not yet completed.

Many reports were written concerning the kaolin and clay deposits of Turkey. They are mainly kept in the archives of the M. T. A. Institute, private industrr, and particularly in the archives of Istanbul Yildiz porcelain factory. Howevey very few books and periodicals are published on kaolin deposits.

References

Eisenhart, Dietrich V. (1960): Porselen ve Türk hammaddelerinden istifade suretiyle imali. M. T. A. Enstitüsü dergisi, No. 55, Ankara.

Erdinç, Ş. Ş. (1962): Ceramic and Refractory in Turkey and their Raw Materials. Cento—Symposium, Lahor—Pakistan.

Orhon, O. (1962): The Kaolin and Clay Deposits in Turkey. Cento—Symposium, Lahor—Pakistan.

Tolun, R. (1960): Feldspatlarin standardizasyonu ve seramik sanayiindeki ehemmiyeti. M. T. A. Enstitüsü dergisi Nl. 54, Ankara.

AFRICA

1969 XXIII INTERNATIONAL GEOLOGICAL CONGRESS (1968) — VOL. 16 PAGE 55—59

Les gîtes de kaolin du Maroc

E. A. HILALI, M. NATAF et L. ORTELLI
Maroc

Le Maroc possède de nombreux indices d'argiles kaoliniques dont la localisation géographique est indiquée sur la figure 1. Mais, seul, le gisement d'halloysite de Maaza (Maroc oriental) présente un intérêt économique et mérite d'être mentionné.

En outre, du kaolin existe en abondance dans les gîtes à pyrophyllite d'âge infracambrien du Haut-Atlas central (Aït Azegrouz, notamment à Azrou-Melloul).

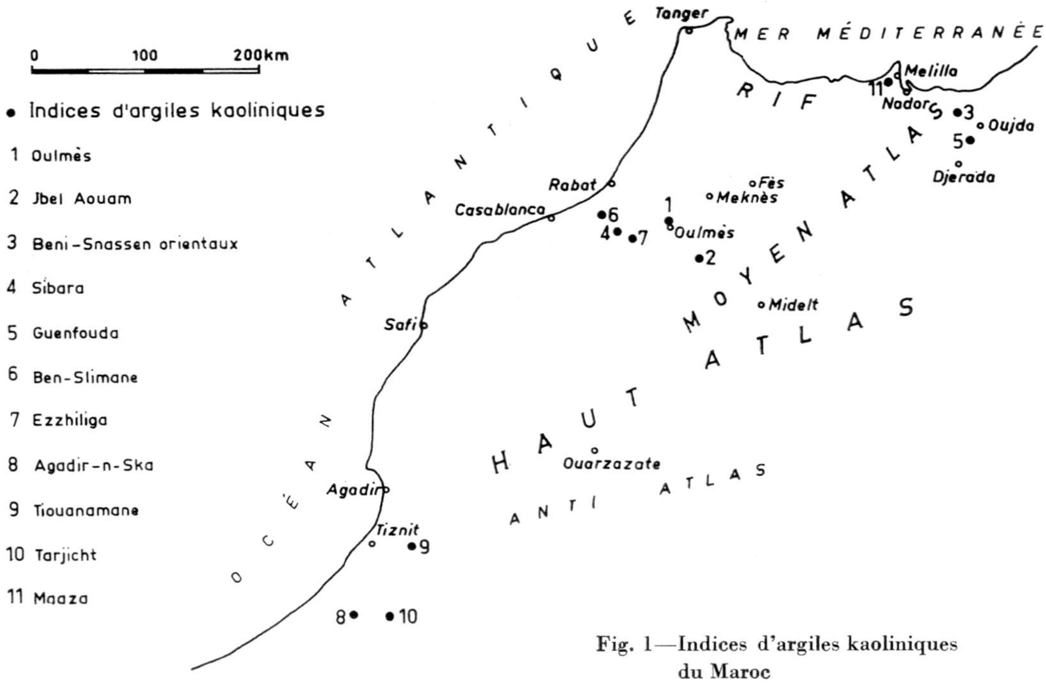

Fig. 1—Indices d'argiles kaoliniques du Maroc

Gisement d'halloysite de Maaza

Situation géographique

Le Jbel Maaza est situé à 7 km à vol d'oiseau de Melilla, près du village de Souk et Had des Beni-Sicar. Il appartient à la face N du massif volcanique du Gurugu et se trouve en bordure de la piste secondaire qui contourne ce massif par le N et l'W (Fig. 2).

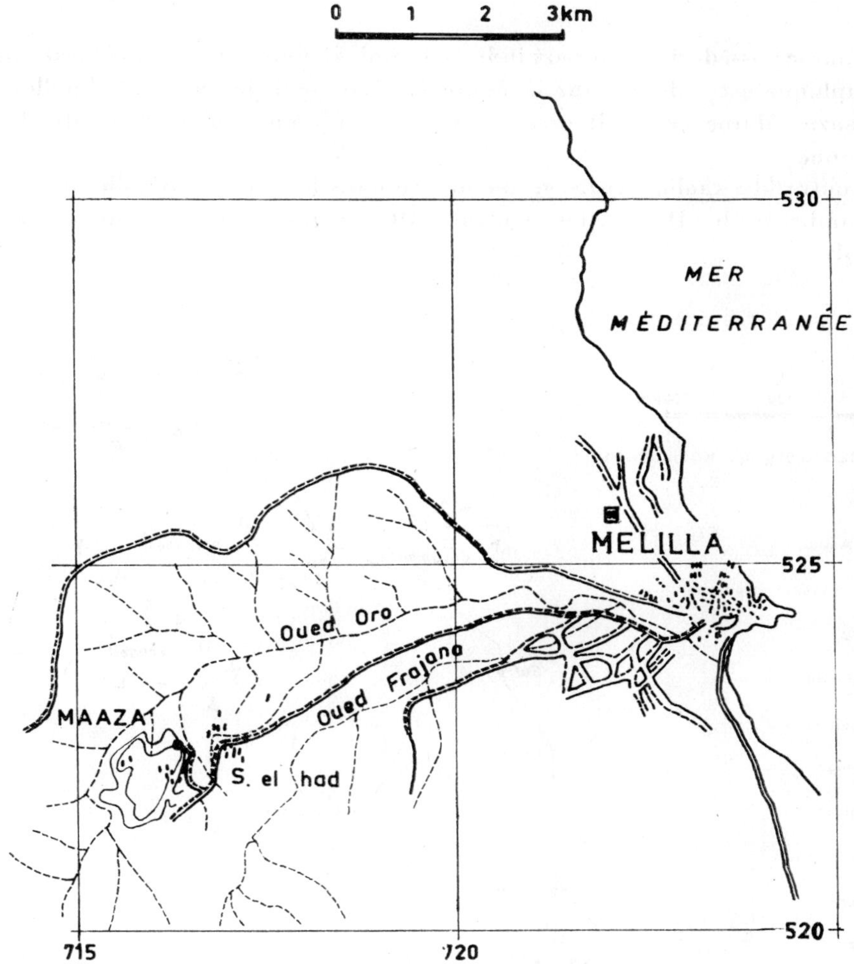

Fig. 2—Gisement d'argile de Maaza (croquis de situation)
● — gisement

Cadre géologique

Le Jbel Maaza est formé de sédiments mio-pliocènes montrant d'importantes variations de faciès et comprenant des intercalations de bancs de cinérites appartenant au complexe volcanique du Gurugu.

La base est un falun calcaire à polypiers qui s'intercale vers le N dans des calcaires rognoneux et travertineux. Ces formations sédimentaires, qui se terminent par un niveau continu de sables blancs dunaires, essentiellement calcaires, ont une épaisseur visible de 20 à 30 m et sont recouvertes par une épaisse série de cinérites et graviers volcaniques rouges, couronnée par une coulée de basalte qui forme les plateaux (Fig. 3).

Les amas et couches d'halloysite apparaissent au sein des formations calcaires de base ou à leur partie supérieure. L'argile est blanche, onctueuse, très pure, à toucher gras, sans éléments clastiques. La couche intéressante d'halloysite est d'une grande régularité, épaisse de 1,50 à 2,50 m, très homogène mais on y trouve parfois des lentilles sableuses ou des bancs de calcaire induré; de plus, l'halloysite est souvent mêlée à de l'alunite.

Si l'on considère une épaisseur moyenne de 2 mètres et une densité en place de l'halloysite égale à 2, les réserves à vue peuvent être chiffrées à environ 20.000 tonnes.

Les caractéristiques moyennes de l'halloysite de Maaza sont groupées dans les tableaux 1 et 2.

Tableau 1—Analyses

Analyse chimique		Rayons X	Analyse thermique différentielle
SiO_2	43,15 %	halloysite, alunite, sulfates doubles, hydroxides d'aluminium, montmorillonite	—pic endothermique vers 220 °C, — 2ème pic endothermique vers 560—580 °C, — pic exothermique vers 980 °C
Al_2O_3	34,61		
Fe_2O_3	1,45		
CaO	1,97		
MgO	1,62		
SO_3	3,70		
H_2O^-	13,30		
Total	99,80		
Humidité	11,50		

Tableau 2—Caractèristiques technologiques

Plasticité	Retrait %		Couleur
	au séchage	à la cuisson vers 1050 °C	après cuisson vers 1050 °C
forte	9,25	21,3	beige très pâle ou légèrement rosé

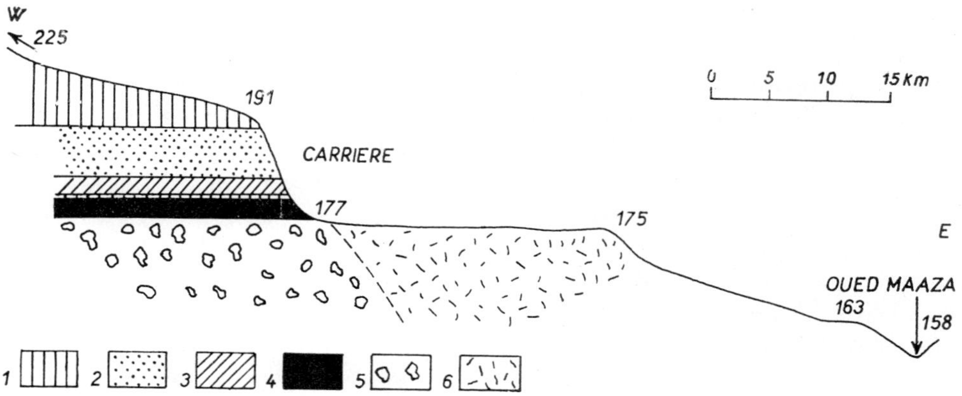

Fig.3 — Coupe schématique du J. Maaza
Mio-pliocène: *1* **- série pyroclastique;** *2* **- niveau sableux;** *3***- montmorillonite;** *4* **- halloysite;**
5 **- travertin calcaire;** *6* **- rembloi**

Ces analyses révèlent l'existence d'impuretés (fer, magnésium, sulfates) qui ont parfois une incidence défavorable sur les qualités technologiques du matériaux.

Production — Exportation — Importation

Le gisement de Maaza n'est plus exploité actuellement. La production, de l'ordre de 1000 tonnes en 1965 a été entièrement exportée.

Le Maroc importe du kaolin brut et traité qui sert presqu'entièrement comme charges dans l'industrie papetière.

Les quantités importées sont groupées dans le tableau 3.

Tableau 3 **(Chiffres arrondis exprimés en tonnes métriques)**

Années	1960	1961	1962	1963	1964	1965	1966
	2 400	1 900	1 500	4 300	1 800	2 700	2 400

Références bibliographiques

Caïa J., Dietrich J. E. et Mazéas J. P. (1967): Etude de gîtes et occurrences de roches à pyrophyllite du Maroc — Notes Serv. Géol. Maroc, t. 28, sous presse, 11 fig., 6 pl. photos, nombreuses références bibliographiques.

Moussu R. (1959): Les gisements de pyrophyllite, kaolinite et alunite du Tifnout. Mines et Géol., Rabat, n° 8, pp. 44—45.

Nataf M. (1967): Possibilités d'utilisations industrielles des roches à pyrophyllite des Aït-Azegrouz et de l'Ougnat. Note Serv. Géol. Maroc, t. 28, sous presse, 11 tableaux, 2 figures, nombreuses références bibliographiques.

„Statistiques du mouvement commercial et maritime du Maroc" années 1960 à 1966 — Ministère du Commerce, de l'Artisanat, de l'Industrie et des Mines-Rabat.

1969 XXIII INTERNATIONAL GEOLOGICAL CONGRESS (1968) — VOL. 16 PAGE 61—65

Kaolin Deposits of the Republic of South Africa

C. B. COETZEE
South Africa

Summary—In South Africa the conditions for the large-scale alteration of granite and pegmatite by hydrothermal or pneumatolytic agents seem to have been generally absent, and there is practically no kaolin formed by endogenic processes. South African kaolins are derived from the weathering of granite and certain shales, consequently all contain to a greater or smaller extent quartz, sericite, (ferrous) colouring matter and occasionally montmorillonite.

The majority of the known deposits are comparatively small: the probable reserves exceed one-quarter of a million tons at each of 4 separate localities or farms only.

Occurrences

The occurrences of kaolin in South Africa may be conveniently classified into 4 groups, based on their host or parent rocks.

Residual kaolins derived from the weathering of Archaean granite occur in the Vanrhynsdorp District of the Cape Province. Those from the following farms are the most promising out of 25 occurrences described: Nieuwhoudts Naauwte (potential kaolin recoverable 283,000 metric tons); De Toemkomst (potential kaolin recoverable from 2 deposits 454,000 metric tons); and Hendriksvlei where there is an estimated reserve of 1,360,000 metric tons of crude, good-quality kaolin and about 6,985,000 metric tons of crude medium-quality kaolin — the reserves of poor-quality kaolin amount to about 8,164,000 metric tons.

The deposit exploited in the Province of Natal in the Ndedwe District falls genetically in the same category; on Witklip in the Nelspruit District of the Transvaal, redeposited kaolin derived from the weathering of Archaean granite has been mined in the past.

Residual kaolins derived from the weathering of Cape granite (Proterozoic) occur in several districts in the south-western part of the Cape Province. The best deposits known lie on the following farms; the approximative reserves of the kaolin which are recoverable are appended in brackets:

Haasendal, Stellenbosch District (363,000 metric tons).
Bakkerskloof, Somerset West District (154,000 metric tons).
De Goede Hoop and Imhoff's Gift, Simonstown District.
Kruispad, Stellenbosch District.

Sedimentary ball clay occurs in addition to kaolin on the farm listed last.

Residual clays derived from the weathering of Bokkeveld shale (Devonian to Carboniferous) occur as small deposits at numerous places in the southern part of the Cape Province and are associated with a silcrete capping in many instances. The deposits are, or have recently been, exploited at Tygerfontein, Gelukshoop and Wiehmansfontein in the Albertinia District, and on Droge Vallei and other farms in the Riversdale District.

Residual clays derived from the weathering of Dwyka or Witteberg shale (Carboniferous) occur in the south-eastern Cape Province. Their formation depended on the development of the Grahamstown peneplain and the total reserves appear to be very large. Since the clay usually contains an appreciable percentage of quartz and feldspar, it can often be used satisfactorily without further additions to the body. The deposits are being exploited on Collingham, Inniskilling, Upper Gletwyn and Bergplaas in the Albany District, and on Strowan in the Grahamstown District.

Inferior kaolins not really suitable for the making of whitewares are associated with or derived from other formations, such as shales in the Pretoria Series and in the Ecca Series.

Chemical data

Analyses of typical samples from each of the four groups described above appear in Table 1.

Utilization

Factors which have so far hampered any substantial exploitation of the deposits in the Vanrhynsdorp District are their great distance from the large industrial areas (250 miles from Cape Town and 900 miles from Johannesburg); the scarcity of water of good quality if the clay were to be beneficiated at the prospective quarries; and the present rail-tariff, which is high for refined products in comparison with crude materials. The clay derived from the shales (cf. groups 3 and 4 below) have a low reflectivity (whitness) and the fine grain of the impurities militates against beneficiation.

Table 1—Chemical and mineralogical data

	1	2	3	4
SiO_2	49.44	46.12	64.90	57.30
Al_2O_3	35.60	36.37	23.80	29.70
Fe_2O_3	0.62	1.58	0.53	0.73
MgO	0.80	0.49	trace	0.53
CaO	0.06	trace	trace	0.16
Na_2O	0.04	0.48	0.46	1.84
K_2O	0.59	1.08	4.16	2.86
TiO_2	trace	0.60	0.92	0.80
Loss on ignition	12.55	12.82	5.43	6.50
Totals	99.70	99.54	100.20	100.42
Fired colour at 1200 °C	white	pale cream	pale cream	white
Linear shrinkage at 1200 °C (%)	8.0	9.0	13.3	10.8
Fraction $>$10 micron (%)	15.0	9.0	18.9	2.1

1 - *Van Rhynsdorp District* (Nieuwhoudts Naauwte)—One-ton sample, washed and bleached. Mineralogical composition: kaolinite 90 %, quartz 2 %, hydrous mica 8 %;

2 - *Stellenbosch District* (Haasendal)—Handsorted unwashed sample. On washing Fe_2O_3 decreased from 1.58 to 0.60 per cent and the sample then fired white. Mineralogical composition: kaolinite 83 %, quartz 2 %, hydrous mica 12 %, montmorillonite 3 %;

3 - *Albertinia District*—Mineralogical composition: coarsely crystalline kaolinite with a little sericite, iron-stained quartz and ilmenite or rutile;

4 - *Grahamstown District*—Mineralogical composition: kaolinite and sericite. This clay probably contains too many fluxes to be used alone satisfactorily but when combined with one or more refractory clays (such as found elsewhere in that area) and a little silica to prevent crazing, it should form the basis of a cheap and satisfactory whiteware body.

The kaolin producers are at present scattered throughout the country. The industry is not yet served by an elaborate beneficiation plant from which standardised and guaranteed grades and blends of kaolin could be obtained.

The kaolin usage in the Republic (expressed in per cent) is as follows according to a survey carried out by the Council for Scientific and Industrial Research in 1961:

	%
Ceramic industry	57.6*
Insecticide manifacture	32.6
Rubber industry	3.7
Paper, cardboard and paper-based fireboard manufacture	3.2*
Paint manufacture	2.6
Plastics and general chemical industries	0.2
Pharmaceutical industry	0.1

* In these cases most of the "kaolin" used consisted of (quarry) mixtures of materials containing a fair percentage of kaolin.

Production, exports and imports of kaolin

The production, local sales, exports and imports of kaolin in the Republic of South Africa are listed in metric tons in Table 2.

Kaolin and kaolin mixtures in the Republic are at present only won from open quarries.

Table 2—Kaolin trade statistics

Year	Production,* metric tons	Local sales,* metric tons	Export,* metric tons	Import,** metric tons
1900	—	—	—	—
1910	36	36	—	—
1920	—	—	—	—
1930	—	—	—	67
1940	1,836	1,836	—	378
1950	1,557	1,532	—	546
1951	10,582	8,592	493	725
1952	7,478	6,545	12	635
1953	7,856	8,060	57	1,280
1954	13,097	12,214	10	1,706
1955	10,229	8,373	—	1,590
1956	10,543	9,990	—	2,532
1957	14,354	11,558	—	2,454
1958	24,124	23,262	—	1,895
1959	18,175	21,776	20	3,801
1960	26,493	26,594	44	3,739
1961	24,017	24,496	—	3,975
1962	29,455	26,908	13	4,470
1963	33,941	34,338	18	5,141
1964	39,459	39,891	105	1,195
1965	41,394	39,342	403	8,390
1966	40,521	42,080	514	—

* These figures, supplied by the Government Mining Engineer, would include the weight of a certain amount of gangue in some of the kaolin batches or quarry mixtures.

** Material listed as Cornish stone, Kaolin and China clay in the records of the Department of Customs and Excise.

— Not available or not estimated separately.

References

Bennetts, K. P. and Van Rensburg, W. C. J. (compilers) (1967): The ceramic raw material resources of the Republic of South Africa: Interim rep. Geol. Surv. S. Afr. (Not published.)

Bosazza, V. L. (1940): The Grahamstown white clays with special reference to the manufacture of porcelain: Bull. min. res. Lab. S. Afr., 5.

Council for Scientific and Industrial Research, Pretoria (1961): An investigation into the present demand for kaolin by each of the main consumer industries in the Union and the particular requirements of each in their kaolin usage. (Confidential, not published.)

Hamburger, J. (1940): The white burning clays at Grahamstown: Bull. min. res. Lab. S. Afr., 5.

Heystek, H., De Jager, D. H., De Weal, P., and Urli, G. P. L. (1961): The kaolin deposits of the area between Bitterfontein and Landplaas, Vanrhynsdorp District: Bull. Geol. Surv. S. Afr., 36.

Jacobs, H., (1935): The clays of the Cape Peninsula: Ph. D. thesis, University of Cape Town. (Not published.)

Mountain, E. D. (1931): The Grahamstown ceramic industry: S. Afr. J. Sci., 28.

1969 XXIII INTERNATIONAL GEOLOGICAL CONGRESS (1968) — VOL. 16 PAGE 67—74

Kaolin Deposits of Swaziland

D. R. HUNTER and J. G. URIE
Swaziland

Summary—Kaolin deposits of economic importance are found in the Mahlangatsha mountains of the Manzini District. In an area of some 13 sq. km. seven basic dykes are known to have suffered complete or partial kaolinization. Reserves of 159,000 metric tonnes have been proved and probable reserves amount to a further 253,000 tonnes.

Geological setting

The kaolinized dykes intrude the arenaceous Mozaan Series which has suffered the following deformations:
Folding into a broad syncline aligned about a northerly to northeasterly axis.
Warping along axes aligned in an east-northeasterly direction.
Folding about northwesterly axes giving rise to a series of isolated centroclinal structures.

Faulting has taken place along both northeast and northwest directions, the majority of the kaolinized dykes having been intruded along the former direction. Fault movement has taken place along, or about 5° off, the northeast direction after the emplacement of the dykes. This has resulted in the pinching out of the dykes in both horizontal and vertical senses.

Two granites intrude the Mozaan Series along its eastern flank. The older granite, dated provisionally at 3070 ± 60 m. y., is medium grained. It is composed of quartz, biotite, microcline and subordinate plagioclase. The emplacement of this granite has resulted in the metamorphism of the Mozaan Series adjacent to its contact to the kyanite and sillimanite grades. The younger granite, which lies within 1.5 km. of the deposits and post-dates the emplacement of the now kaolinized dykes, is a coarse-grained, porphyritic rock with sharp contacts. The intrusion of this granite is related to the warping of the Mozaan Series along the northwesterly aligned axes. The contact metamorphic effects of this granite are confined to limited recrystallization of the quartzites of the Mozaan Series.

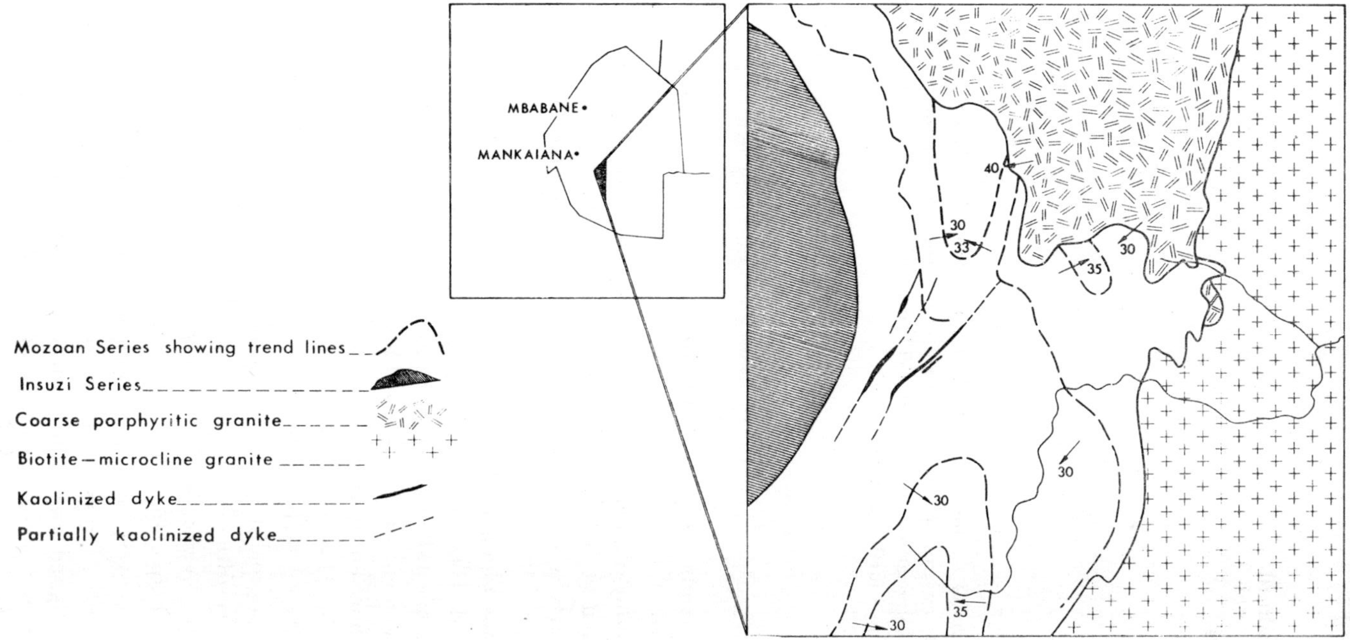

SWAZILAND

LOCALITY MAP OF KAOLIN DEPOSITS

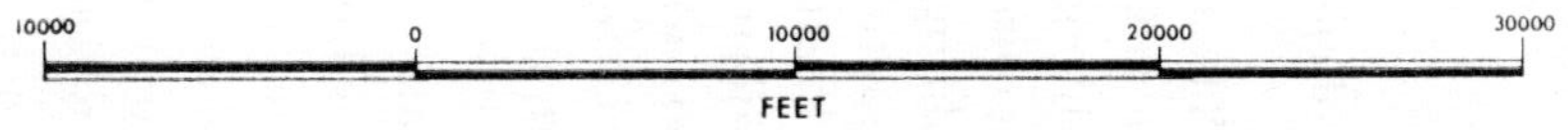

The kaolinized dykes can be traced for distances of 2000 metres. The larger dykes have a width of 10 metres and the smaller dykes average about 3 metres in width. The dykes dip at about 80° at surface but deep drilling on one dyke suggests that the dip becomes less steep in depth. The dykes fail to give rise to any outcrops and are marked by grassy depressions transecting almost continuous outcrops of conglomerate, quartzite and quartz-schist. The dykes are not completely kaolinized throughout their length. Areas of complete kaolinization occur where there is a cover of black, marshy, turf clay, which has an average thickness of 1 m. but locally reaches thicknesses of 5 metres. The intensity of kaolinization decreases as the dykes are traced into areas where the soil cover in the depressions becomes less charged with organic matter and the moisture content of the soil is reduced. The greatest strike length of complete kaolinization revealed to-date is 620 metres. Kaolinization in depth has been found to be highly variable ranging from 20 to 75 metres below surface. It is not possible to relate this erratic distribution with the presence of erosion surfaces.

The kaolin varies in colour from white through pinkish white to pale red, the different colour varieties being intimately intermixed. A delicate banding may sometimes be observed as a result of the alternation of layers of different colours, individual layers varying from paper thickness to 2 mm. A porphyritic texture is often retained. Cream or white blebs, either spherical or elliptical in shape, are set in both the banded and unbanded kaolin. Along the footwall side of the larger dykes the kaolin is red-brown in colour and is identical to the kaolinitic clay found underlying the drier portions of the depressions along strike from the completely kaolinized areas. In this kaolinitic clay, which is found both along strike and down dip, nodules of the original dyke rock are preserved. Very rarely traces of the spheroidal weathering, typical of basic dykes, are observed in the completely kaolinized areas. The trace of the successive skins of the spheroidal weathering is picked out in some cases by curved layers of limonite up to 10 mm. thick.

The variation in colour of the kaolin can be attributed to the iron content. The presence of titanium is of no significance in this regard, as shown by the analyses on raw kaolin in Table 1.

Table 1

	1	2	3	4
Fe_2O_3	8.30 %	3.03 %	2.82 %	0.78 %
TiO_2	2.66 %	2.56 %	2.28 %	2.54 %

1 - Brown kaolin; *2* - Pale brown kaolin; *3* - Pink kaolin; *4* - White kaolin

Firing tests on pink kaolin show that when fired at 1300 °C. for two hours it becomes cream in colour. It seems probable, therefore, that some of the

pinkish discoloration of kaolin near the surface may be due to organic material.

The parent rock of the kaolin is grey or pale olive-green in colour. It is composed of a fine-grained aggregate of laths of green amphibole, some chlorite, feldspar and epidote. Highly saussuritized phenocrysts of andesine are set in this groundmass. Minor amounts of interstitial quartz are sometimes present and ilmenite, altering to leucoxene, is common. In some places the amount of feldspar is reduced and the rock consists of ragged laths of green amphibole set in a groundmass of zoisite and epidote.

The dykes are intruded by the pluton of coarse-grained porphyritic granite. Age determinations (Sr/Rb whole rock) on a similar pluton 50 km. to the north gave an age of 2650 $\pm$ 60 m. y. Preliminary results from the Mahlangatsha granite indicate that this rock has a similar age.

These occurrences of kaolin are difficult to classify from a genetic point of view. The deposits are clearly in situ and have not suffered any re-washing. However, it is apparent that the origin of these deposits cannot be accounted for by simple weathering or hydrothermal processes. Nor is it valid to postulate the presence of different parent rocks to explain the presence of white, iron-free and brown, iron-rich kaolin, because of the intimate intermixture of these different types. Simple surface weathering is not adequate in itself because the surrounding quartzites, often granular and rich in sericite, and quartz-sericite schists show no signs of weathering. Furthermore a quartz vein carrying unoxidized galena was observed in the vicinity of the kaolin deposits. Surface weathering has played a part but an additional factor is required to explain the localization of kaolinization. The shallow depths to which kaolinization has proceeded in many places suggests that hydrothermal activity is not wholly responsible.

It is concluded that a combination of various processes has been responsible for the formation of these deposits, these processes being concentrated along lines of structural weakness occupied by basic dykes. Basic dykes were emplaced along northeasterly aligned faults in the Mozaan Series. Possibly as a result of the emplacement of the pluton of coarse-grained porphyritic granite further movement took place along these faults shearing the dykes. Hydrothermal solutions permeated the Mozaan Series giving rise to the quartz veins which cut the dykes and at the same time altering the minerals of the dykes. Normal weathering processes then operated breaking down the hydrothermally altered dykes to stiff brown clay. Where the shearing was more intense or where the dykes had originally intruded softer country rocks, larger depressions would be formed. Water collecting in these marshy depressions would become charged with organic matter forming dilute humic and carbonic acids capable of reducing the iron to a soluble ferrous state. Surface waters percolating downwards would gradually dissolve the iron in the brown clay. The level to which such a process is operative would depend on the intensity of earlier hydrothermal alteration and the depth to which surface weathering had proceeded.

The changes involved in this process can be demonstrated by a study of the chemical analysis (Table 2).

Table 2

	1	2	3	4
SiO_2	49.48	50.81	43.48	47.46
TiO_2	0.98	1.42	2.66	2.54
Al_2O_3	17.64	20.12	32.26	35.00
Fe_2O_3	12.98	12.08	8.30	0.78
MgO	5.63	5.22	0.01	0.17
CaO	10.34	2.83	0.64	0.68
Na_2O	0.67	0.24	0.27	0.12
K_2O	0.13	0.22	0.11	0.27
Loss on ignition	2.08	7.10	12.32	13.02
Σ	99.93	100.04	100.05	100.04

1 - Dyke, hydrothermally altered; *2* - Partially altered dyke; *3* - Brown kaolinic clay; *4* - White kaolin (raw)

The process of kaolinization involved firstly the removal of much of the lime (compare analyses 1 and 2) followed by the removal of the remainder of the lime and the magnesium (compare analyses 2 and 3). When the dilute humic and carbonic acids were formed the iron was leached leaving white kaolin.

Composition and properties

The Mineral Resources Division, Overseas Geological Surveys, London, carried out a minerological examination, using X-ray diffraction and differential thermal analysis. Their report showed that kaolinite is the major constituent of the white kaolin. Examination by differential thermal analysis produced a curve showing characteristic thermal peaks for well-crystallized kaolinite. The weight loss on decomposition of the kaolinite between 500 °C. and 800 °C. measured by a thermobalance, indicated a kaolinite content of 90 per cent, when compared with a standard of "Supreme kaolin" (English clays).

The particle size distribution of the raw kaolin shows that between 95 and 98 per cent passes through a 325-mesh seive, the coarse residue being composed of grains of quartz and some mica. The Mineral Resources Division obtained the following particle size distribution on white kaolin of the —240 mesh fraction (in weight %):

—240 mesh + 20 microns	3
—20 microns + 6 microns	29
—6 microns + 2 microns	30

The reflectivity of various grades of kaolin is given below (in %):

Raw kaolin (buff coloured)	77
Raw kaolin (pinkish)	73
Raw kaolin (off white)	78
Kaolin —325 mesh	76
Kaolin —10 micron	79
Kaolin —2 micron	80

The determination of the Atterberg plastic limits by the Mineral Resources Division indicated that the clay samples, although slightly sticky when wet, are non-plastic.

Ceramic tests on the clay show that the white clay has a linear drying shrinkage of between 2 and 3 per cent, while for the pink clay the percentage is nearly 6. The dry strength of white clay ranges from 2 to 4 kg/sq. cm. and of the pink clay is 6.6 kg/sq. cm.

Results of firing tests are listed in Table 3.

Table 3

Temperature in °C	Linear shrinkage in %	Colour
1100	3	merest cream
1150	5	merest grey
1200	5	merest grey

The trial slabs were fired at each temperature for 30 minutes.

	Parent rock	Raw kaolin	Marketable product (less than 0.074 mm)	Clay substance (less than 0.002 mm)
SiO_2	49.48	51.11	44.55	45.06
TiO_2	0.98	2.31	2.92	0.67
Al_2O_3	17.64	32.39	37.69	39.02
Fe_2O_3	12.98	1.33	0.39	0.54
MgO	5.63	0.33	Tr	Tr
CaO	10.34	0.40	0.52	0.34
Na_2O	0.67	0.17	0.05	0.05
K_2O	0.13	0.21	0.11	0.02
Loss on ignition	2.08	11.77	13.80	14.10
Total	99.93	100.02	100.03	99.80

In the above table the composition of the raw kaolin is the mean of eleven samples ranging from white to pale brown or buff in colour. At present the deposits are still in the exploratory stages of development so that the analysis of

the —200 mesh (0.074 mm.) product is based on one analysis obtained under laboratory conditions.

The titanium content is highly variable in the kaolin ranging from 1.18 to 3.56 per cent. Over thirty samples were analysed for TiO_2 but no systematic distribution could be established either as regards to colour or position within the dyke.

Utilization and economic evalution

Although small quantities of kaolin have been produced since 1961 the deposits are not yet being fully exploited. Investigations are being conducted in an effort to find a method of beneficiating the kaolin to a standard acceptable for the paper industry. The majority of the output to-date has been used in the pottery industry.

The reserves of kaolin can be summarized as follows:

Proven reserves	159,000 tonnes
Probable reserves	161,000 tonnes
Possible reserves	92,000 tonnes
Total	412,000 tonnes

The disproportionate ratio of length to width is a considerable handicap to the mining of the dykes in depth. Considerable tonnages of waste rock will have to be removed if conventional methods of quarrying are employed. It is considered that, on present information, it will not be economically feasible to mine the widest dykes to depths greater than 15 metres. In the case of the narrower dykes the limit of mining will vary from 7 to 10 metres, depending on the width of the dyke. Proven reserves are, therefore, given to a depth of 15 metres. Although one dyke has reserves proved to a depth of 40 metres, these reserves are regarded as probable as a result of the mining problems.

Exploitation is taking place on one dyke. Quarrying operations on a small scale are confined to a strike length of 200 metres where the lowest bench is only 4 metres below surface. On the hanging wall it is necessary to bench the friable quartzites which crumble on exposure. Drainage is a serious problem as the deposits are situated in an area which receives an average of 150 cm. of rain, mainly during the summer months. Fortunately the topography is such that adequate drains can be dug without resorting to pumping.

The kaolin is sun dried during the winter months and is then transported over 40 miles of gravel road to the nearest railhead at Piet Retief in the Republic of South Africa, from where it is railed to consumers in Johannesburg.

The major aspect influencing production is that the reflectivity of the kaolin

except for the micron sizes does not reach the minimum specifications of the paper industry. In view of the fact that the deposits are isolated from large industrial centres, the economic viability of the deposits depends on obtaining sales in that industry, where the highest prices are offered.

Annual production

1961	52.71 tonnes
1962	2487.90 tonnes
1963	2006.00 tonnes
1964	311.80 tonnes
1965	752.36 tonnes
1966	586.78 tonnes

The entire production was exported to the Republic of South Africa, there being no internal demand in Swaziland for kaolin.

The kaolin deposits have been known for many years. They were investigated in reconnaissance by a South African mining house in 1950 but it was not until 1962 that detailed prospecting by means of trenches and boreholes was started by the Geological Survey and Mines Department. Late in 1961 a mining location was granted, with the result that preliminary exploitation of the deposit began. Until suitable markets can be found development is likely to be limited.

References

Bain, J. A. (1965): Kaolin from the Mahlangatsha deposit, Swaziland.—Special Report No. 209, Min. Res. Div., Overseas Geol. Survey, London.

Hunter, D. R. and Urie, H. G. (1964): The kaolin deposits, Mahlangatsha, Manzini District.—Bull. No. 4, Geol. Survey and Mines Department, Swaziland.

1969 XXIII INTERNATIONAL GEOLOGICAL CONGRESS (1968) — VOL. 16 PAGE 75—78

Kaolin Deposits of Tanzania

J. F. HARRIS
Tanzania

Kaolin is a common raw material in Tanganyika. It seems to be the characteristic product of rock decay under the climatic conditions affecting much of the territory and it is widely distributed in soils and clayey superficial deposits, particularly in the more elevated areas. In such deposits it is usually mixed with sand and other impurities and is of no commercial value.

Kaolin of hydrothermal origin does not form important deposits in Tanganyika. Feldspar in the tin veins of the Karagwe area and in certain mica pegmatites, notably those of the Bundali Hills, is commonly converted to kaolin, but such occurrences are always small.

Kaolin deposits of possible economic interest are known in Tanganyika in sedimentary rocks, and in areas of deep surface weathering of feldspathic rocks.

Kaolin in sedimentary rocks

Pugu Hills, near Dar es Salaam.—The Pugu Hills form a well-defined escarpment running north-north-east for six and a half miles, about seventeen miles west of Dar es Salaam. The Neogene sediments which form the Pugu Hills dip at about 3° to the west. Part of the succession is made up of soft, white kaolin sandstones, the total thickness of which is probably 600 feet. Kaolin makes up about 30 per cent. of these rocks and the total reserves are estimated at about 2,000 million tons.

About 150 tons of kaolin per year were produced from claims in the Pugu area from 1942 to 1955 by a local company, the East African Mining and Development Co. Ltd. In recent years the output has been smaller and irregular.

In 1947, New Consolidated Gold Fields Ltd. started to develop the deposit for underground mining and to build a pilot mill. By 1951 the pilot mill was in operation and by 1953 the testing period was complete. About 2,500 tons of kaolin had been produced, overseas markets had been developed, and consider-

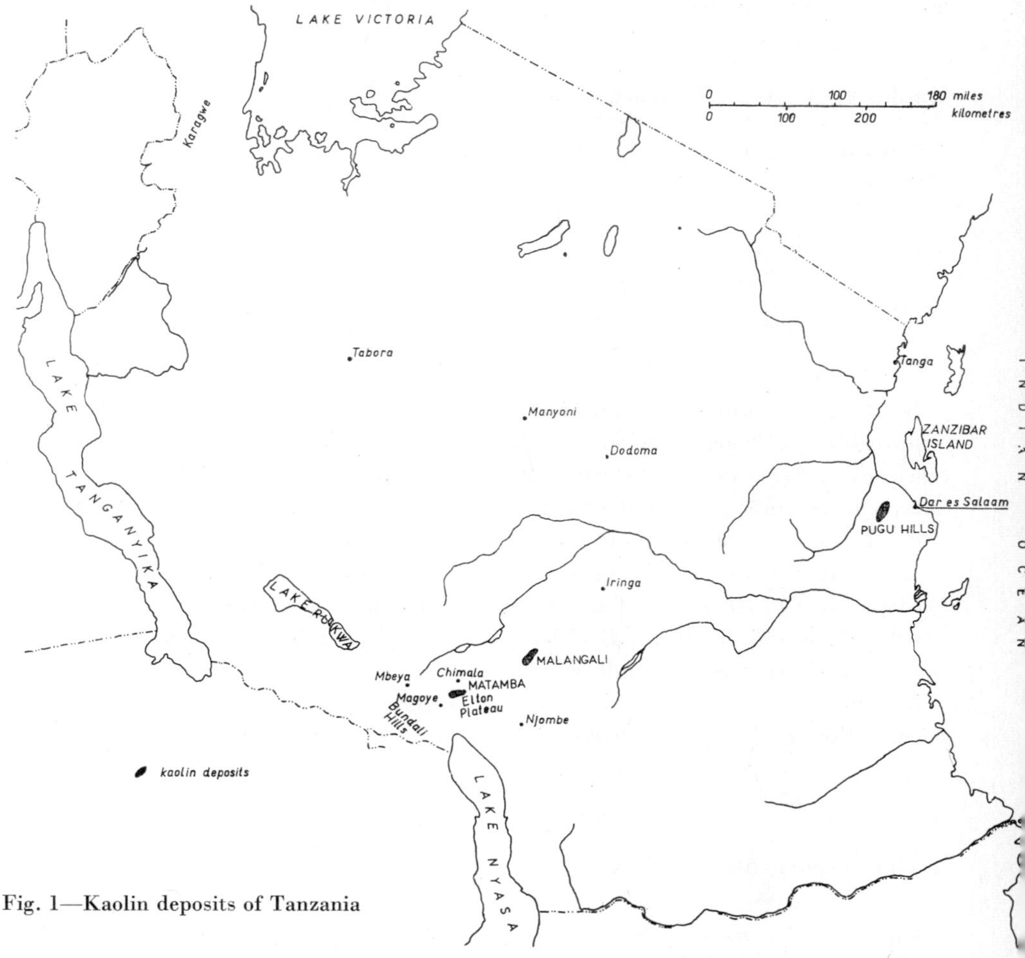

Fig. 1—Kaolin deposits of Tanzania

ation was being given to advancing to the production stage. However, it was decided to abandon the prospect, and the Pugu China Clay Company Ltd., which had been formed in 1952 to exploit the kaolin, went into liquidation at the end of 1953.

It was found that by a fairly complicated process of beneficiation to remove the kaolin from the sand, a product suitable for use as a filler for paper and rubber could be made and there was a possibility of producing a grade suitable for surfacing paper. Some overseas markets were developed during the testing period but definite large-scale contracts for a regular supply could not be obtained. The local demand for kaolin is small.

The clean sand separated from the kaolin was found to be suitable for glass-making.

Kaolin as surface deposits formed by weathering

Matamba area, north-western Njombe District.—Large areas of the gentle valley slopes of the Ndumbi and Ruaha rivers east and north-east of Matamba are covered with a thick blanket of kaolin. The kaolin is found over both leuco-gabbro and granite and was probably formed by accelerated weathering of these feldspathic rocks under acidic conditions when the area was covered by a large swamp in Pleistocene times.

In the best area, east of Magoye Mission, the kaolin layer probably averages 20 feet in thickness; near the top of the valley slopes it may be 50 feet or more in thickness, but in the bottoms of the valleys it has been removed by erosion. The kaolin seems generally free from gritty impurities but it is speckled with pink iron-staining.

Of five samples from these deposits sent for testing to the East African Industrial Research Organization, Nairobi, four were reported as being low-grade china clays which could be used in a pottery body, and the remaining one (which has a very high contraction at 1,200 °C.) as possibly being suitable for use in a stoneware body. The clays fire pink at 1,050 °C., but are creamy-white to white at 1,200 °C. The sample discs fired at the latter temperature are hard and free from cracks.

Reserves of the order of several million tons are available for quarrying, and water supplies for washing the kaolin are plentiful.

The region is not well served by communications as, although it is only about ten miles south of the main Iringa-Mbeya Road at Chimala, it is some 2,000 feet higher and the construction and maintenance of a road suitable for heavy traffic up the steep escarpment would not be easy. Alternative access roads across the Elton Plateau from Mbeya or through Ukinga from Njombe are both long and hilly.

Kaolin is also found nearer the edge of the Chimala Escarpment, as, for instance, in the headstreams of the Chimala River, but only as isolated pockets on the steep slopes of the valleys.

Malangali, south-west Iringa District.—Large deposits of impure kaolin occur near Malangali and are accessible from the Iringa-Malangali Road. They consist of rotted granite, which in certain localities has been almost completely kaolinized down to depths of fifty to one hundred feet. The material is soft and readily excavated, but is subject to very rapid erosion in cuttings. Iron-staining is common and all shades ranging from pink to reddish-brown are seen, but some of the material is white and comparatively pure. The quantity of mica present is thought to make the kaolin useless as a fireclay.

Domestic uses include the manufacture of bricks and tiles, and various articles for the building industry. There is as yet no local ceramic industry but interest is being shown in this development. Virtually all of the kaolin presently being

produced is obtained as a by-product in the dressing of silica-sand for a recently established glass-making industry.

Tanzanian production and exports of kaolin (in metric tons)

	1961	1962	1963	1964	1965	1966
Production	157	159	182	111	164	310
Exports	75	91	52	10	—	—
Domestic use	82	68	130	101	164	310

References

Harris, J. (1961): Summary of the geology of Tanganyika Part IV—Economic geology. Mem. geol. Surv. Tanganyika, 1, pp. 18, 20, 21, 91—93.

Robertson, R. H. S., and others (1954): Mineralogy of kaolin clays from Pugu, Tanganyika. Am. Miner., v. 39 (1/2), pp. 118—138.

Sampson, D. N. (1956): Forrest-cap clays of the Western Uluguru Mountains. Rec. geol. Surv. Tanganyika, v. 4, pp. 64—69.

Teale, E. O. and Oates, F. (1946): China clay (Kaolin) in mineral resources of T. T. Bull. geol. Surv. Tanganyika, 16, pp. 82—85.

1969 XXIII INTERNATIONAL GEOLOGICAL CONGRESS (1968) — VOL. 16 PAGE 79—86

Gîtes de la kaolinite en Tunisie*

JÁN ILAVSKÝ
Tchécoslovaquie

Contenu—Les gîtes de la kaolinite en Tunisie appartiennent à des différents types génétiques, dont les plus importants sont réprésentés par les gîtes sédimentaires dans la zone du flysch d'âge oligocène. Les gîtes volcano-sedimentaires de halloysite-kaolinite, liés dans les formations du Pliocène n'ont qu'une importance théorique. Les sédiments kaoliniques dans le Pliocène, ou dans le Quaternaire, comme aussi les indices résiduels dans le Trias, Eocène etc., ne sont pas explorés suffisamment pour une éstimation satisfaisante.

Géologie des gîtes

Tous les gîtes de la kaolinite, possédant l'importance économique, se trouvent dans le Nord Tunisien. Il s'agit des gîtes de la kaolinite sédimentaire, liés dans la zone du flysch d'âge oligocène inférieur ou eocène supérieur. Cette zone lithofaciale se compose de grès en bancs à intercalations argileuses ou marneuses, qui affectent une direction NE-SO et les pendages différentes, indiquant un plissement intense.

Les argiles blanches, kaoliniques, sont associées à des grès friables, blancs. Elles se présentent sous forme de couches lenticulaires de dimension variée. Les longeurs atteignent 250—300 m, quant à l'épaisseur elle varie entre 20 cm et 4—6 m. La série est bien plissée possédant les pendages assez importants.

Dans la composition d'argile kaolinique prennent part : kaolinite, en partie montmorillonite, illite et oxydes de fer. Les grains de quartz et micas blancs sont assez abondants dans les parties marginales du gîte. Rutile et titanomagnétite sont souvent inclus à l'intérieur. Les feldspathes potassiques ont été trouvés comme accessoires.

L'origine sédimentaire des gîtes est impressionant et bien documentée par tous les phénomèns géologiques. Il s'agit ici de deux gîtes exploités: Djebel Touila I

*) Cette contribution est publiée avec la permission du Gouvernement de la Tunisie et des Nations Unies. Les idées exprimées dans cet article n'ont pas besoin d'être accordées par ces deux organisations.

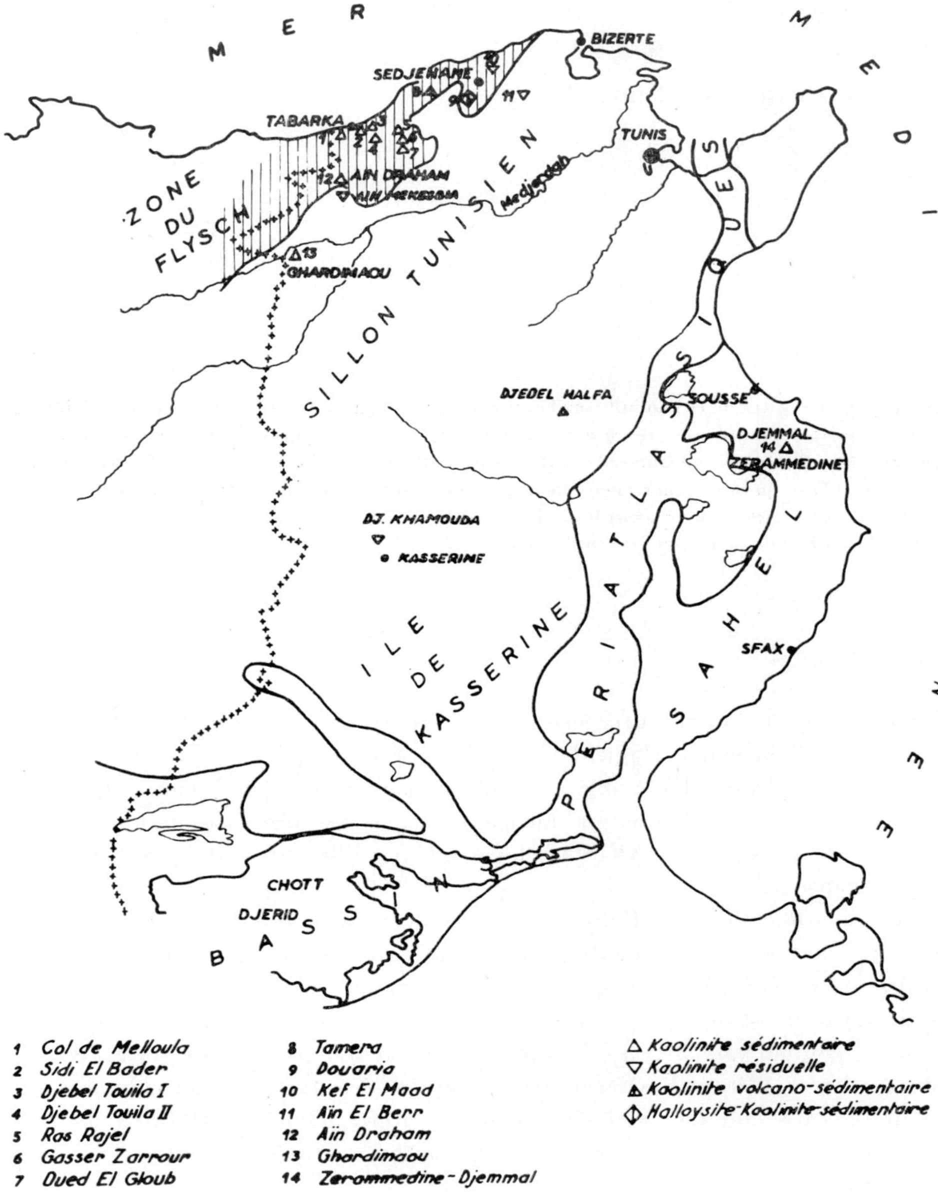

Fig. 1—Gîtes de kaolinite de la Tunisie

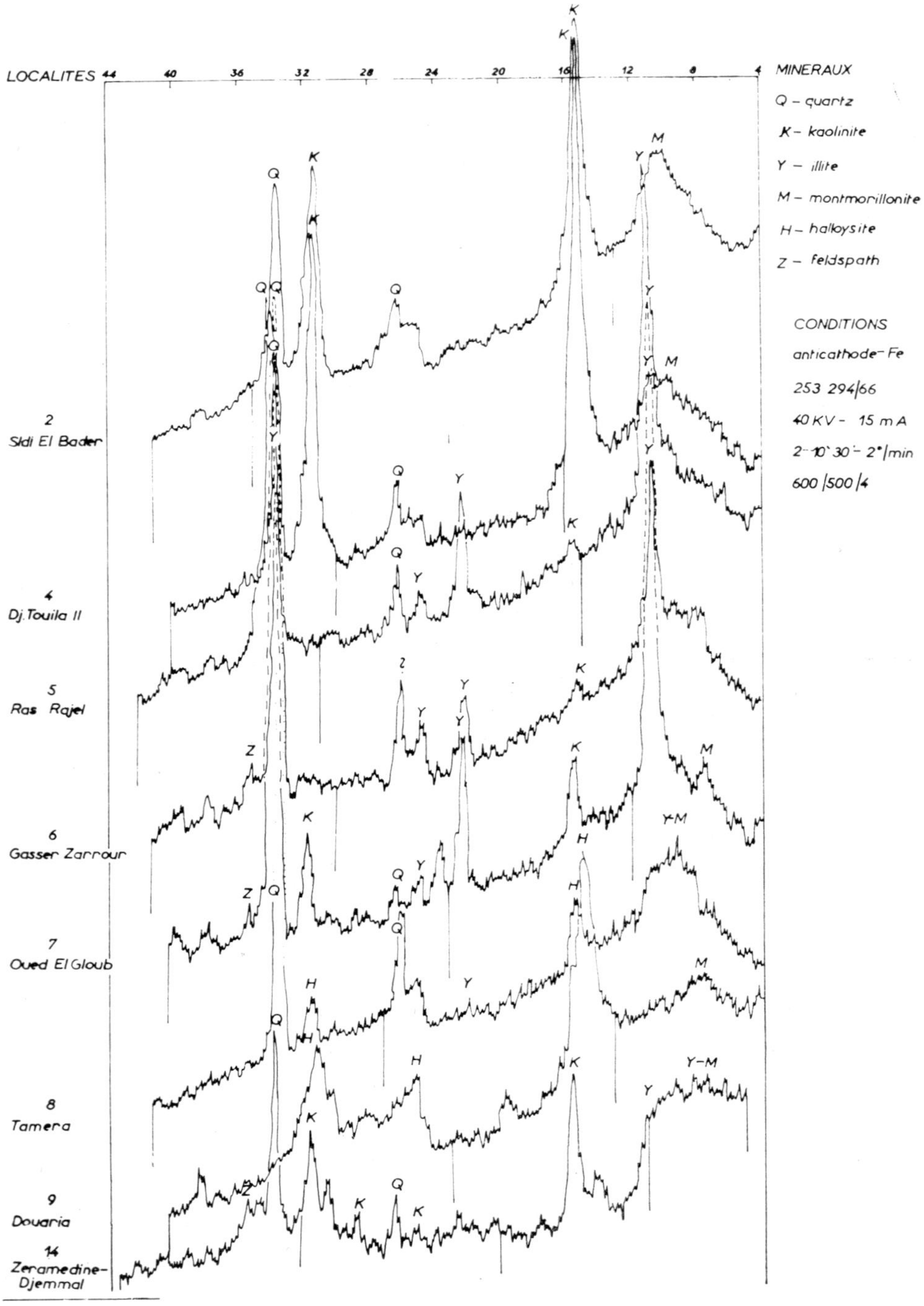

Fig. 2—Roentgenogrammes des argiles kaoliniques

et celui de Sidi El Bader à une petite distance de Tabarka. Col de Melloula, Djebel Touila II, Djebel Handala etc. sont moins importants. (Voir également chez Sassi M. Sassi—1963, 1964).

Un autre type génétique se présente comme couches de halloysite-kaolinite d'origine volcano-sédimentaire, liés dans une formation ferrifère de Douaria-Taméra, appartenant au Miocène (Pontien), qui est déposé par le flysch d'Oligocène et d'Eocène dans la Tunisie septentrionale.

Ce type d'argile semble être en rapport avec celui des roches volcaniques contemporaines. Les couches sont interstratifiées sous forme des lentilles dans le minérais de fer à intercalations de tuffs dacitiques et tuffites vertes. Leurs puissances varient de 0,20 à 0,4 m, exceptionnellement d'un mètre.

La composition d'argile à Douaria est uniquement d'halloysite, étant d'une texture massive ou très finement stratifiée, elle est très pure, blanche, et en présence de l'eau un peu élastique. Un certain mélange de la kaolinite n'est pas exclu.

A Taméra il s'agit d'une argile mélangée de la kaolinite et de la halloysite, à un certain pourcentage d'illite, chlorite et aussi de montmorillonite. L'oxyde de fer, carbonates et feldspathes, sont également présents. Comme accessoires il faudrait signaler de rutile, titanomagnétite, barytine, apatite, quartz et probablement des minéraux du Plomb-Zinc-Bismouth (voir les analyses spéctrales).

Un troisième type se présente sous forme de kaolinite résiduelle, située dans les milieux différents:

Une première catégorie est liée aux formations du Trias au sein de schistes argileux, verdâtres, dont elles sont issues par décomposition. Exemples: Ain El Berr, Kef El Maad (Route de Sédjenane-Nord Tunisien).

Une seconde catégorie se présente sous forme d'amas ou de fissures irrégulières soit dans des schistes bruns, verdâtres ou grisâtres appartenant à l'Eocène moyen, ou supérieur, soit dans les éboulis du Quaternaire. Exemples: Ain Draham-sud etc. Ces types d'argile kaolinique n'ont évidement qu'un interrêt théorique.

Les argiles kaoliniques résédimentées sont un dernière type, qui représente de gîte le plus typique, celui de Ain Mekesbia (Sud d'Ain Draham). Elles sont liées au Pliocène continental, composé de graviers polygéniques, cimentés par de l'argile kaolinique-blanche. Ce gîte est en rélation directe avec les gîtes de kaolinite primaire et sédimentaire de la région d'Ain Draham.

Caractère chimique et technologique

Toute les argiles kaoliniques contiennent un faible pourcentage de Fe_2O_3, sous forme soluble, et elles possèdent un certain caractère arkosique (K, Na, Ca, TiO_2), dont il faudrait rechercher l'origine dans les feldspathes décomposés.

L'étude de l'intensité, de la localisation et de la forme des courbes thermo-différentielles, comme telle de technologie (voir les tableaux intéressés), montrent que trois groupes d'argile kaolinique existent:

Tableau 1—Analyses chimiques-silicates d'argiles kaoliniques

	SiO_2	Al_2O_3	Fe_2O_3	CaO	MgO	MnO	K_2O	Na_2O	TiO_2	P_2O_5	SO_4	CO_2	Pert. au feu	Total
1 Col de Melloula	58,72	23,68	2,85	0,56	0,71	0,01	0,94	0,16	1,55	0,13	0,21	—	10,36	99,88
Col de Melloula	66,56	15,83	3,84	0,70	0,80	0,02	2,28	0,19	1,09	0,13	0,13	—	7,44	99,01
2 Sidi El Bader	55,67	26,13	2,60	0,72	0,72	0,02	1,80	0,28	1,59	0,72	0,16	—	9,78	99,74
Sidi El Bader	63,70	21,31	1,82	0,77	0,52	0,01	1,21	0,24	1,44	0,29	0,14	—	8,00	99,40
3 Djebel Touila I	56,59	26,82	1,95	0,84	0,50	0,01	1,06	0,36	1,31	0,27	0,09	—	10,20	99,92
4 Djebel Touila 2	57,14	27,02	1,12	0,56	0,20	0,01	1,43	0,21	1,68	0,19	0,15	—	10,04	99,75
5 Ras Rajel	57,87	19,47	5,58	0,84	3,12	0,01	6,20	0,25	0,84	0,25	0,15	0,11	4,92	99,50
6 Gasser Zarrour	71,22	13,51	3,75	0,84	1,31	0,01	3,62	0,84	0,81	0,20	0,12	—	3,49	99,36
7 Oued El Gloub	60,07	16,54	4,50	2,24	3,53	0,01	5,44	0,24	0,84	0,27	0,09	1,01	5,25	99,92
8 Tamera	44,03	32,29	4,12	0,84	0,90	0,02	0,17	0,31	1,56	0,73	0,12	—	14,51	99,60
9 Douaria	44,22	34,61	1,28	0,84	0,40	0,02	0,17	0,18	0,34	0,58	0,49	—	16,54	99,67
Douaria	44,69	21,26	17,85	0,84	0,60	0,02	1,31	0,18	0,78	0,39	0,13	—	11,99	99,98
10 Kef El Maad	57,48	19,17	7,04	1,12	1,01	0,06	1,60	0,33	1,09	0,26	1,05	—	10,02	100,23
11 Ain El Berr	51,96	18,30	4,21	3,35	7,01	0,02	4,32	0,33	1,04	0,22	0,08	—	9,13	99,98
12 Ain Draham	57,24	24,11	3,00	1,40	0,20	0,02	1,22	0,15	1,37	0,25	0,12	—	9,89	99,75
14 Djemmal - Zeramedine	53,00	18,80	6,90	0,76	2,40	0,04	3,00	0,76	1,18	0,07	4,15	—	9,50	100,48
Djemmal - Zeramedine	60,60	18,17	3,94	0,45	1,87	0,03	2,40	0,80	1,37	0,10	0,74	0,15	9,51	99,98

Tableau 2—Analyses semiquantitative-spectrales des argiles kaoliniques

Localité	Ag B Ba Be Bi Co Cu Cr Ga Ge In La Li Mn Mo Nd Ni P Pb Sc Sn Sr Y Yb Ti V Zn Zr	K Na	L'âge	Type génétique
2 Sidi El Bader	× ○ ● ○ × ○ ◐ ◐ ○ × × × ○ ● × × ○ × ○ ◐ ○ ● ◐ ◐ ▪ ○ ○ ◐	● ●	oligocène	sédimentaire
3 Djebel Touila I.	○ ○ ● × × ○ ○ ◐ ○ – – × × ● – × ○ – ○ ○ ○ ◐ ○ ○ ● ○ × ○	▪ ▪	"	"
5 Ras Rajel	○ ○ ● ○ – ○ ○ ◐ ○ – – – – ● ○ – ○ – ○ ○ × ◐ ○ × ▪ ○ ○ ○	■ ■	"	"
6 Gasser Zarrour	○ ○ ◐ × – ○ ○ ◐ ○ – – – × ● × – ○ – ○ ○ ○ ◐ ○ × ● ○ × ○	■ ■	"	"
7 Oued El Gloub	○ ○ ● × – ○ ○ ● ○ – – – × ▪ × – ○ – ○ ○ ○ ◐ ○ × ▪ ○ × ○	■ ■	"	"
10 Kef El Maad	– – – × – ○ ○ ○ × – – – – ○ – – ○ – – – – ○ – – ▪ ○ ○ –	■ ▪	trias.quaternre	résiduel
14 Zeramedine Djemmal	× ○ ◐ ○ – ● ○ ◐ ○ – ○ – ○ ■ – – ● – ○ ○ × ○ × × ▪ ○ ● ○	■ ■	miocène	sédimentaire
8 Tamera	◐ ○ ○ × ○ ○ ◐ ◐ ○ ○ × – × ▪ ○ – ◐ – ● ○ ○ ◐ × × ▪ ○ ○ ○	● ●	pliocène	valcano-sédimentre
9 Douaria	◐ ○ ● ○ ● ○ ◐ ◐ ○ – × ○ × ▪ × – ○ ○ ● ○ ○ ○ × ○ ▪ ○ ○ ○	● ●	"	"

■ 100–10% ▪ 10–1% ● 1–0,1% ◐ 0,1–0,01% ○ 0,01–0,001% × 0,001–0,0001%

Kaolinites à structure cristallographique bien ordonnée (well-ordered kaolinites)

Elles ne présentent pas la déviation endothérmique, mais au contraire, suivant la déshydratation, elles montrent une endotherme symétrique bien dévéloppée,

ainsi qu'une déviation exothermique complétée de 900 à 1000 °C. Composition minéralogique est présentée dans le tableau 2. La couleur des produits cuits est beige ou blanchâtre, les propriétés réfractaires: 31—32 degrés, cela signifie que ces argiles sont utilisables pour la fabrication des produits de céramique précieuse: carreaux de faïence, céramique sanitaire, isolateurs, briques réfractaires etc. Quoiqu'elles contiennent des % de Fe_2O_3 assez élevés, ces argiles peuvent être néanmoins traitées et leur teneurs en Fe_2O_3 abaissées à 1—2 %.

On pourrait ranger dans cette catégorie le gîte du Djebel Touila II.

Kaolinites à structure cristallinée desordonnée (dis-ordered kaolinites)

D'après la forme de la courbe thermo-differentielle, on peut définir que ces argiles présentant une certaine instabilité électrostatique des couches d'unités cristallographiques et structurales. Les contenus de Fe_2O_3, étant assez élevés (tableau 3), peuvent être traités soit par dissolution de HCl à 10 %, soit par flottation à l'hydrocyclone, ce qui permet d'abaisser ces teneurs à 2—2,5 %. Cette teneur correspond déjà à celle des kaolinites utilisables en générale dans l'industrie céramique vrai sens du mot. Quelques exceptions existent néanmoins: indice de retraits par séchage un peu élevés (Tamera) ou redressés (Ain El Berr).

Nous rangeons dans ce groupe les gîtes suivants: Sidi El Bader, Djebel Touila I, Tamera, Col de Melloula, Ain Mekesbia, Ras Rajel, Gasser Zarrour, Oued El Gloub, Ain Draham.

Halloysites contenant des proportions variables de kaolinite

Il s'agit ici soit des kaolinites-halloysites contenant de Fe_2O_3 soit d'halloysites très pures et de bonne qualité, capable de fournir des produits réfractaires, la porcelaine etc. C'est la localité Douaria ou ces argiles sont assez courrantes.

Il faudrait noter qu'avec l'acide chlorhydrique, presque toutes les impuretés telles que la limonite et carbonates, ont disparus. Les teneurs de potass sont diminuées également, tandis que Na demeure presque inchangé.

Et enfin comme conclusion, quelques remarques d'ordre géochimique. Comme on voit dans le tableau 4, les différents types génétiques d'argiles kaoliniques contiennent les oligoélements variés. Les kaolinites sédimentaires dans l'oligocène sont très bien remarquables par les oligoélements Bi, Ge, In, La, Li, Nd, P, Sc, Sr, Y, Yb, Zr. Kaolinites-halloysites d'origines volcano-sédimentaires sont abérrantes par leur contenus en oligoélements (Ag, Bi, Cu, Ge, In, La, P, Pb), tandis que la kaolinite résiduelle possède au contraire une pauvreté rélative en oligoélements.

Tableau 3—Analyses technologiques d'argiles kaoliniques

	Gonflement	L'eau de gachôge en %	Retraits par Sechage en %	Retraits par Cuisson en %	Pertes totales par Cuisson en %	Adsorbtivité pondérale en %	Capilarité près de 1350 °C	Resistence a la flexion sur la surface tendue en Kp/cm²		Propriétés refractiores en degrés
								après séchage	après cuisson	
1 Col de Melloula	2,6	> 40	4–7	12,50	—	—	—	—	—	—
2 Sidi El Bader	1,8–2,4	> 40	4–7	8,3–17,14	6–24,45	14,39-24,8	30,4–51,9	28,5	94,4	30–32
3 Djebel Touila 1	3	< 40	4–7	11,8–12,6	9,72	10,14	9,0	23,7	81,3	32
4 Djebel Touila 2	2,3	< 40	4–7	10,3	—	—	—	—	—	31—32
5 Ras Rajel	1,8	> 40	4–7	5,4	3,92	15,38	4,50	21,8	53,7	> 26
6 Gasser Zarrour	1,6	> 40	4–7	5,7–7,0	13,24	30,42	27,0	22,6	44,7	> 26
7 Oued El Gloub	1,8	> 40	4–7	6,7–11,6	10,38	19,40	20,0	28,5	68,0	> 26
8 Tamera	1,6–2,8	< 40	4–7	11,8–12,6	14,30	24,6-30,2	20–37,0	16,7	46,4	30
9 Douaria	2,6–3,2	24,8–44,7	4,9–8	13,4–28,9	7,9–14,30	14,45–24,6	32–51,0	18,1	125,6	< 35
10 Kef El Maad	2,6	> 40	4–7	11,5–41,6	11,22	8,08	43,0	—	—	—
11 Ain El Berr	2	> 40	4–7	13,3	7,97	14,45	51,0	—	—	—
12 Ain Draham	2,2	> 40	4–7	11,4–12,8	—	—	—	—	—	—
13 Ghar dimaou	1,4–2,4	24–29	7,2–7,4	12,7	—	—	—	41,6–47,5	—	—
14 Zeramedine	2,6–3,2	28,3	6,97	14,2–15,2	—	--	—	36,4	—	25

Tableau 4—Essais d'activation des argiles kaoliniques dans une solution de HCl à 10 %

	Fe_2O_3 Total dissous	MgO Total dissous	K_2O Total dissous	Na_2O Total dissous	CaO Total dissous
1 Col de Melloula	3,84	0,80	2,28	0,19	0,70
	1,70	0,10	0,07	0,13	0,42
4 Djebel Touila II	1,12	0,20	1,43	0,21	0,56
	0,25	0,10	0,11	0,12	0,28
8 Tamera	4,12	0,90	0,17	0,31	0,84
	2,65	0,10	0,09	0,14	0,42
9 Douaria	1,28	0,40	0,17	0,18	0,84
	0,30	0,10	0,06	0,14	0,42
	17,85	0,60	1,31	0,18	0,84
	10,84	0,10	0,07	0,12	0,42
10 Kef El Maad	7,04	1,01	1,60	0,37	1,12
	5,15	0,40	0,09	0,33	0,80
11 Ain El Berr	4,21	7.02	4,32	0,33	3,35
	1,25	0,30	0,07	0,20	3,21
12 Ain Draham	6,82	1,31	1,30	0,18	3,51
	5,11	1,01	0,13	0,13	3,22
	3,00	0,20	1,82	0,15	1,40
	0,72	0,10	0,07	0,15	0,28

Utilisation-possibilités

Il y a en Tunisie deux gîtes uniquement (Djebel Touila I, Sidi El Bader), dont les reserves atteignent l'interêt économique (1 milion de tonnes environ) et qui sont exploités actuellement. Une usine est construite sur le gîte de Sidi El Bader (carreaux de faïence), une autre à Nabeul (vaisselle) est éloignée à 230 km. En même temps une troisième usine à Bizerte (céramique sanitaire), éloignée 150 km de Tabarka, consomme la kaolinite importée de l'étranger.

En estimant qu'une augmentation de la demande en kaolinite indigène devrait se manifester en Tunisie dans l'avenir, il y a une raison réelle pour la construction d'une laverie, déstinée à produire de la kaolinite pure ou flottée, utilisable en différentes branches de l'industrie tunisienne. Cette laverie devra être installée au Nord Tunisien près de Tabarka. Les autres petites localités dans la Tunisie centrale pourront être récupérées par la poterie locale.

Conclusion

En conclusion nous aimerions remercier tous les organismes et toutes les personnes qui nous ont aidé dans l'accomplissement de notre mission. Nos remerciements s'adressent aux collaborateurs de Recherches Géologiques de Turčianske Teplice en Tchécoslovaquie, à M. Ivan Horvath surtout, qui ont éxécuté, suivis et en partie évalués les analyses minéralogiques, chimiques et spéctrales et technologiques.

Au Bureau pour l'Assistance Technique des Nations Unies à Tunis et à New York, comme aussi à l'Office National des Mines à Tunis, j'adresse mes remerciements particuliers pour m'avoir facilité tous les travaux sur le terrain et dans les laboratoires et en même temps pour m'avoir accordé la permission de publier cette contribution.

Bibliographie

Jauzein A. - Rouvier E. (1964): Sur les formations allochtonnes de la Kroumirie. Inédit. Faculté des Sciences., Lab. Géol. Appl., Paris.

Sassi M. Sassi (1963): Etude préliminaire des argiles éocènes, mio-pliocènes et actuelles de la région Métlaoiu-M'Dilla et du Cap Bôn (Tunisie). Bull. Soc. Géol. France, VII série, T. V., Paris.

— (1964): Associations minéralogiques argileuses dans certaines formations triassiques et jurassiques en Tunisie. Compte Rend. Somm. Séanc. Soc. Géol. France, facs. 7, Paris.

Vanderwayden J. M. (1963): Reports of activities. Manuscrits inédits, Office Nationale des Mines-Bureau pour l'Assist. Techn. des Nations Unies, Tunis.

AMERICA

1969 XXIII INTERNATIONAL GEOLOGICAL CONGRESS (1968) — VOL. 16 PAGE 89—96

Kaolin Deposits of Chile

MAURICIO TABAK B.
Chile

Summary—The most important Chilean clay deposits are located in Central Chile between parallels 35°35'S lat. From north to south, four different geographic zones may be separated, in which different clay mineral assemblages occur. They have been individualized as zones A, B, C and D of the following characteristic:

Zone A: Contains clay deposits mostly constituted by kaolinite, quartz and alunite.

Zone B: Includes highly aluminous clay (flint clay), mainly constituted by kaolinite, boehmite and gibbsite.

Zone C: Contains deposits mainly constituted by kaolinite, halloysite and pyrophyllite.

Zone D: Includes clay deposits mostly constituted by kaolinite and traces of halloysite. "Under clays" and "plastic clays" are distinguished separately in different deposits.

Geographical data

Geographic distribution of the clay deposits in Chile permits individualization of the zones A to D, shown on Fig. 1. Listed on this map are either single deposits or groups of working which do not necessarily represent one single type of deposit. Most of them fall within the category of generally undeveloped deposits and, as far as it is known, of less than 1 million tons.

Geological conditions

Investigations conducted so far have provided geological information only on those deposits of zones C and D (Fig. 1) which are listed in Table 1.

Table 1—Geological conditions

Zone C deposit	Geological position of deposits	Shape of deposits, area and proved depth of kaolinization	Exploitable thickness of deposits	Parent rock of kaolin, its age and petrographical characteristics	Overburden of kaolin, its age and petrography; relation to kaolinization	Genetic type of deposits	Morphological type	Relations of kaolinization to tectonic and geomorphologic development of the area	Are the deposits in situ or are they partially rewashed
Blanquita C-5	Terrace level developed on Paleozoic granitic rocks.	Irregular, bolson-like filling depressions of the plain surface, 1 km^2. Kaolinization to a depth of 14 m.	± 12 m.	Granite.	Silt and pebbly sand, 2 m. and soil, 0.5–1.0 m., thick. Quaternary. No relation to kaolinization.	Kaolinization is mostly result of weathering. Halloysite occurs in dikes.	Areal and irregular.	Kaolinization was probably related to the formation of a coastal plain in Early Tertiary time.	Mostly in situ.
Audelina C-10	Idem.	Same as above. Kaolinization to a depth of 18 m.	± 15 m.	Idem.	Idem.	Idem.	Idem.	Idem.	Idem.
Mercedes C-7	Hydrothermal zone.	Very irregular. Known alteration to a depth of 60 m.	Variable within the alteration zone.	Volcanic rock.	None.	Hydrothermal.	Irregular.	Pyrophyllite concentrated mostly within an intensively fractured and sheared zone.	In situ.
Mary C-9	Idem.	Irregular local concentrations in roughly tabular bodies at least 5 m. thick, dipping about 45 °.	4 to 5 m. to known depth of approximately 6 m.	Idem.	None.	Idem.	Irregular roughly tabular.	Idem.	Idem.
San José C-1	Dissected gravel terrace level.	Irregular lens-like. Kaolinization to a depth of 6.5 m.	4 to 5 m.	Probably a granitic rock.	Fine alluvial sediments and soil, and weathered zone with clay containing abundant limonite. Total overburden 1.5 to 3.0 m. Quaternary. No relation to kaolinization.	Weathering.	Lenticular.	?	Mostly in situ.
San Miguel C-2	Idem.	Stratiform, large areal extent. Kaolinization to a depth of 9 m.	2 m., average.	Probably a granitic rock.	Soil, silt and pebbly sand, 2 to 3 m. thick. Quaternary. No relation to kaolinization.	Idem.	Stratiform, filling large, nearly flat depressions.	?	Idem.

Table 1 (continued)

Zone D deposit	Geological position of deposits	Shape of deposits, area and proved depth of kaolinization	Exploitable thickness of deposits	Parent rock of kaolin, its age and petrographical characteristics	Overburden of kaolin, its age and petrography; relation of kaolinization	Genetic type of deposits	Morphological type	Relations of kaolinization to tectonic and geomorphologic development of the area	Are the deposits in situ or are they partially re-washed
La Chica	Fault zone with hydrothermal alteration.	Roughly lenticular, nearly vertical, irregular outline. Kaolinization within a width of 25 m. but with high concentration in center.	—	Tonalite.	Not overburden in present working.	Primarily hydrothermal; subsequent weathering.	Tabular, roughly lenticular, nearly vertical.	Hydrothermal alteration in close association with movement and brecciation in a fault zone.	In situ.
Lo Pequén D-12	Dissected rolling surface.	Wedge-shaped, 85 m. long, 25 m. wide. Kaolinization to a depth of 15 m.	In sectors to be exploited, up to 6.5 m.	Most probably granitic rocks.	Soil and colluvial material, 8.5 m. Quaternary. No relation to kaolinization.	Weathering.	Wedge-shaped.	?	Partly in situ, partly transported gravitationally.
Landa D-11	Weathered surface.	Roughly stratiform, wedge-shaped, 40 m. long, 15 m. wide. Kaolinization to a depth of 11 m.	About 3 m.	Granitic rocks with inclusions of metamorphic rocks.	Soil gravel deposits, sand and mechanically desintegrated granitic rocks. 7 to 8 m. Quaternary. Lower part of overburden belongs to the kaolinization zone.	Idem.	Roughly stratiform.	?	Mostly in situ; in part transported gravitationally.
Yovilo D-8	Interstratified with coal-bearing. Tertiary sedimentary rocks.	Stratiform, large areal extent.	Approximately 0.3 to 0.5 m.	Sedimentary rock. Argillaceous siltstone.	Underground exploitation, by-product in coal exploitation.	Probably represents material transported to the environment of deposition of the sediments.	Tabular, stratiform.	?	In situ.
La Puntilla D-15 Eucaliptu Grande D-14	Idem.	Idem.	Two beds, one up to 2 m. thick; the other 0.7 m. thick.	Sedimentary rocks. Claystone.	Workings being opened in one bed. The other has an overburden up to 5 m. of sedimentary rocks. Tertiary. No relation to kaolinization.	Represent material transported to the environment of deposition.	Idem.	?	Idem.
Maño Plástico D-6	Near top of a rolling, dissected surface.	Wedge-shaped irregular, 35×20 m. approximately. Kaolinization to a depth of 5 m.	Variable between 0.10 and 2 m.	Slate.	Soil, argillaceous soil, and impure clay. Up to 3 m. thick. Partly belongs to the kaolin deposits.	Weathering.	Wedge-shaped irregular.	?	Probably partly removed.

Table 2—Average chemical composition (raw kaolin) (in %)

Zone	Deposit	SiO_2	Al_2O_3	Fe_2O_3	FeO	Na_2O	K_2O	CaO	MgO	MnO	TiO_2	P_2O_5	H_2O^+	H_2O^-	SO_3	BaO
Zone A																
A-1	Encarnación ≠ 1	58.63	23.67	1.12	0.47	0.07	2.09	0.48	0.26	0.006	1.03	0.41	7.39	2.95	0.21	
A-2	Silvia	54.46	26.71	2.11	0.25	0.17	1.21	0.39	0.18	0.009	0.86	0.25	9.01	3.67	0.13	
A-3	Perico	54.49	26.38	1.01	0.45	0.14	2.09	0.29	0.20	0.005	0.89	0.25	8.95	2.38	1.54	
A-4	Socorro	49.26	28.32	2.03	0.54	0.65	0.57	0.21	0.27	0.003	0.78	0.29	10.92	3.36	1.75	
A-5	Magnesia	64.73	24.10	0.00	0.32	0.03	0.08	0.21	0.07	0.004	0.93	0.09	8.65	0.22	0.08	
A-6	Sta. Filomena	58.96	25.15	0.68	0.49	0.14	1.50	0.18	0.09	0.006	0.60	0.22	9.07	1.92	0.16	
A-7	Lo Gallardo 1	41.95	30.08	0.19	0.45	1.40	0.33	0.24	0.13	0.006	0.83	0.14	12.41	2.10	9.20	
A-9	Empalme 1	51.18	28.54	2.41	0.63	0.60	0.37	0.14	0.17	0.034	0.89	0.22	11.20	2.46	0.41	5.95
A-10	Mina Vieja	45.52	28.63	0.10	0.32	0.18	0.13	0.25	0.05	0.002	2.04	0.08	9.08	0.24	0.18	0.41
Zone B																
B-1	La Nueva Paloma	43.38	36.12	0.46	0.58	0.17	0.04	0.39	0.07	0.012	2.14	0.03	13.47	1.74	0.09	0.39
B-5	La Codorniz	41.76	33.58	4.39	0.61	0.06	0.04	0.64	0.46	0.034	1.20	0.19	12.71	3.01	0.07	
B-10	El Guindo	13.27	43.95	13.40	1.09	0.07	0.04	0.31	0.06	0.021	2.53	0.18	23.02	0.71	0.05	
B-15	El Quisco	42.37	38.49	0.15	0.75	0.23	0.04	0.34	0.11	0.003	2.51	0.05	13.34	1.60	0.03	
B-20	Los Colitos	43.50	36.59	1.03	0.48	0.23	0.04	0.28	0.37	0.014	1.92	0.06	13.44	1.91	0.05	
B-21	El Andacollo	41.66	39.52	0.78	0.41	0.21	0.04	0.25	0.04	0.010	2.03	0.05	13.86	1.26	0.03	
B-25	El Yal	41.24	39.19	0.03	0.71	0.18	0.04	0.36	0.04	0.008	2.63	0.16	13.91	1.17	0.07	
Zone C																
C-1	San José	49.71	29.14	0.47	2.20	0.14	2.29	0.42	0.74	0.006	1.00	0.10	10.56	2.10	0.43	
C-2	San Miguel	48.59	29.65	2.50	0.56	0.20	2.29	0.31	0.88	0.015	0.96	0.10	10.34	2.67	0.06	
C-5	Blanquita	63.70	24.35	0.20	0.75	0.03	0.12	0.17	0.03	0.009	0.15	0.05	9.44	0.94	0.05	
C-7	Mercedes	54.10	32.87	1.59	0.24	0.09	0.04	0.14	0.00	0.004	0.46	0.16	9.53	0.33	0.28	
C-9	Mari	70.74	16.99	0.21	1.12	0.38	4.12	0.19	0.11	0.009	0.21	0.03	4.90	0.67	0.02	
Zone D																
D-6	Mañio Plástico	54.16	28.58	0.63	0.92	0.26	1.03	0.29	0.15	0.033	2.38	0.16	9.01	0.95	0.06	
D-8	Yovilo	49.80	27.18	0.47	2.22	0.15	1.20	1.36	0.64	0.009	1.13	0.09	10.17	4.49	0.50	
D-11	Landa	64.07	21.24	0.05	1.28	0.31	2.02	0.52	0.37	0.030	1.33	0.03	6.18	1.81	0.04	
D-12	Lo Pequén	64.89	20.99	0.00	1.26	0.36	2.18	0.70	0.38	0.015	1.33	0.09	6.04	1.43	0.14	0.13
D-13	La Chica	55.55	28.67	0.13	0.41	0.32	0.20	1.04	0.11	0.011	0.17	0.02	10.85	2.29	0.03	
D-15	La Puntilla	43.78	32.84	3.29	0.45	0.27	0.93	0.27	0.16	0.002	1.58	0.08	13.55	2.50	0.03	

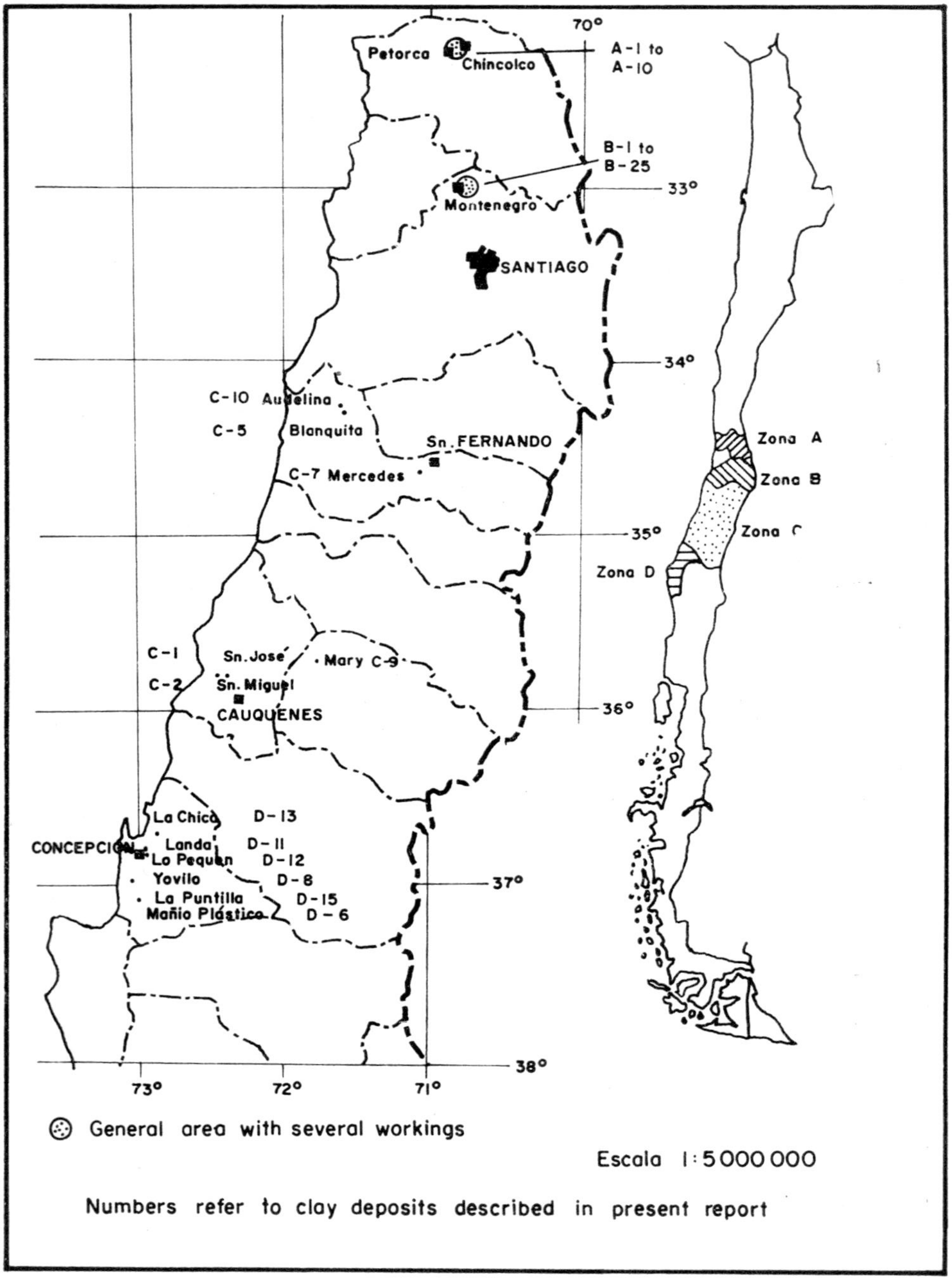

Fig. 1—Map showing location of clay deposits of Central Chile

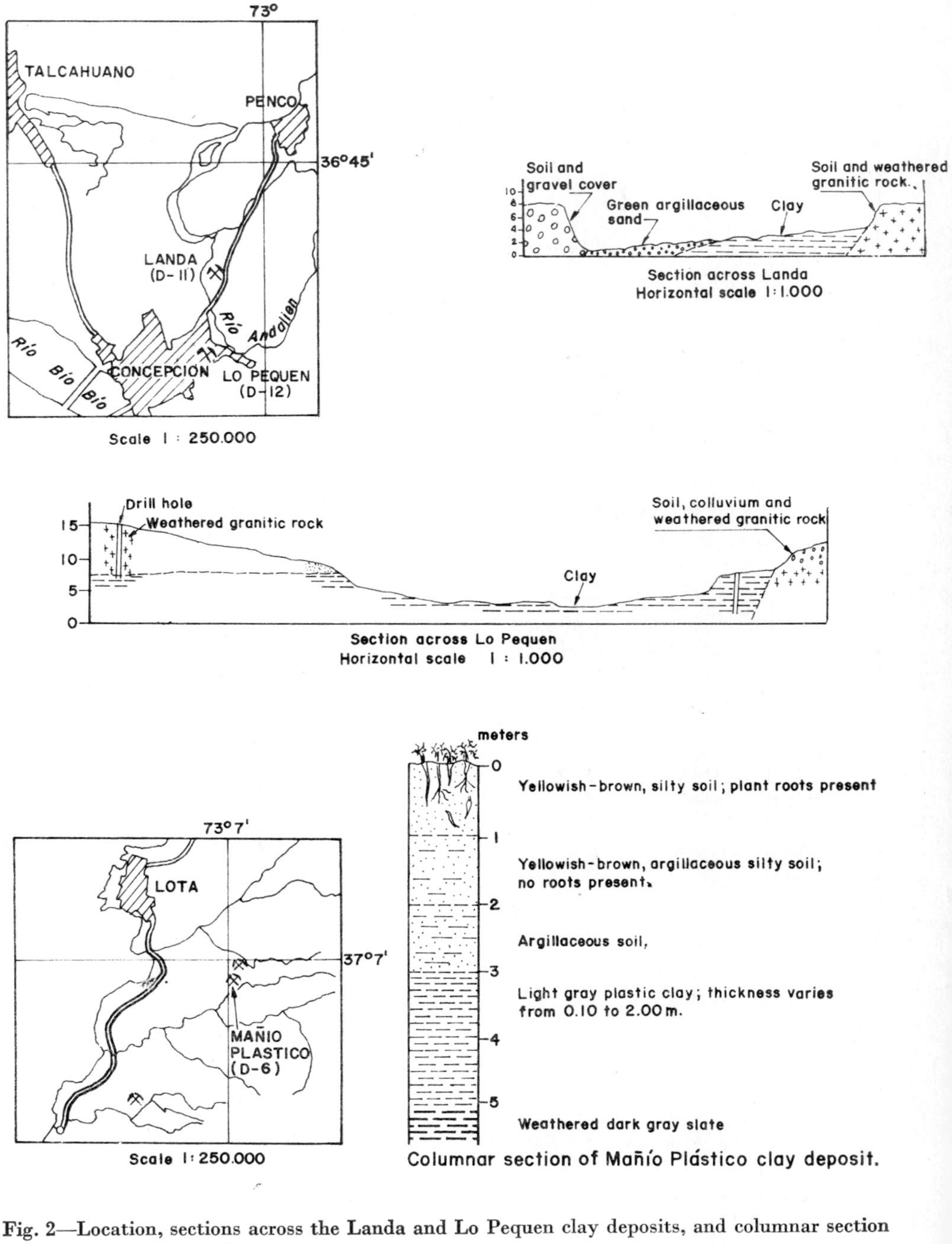

Fig. 2—Location, sections across the Landa and Lo Pequen clay deposits, and columnar section of Mañio Plástico clay deposit

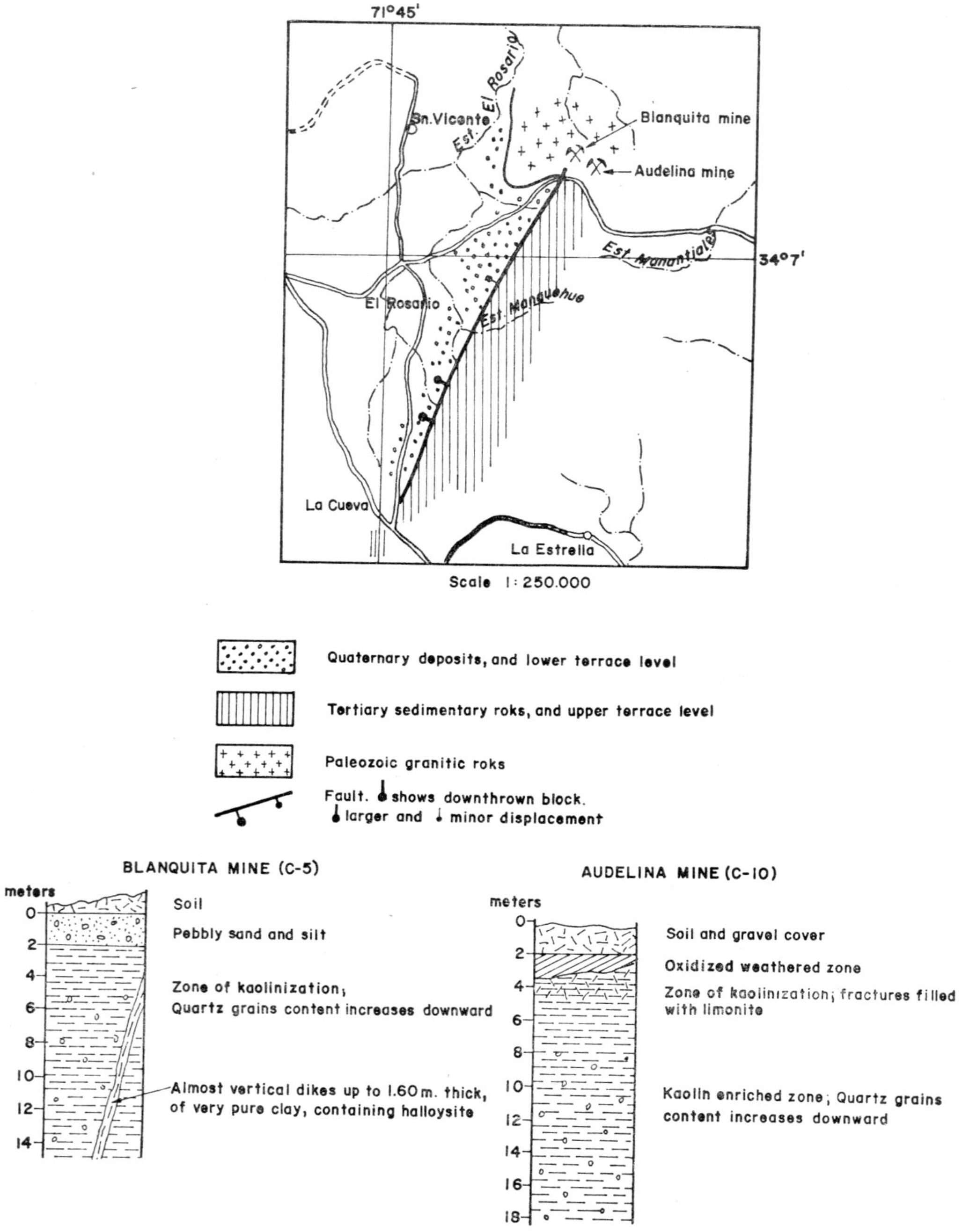

Fig. 3—Location and columnar sections of Blanquita and Audelina clay deposits

Utilization of kaolin

Table 3—Technology of washed kaolin

	1	2	3	4	5	6
Zone A						
A-1	22.10	7.50	7.60	29	4.20	1.79
A-2	16.68	8.40	1.82	— 26	0.80	2.28
A-3	27.84	—	0.97	29	0.42	2.30
A-4	38.04	7.69	5.68	+ 28	2.78	2.05
A-5	4.43	—	24.90	32	12.20	2.05
A-6	25.69	10.18	4.26	+ 32	1.90	2.25
A-7	25.72	9.35	18.64	32—33	10.15	1.84
A-9	36.93	9.52	8.67	— 32	4.02	2.15
A-10	1.82	—	23.60	32—33	11.60	2.06
Zone C						
C-1	41.75	6.61	1.92	+ 31	1.00	1.97
C-2	37.55	8.52	2.29	—	0.98	2.35
C-5	16.65	4.80	25.03	+ 33	12.39	2.02
C-7	7.02	2.77	26.86	+ 30	13.94	1.92
Zone D						
D-6	18.78	4.79	11.52	31	5.90	1.96
D-8	30.25	2.89	4.04	30—31	2.06	1.99
D-11	29.44	— 0.41	13.47	26	7.70	1.75

1 - Bending compression strength after drying [kg/cm^2] (110 °C)
2 - Shrinkage after burning to 1,450 °C [%]
3 - Porosity after burning to 1,450 °C [%]
4 - Refractoriness
5 - Sorption capacity [%]
6 - Specific surface area [%]

Note: All of the clay minerals from zone B are "flint-clays".

Economic evaluation of deposits

Mining conditions:

Quarrying in most of the deposits, except for a few, no more than five, which are exploited underground.

Economic aspects influencing the production:

Mostly transport, which is expensive and difficult because of bad roads, especially in zones A and C.

References

Corvalán J., A. Dávila, M. Tabak y A. Aguilar (1967): Estudio Geológico de Yacimientos de arcilla en las provincias de Concepción, Maule, Linares y Colchagua. Inst. de Invest. Geol., Santiago, Chile.

Ruiz F. Carlos (1947): Informe sobre yacimientos de materiales refractarios. Compañía de Aceros del Pacífico, Santiago, Chile.

Vila Tomás (1953): Recursos Minerales No-Metálicos de Chile. 3â ed. Editorial Universitaria, Santiago de Chile, p. 25—52.

1969 XXIII INTERNATIONAL GEOLOGICAL CONGRESS (1968) — VOL. 16 PAGE 97—100

Kaolin Deposits in Colombia

FERNANDO CALVACHE C.
Colombia

A general description is presented and the different kaolin deposits in Colombia are plotted on a map.

1—Antioquia Department—In the locality of La Unión, kaolin of residual type is found. This deposit occurs in gently sloping domes and the host rock is granite (Eduardo Nichols, Report No. 1349, S. G. N.)

2—Medellín district—Along the road from Medellín to Rionegro on km 20 exists a pegmatite dike. Feldspars of this locality are completely kaolinized. Two km further on the same road an outcrop of kaolin is also found. (Royo y Gómez, Report No. 309, S. G. N.)

3—Santuario district—In "Palmar" Creek valley is a pegmatite dike of large volume which is under exploitation. (Royo y Gómez, Report No. 309, S. G. N.)

4—Envigado district—In "Cienpesos" Creek valley kaolin is found in a vein about 5 m thick. It is of residual type. (Royo y Gómez, Report No. 309, S. G. N.)

5—Boyacá Department—In the hill of "El Salvador" or "El Volcán" between the towns of Paipa and Tuta, kaolin outcrops in the following way: it is formed by a rhyolitic porphyry which has been considerably weathered on the surface. (Sarmiento, Soto, Report No. 353, S. G. N.).

6—Tunja district—In the hamlet of Chivatá, Hacienda of "Rumba" and Portezuela village, is a deposit of sedimentary type; this material is mined for the factory "Caolines Boyacá". It is considered to be found in the cherts of the Guadalupe Formation. (Upper Cret.). (Report No. 1504, S. G. N.)

7—Cundinamarca Department—Bogotá plateaux—Small kaolin outcrops were found by the geologist T. v. d. Hammen in the Santa Sofía hill at the end of the "Regadera" ridge. The deposit is of sedimentary type and is located within the Tilatá Formation. (Pliocene).

8—Mosquera district—In the farm "Mondoňedo" a sedimentary kaolin deposit was found in the Guadalupe Formation (Upper Cret.). Analyses of this kaolin were made with following results:

SiO_2	Al_2O_3	TiO_2	CaO	MgO	Per cal	P. C. E.
48.20 %	33.58 %	0.55 %	1.12 %	0.24 %	14.10 %	33.00 %

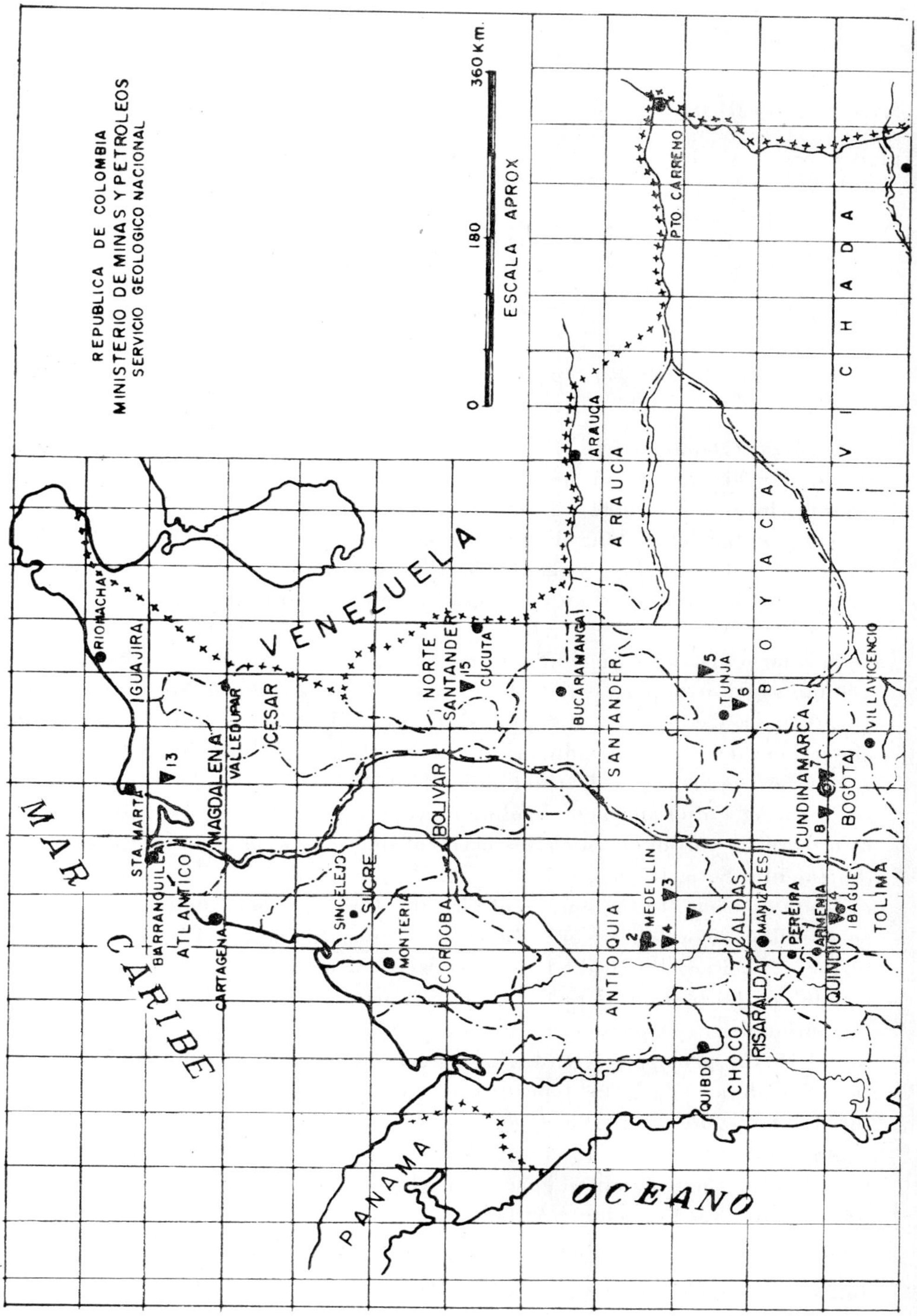
REPUBLICA DE COLOMBIA
MINISTERIO DE MINAS Y PETROLEOS
SERVICIO GEOLOGICO NACIONAL
0
180
360 Km.
ESCALA APROX
MAR CARIBE
OCEANO
VENEZUELA
PANAMA
RIOHACHA
GUAJIRA
STA. MARTA
BARRANQUILLA
ATLANTICO
CARTAGENA
MAGDALENA
VALLEDUPAR
CESAR
SINCELEJO
SUCRE
MONTERIA
CORDOBA
BOLIVAR
NORTE SANTANDER
CUCUTA
BUCARAMANGA
SANTANDER
ARAUCA
ANTIOQUIA
MEDELLIN
CHOCO
QUIBDO
CALDAS
MANIZALES
RISARALDA
PEREIRA
ARMENIA
QUINDIO
IBAGUE
TOLIMA
CUNDINAMARCA
BOGOTA
BOYACA
TUNJA
VILLAVICENCIO
VICHADA
PTO. CARREÑO
OBANDO

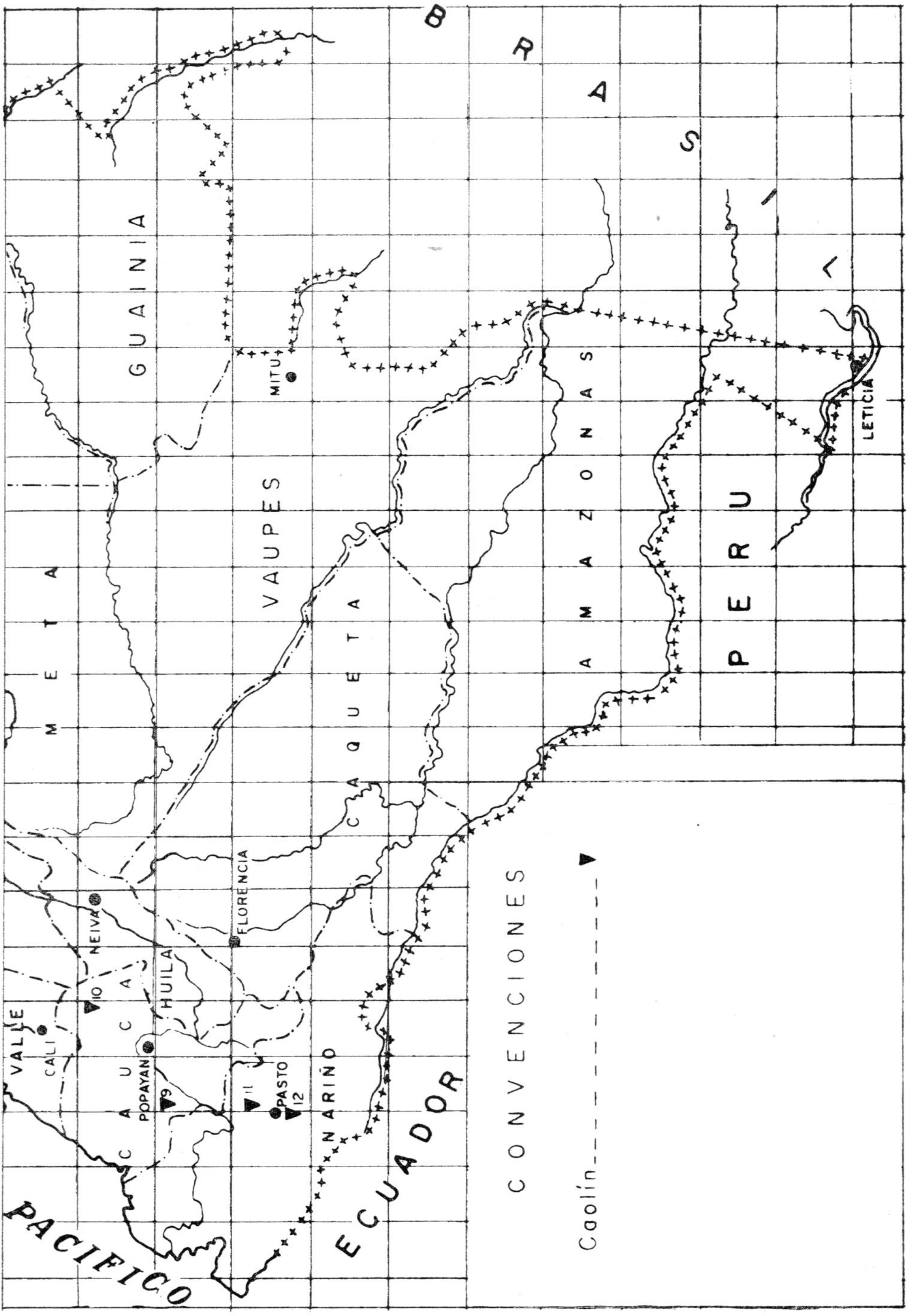
BRASIL
GUAINIA
META
VAUPES
MITU
CAQUETA
AMAZONAS
PERU
LETICIA
FLORENCIA
NEIVA
HUILA
VALLE
CALI
CAUCA
POPAYAN
NARIÑO
PASTO
9
10
11
12
ECUADOR
PACIFICO
CONVENCIONES
Caolín

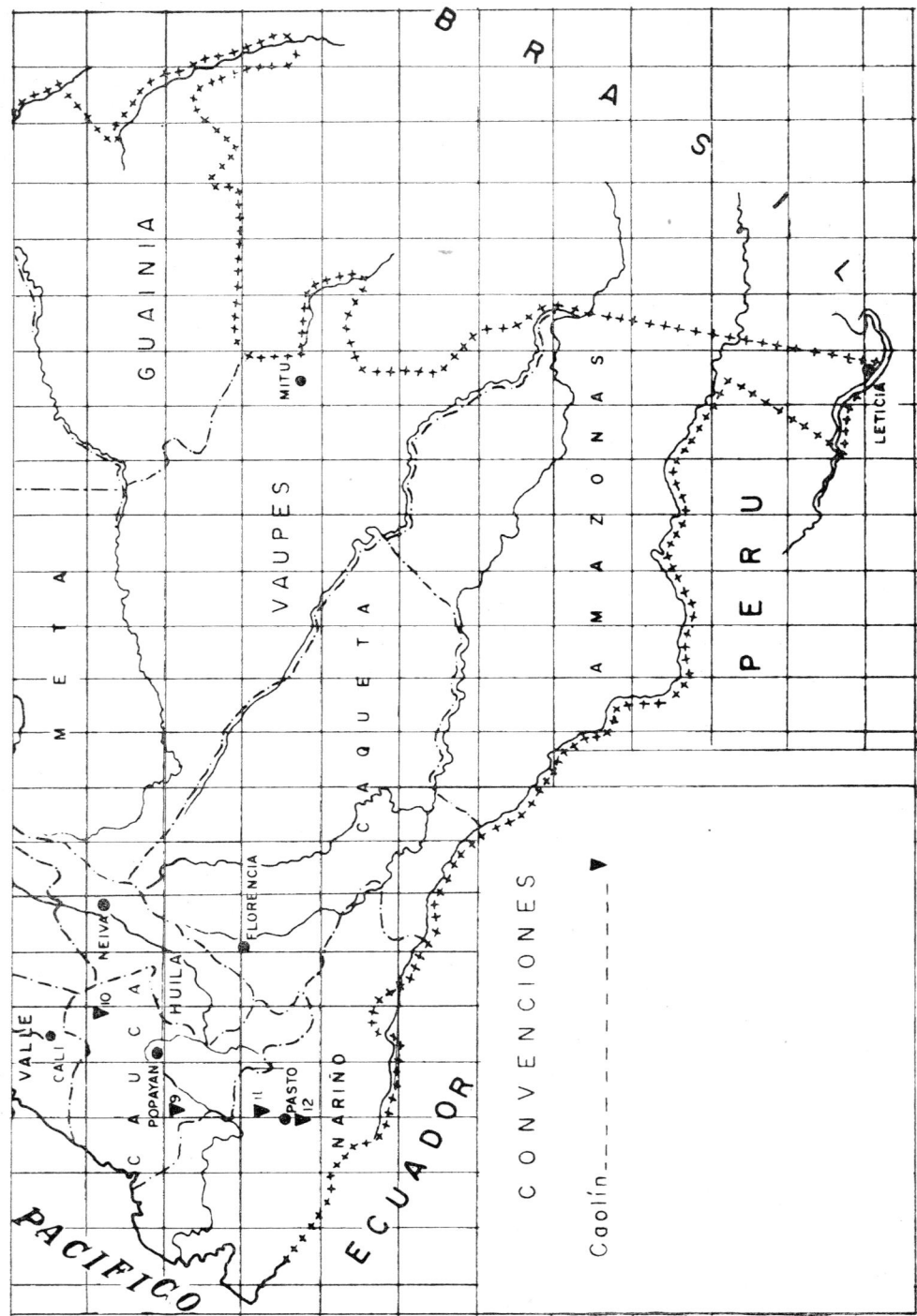

9—Cauca Department—A deposit of almost pure kaolin suitable for the manufacture of porcelain exists at the place named "Tierras Blancas" in the Guavito Creek tributary of "Guschicano River". This deposit was found by Dr. Grosse. (Compilación de Estudios Geologicos en Colombia, pp. 227—229).

10—In Santander the Quilichao in the northern part of the Cauca Department are kaolins of residual type. (Verbal communication of the geologist Ignacio Cucalón).

11—Nariño Department—San Lorenzo district.—In "La Caratosa" hill exists residual kaolin in large amounts.

12—Pasto District—In the Páramo of "Tescual" a kaolin deposit resulting from volcanic tuffs or alteration of a porphyritic dike is found (Report 313, Royo y Gómez, S. G. N.).

13—Magdalena Department—In the "San Pedro de la Sierra" and "Cuatro Caminos" districts a primary deposit is found in the alteration zone of granites (Report 1399, S. G. N.)

According to the Mineral Resources Survey, besides this deposit there is another occurrence near the Sevilla locality.

14—Tolima Department—In "Iman" hill, there is residual clay and NW of the hamlet of "Salado", Ibagué, there is also a deposit of residual kaolin (Dr. Dario Suescún Gómez, Report No. 872, S. G. N.)

15—N. of Santander Department—In Ocaña district in the place named "La Labranza" there is a deposit of residual kaolin. (Carlos Cáceres and Inra Megyesin, Report No. 1524, S. G. N.)

There are some other kaolin outcrops in Colombia, but owing to their insignificance they are not worth to be mentioned.

Kaolin Deposits of Jamaica

H. R. VERSEY
Jamaica

There are only three significant deposits of kaolinitic clay known in Jamaica one of which exceeds several million tons. They each have a different mode of origin the largest being alluvial, the second largest residual and the smallest hydrothermal. Studies from different technological standpoints have been made on each but none is as yet exploited. Their location is shown on the accompanying map (Fig. 1).

The Black River deposits

In and around the morasses in the lower reaches of the Black River, the alluvium consists of a mixture of kaolin and quartz sand. The parent rocks of this material are the Cretaceous tuffs in the western end of the island's central inlier. Some of this alluvium is represented by terrace levels but the major part is with-

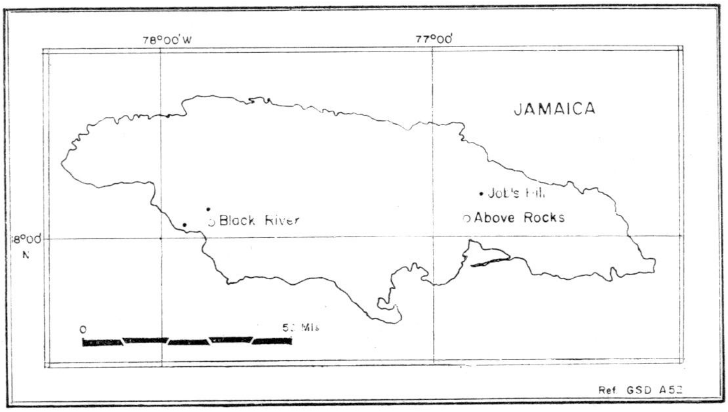

Fig. 1—Location of the kaolin deposits of Jamaica

in the present flood plain of the river. In this latter situation it appears that the swampy and sulphurous conditions have allowed the removal of much of the iron oxide which discolours the clay in the upper terrace levels and in the clay deposits in higher reaches of the river.

The Holland deposit to the north of the Black River has been prospected and in two selected areas a total of 200,000 tons (bulk) has been proven. The whole deposit must be considerably bigger. Ceramic testing showed a satisfactory plasticity which was improved by grinding (necessary on account of the silica content) and a good firing colour. Water absorption of the fired products is high but reduces with the addition of normal fluxes. The clay was thought quite suitable for off-white earthenware.

The Frenchman deposit southeast of the river opposite Holland is similar in character but more consistently light-coloured. It has been estimated that the deposit contains $2^1/_2$ million tons (bulk) or 1 million tons of clay. Chemical analysis showed that the 10 micron fraction contained a silica content of 48 % as against 60 % for the bulk material, representing roughly a 75 % reduction in the free quartz content. An X-ray diffraction pattern showed the clay fraction to be a disordered kaolinite. It has been shown that the iron and titania are contained in the quartz fraction and that removal of most of this fraction leaves a clay purer and whiter than many of the conventional ball clays or chine clays currently marketed. This benefication cannot be done by sieving however as the significant grain size is too fine. Hydro-cycloning has been used in preliminary testing and the economics of its commercial use is now being investigated.

The Cowmarket deposit lies about two miles west of the mouth of the Black River. It appears to be of the same type as the Holland and Frenchman clay and is thought to be extensive. No tests have been carried out however. It is overlain by the major silica sand deposit of the island which is currently being mined.

The Above Rocks deposit

Weathering of the adamellitic rocks around the village of Above Rocks has produced extensive residual clay deposits. The weathered material consists of clay together with quartz and feldspar sand, and is soft enough to be easily dug by hand. The clay content varies between 9% and 58 % by weight with the average grade between 20 % and 25 %. When separated by washing, the clay is light buff in colour. This weathered material has been proved to a depth of 20 feet in trenches, and within a circle of one mile radius centred in Above Rocks village it covers a surface area of a million square yards. If the top 15 feet of this material were to be extracted and refined by waterwashing it would yield a million tons of light buff clay suitable for ceramic use.

Within the same area small deposits of a very high quality white china clay occur at Long Coconut (G. R. 59534335) and at Golden River (G. R. 59174381) as a result of the weathering of dykes of leucocratic adamellite. This material consists of brilliant white clay with grains of quartz and white feldspar. The clay content ranges from 29 % to 50 % by weight with an average grade of about 35 %. It is estimated that refinement by waterwashing would yield about 2000 long tons of the pure white clay.

A similar white clay of Harkers Hall would probably yield a further 500—1000 tons after water-washing.

Jobs Hill dickite

This deposit is best exposed to the northeast of the bridle road from Jobs Hill to Mount Cheerful. It has the form of a wall rock to a large granodiorite intrusion and is thought to be product of hydrothermal alteration of the Wagwater formation, a series of red sands and conglomerates, which consitutes the country rock. Brecciation and widespread slickensiding suggest the deposit is faulted, probably against the granodiorite. The outcrop is a wedge some 200 feet long by 50 feet wide.

The clay is hard and compact, white when pure but grading to purple or green where the alteration has been incomplete. The available reserves have been estimated at 7,000 tons.

No tests have been carried out on this material.

References

In compiling this note I have drawn on material in the field of the Jamaica Industrial Development Co., Development Finance Corporation, and the Geological Survey Department.

1969 XXIII INTERNATIONAL GEOLOGICAL CONGRESS (1968) — VOL. 16 PAGE 105—110

Kaolin Deposits of Mexico

R. PASQUERA, L. DE PABLO and M. CARBONELL
Mexico

The largest part of the Mexican kaolin deposits are of the primary type which originated by hydrothermal alteration of Tertiary rhyolitic rocks. Exploitation is usually effected in open pits; in the country there are large reserves, available amounting to more than 500,000,000 tons. There are no deposits of ballclays or china clays for paper industry, so that 14,000 tons are to be annually imported.

Kaolins can be found in almost every state of the country. The main producing areas are indicated in the enclosed table and schematic map.

The kaolin deposits of Mexico are associated mainly with volcanism or with ore veins. They occur also in karst areas where pyroclastic material had filled cavities, and in a lesser degree in sedimentary basins. Volcanism has been the most usual process, during the final stages of which kaolinization had taken place; a good many deposits of this type are found in the central part of the country, following the so called Neo-volcanic axis or Clarion fault, with numerous geysers and high hydrothermal activity.

The genesis includes (1) hydrothermal alteration of Tertiary rhyolites and rhyolitic tuffs, (2) weathering of rhyolitic tuffs, and (3) sedimentary deposits. Parent rocks are usually rhyolitic and the overburden are usually recent soils, Quaternary basalts, or Tertiary rhyolites. Most deposits are primary, in situ, and a few are sedimentary.

Kaolinization varies in depth, form and mineralogy. In clearly hydrothermal deposits, the kaolins are whiter and cleaner than those originated by weathering and are also more variable, due to the selectivity of the hydrothermal process and the nature of the solutions. Variations in chemical composition, physical characteristics, and mineralogy are common. The kaolins contain frequent chalcedony, quartz, fluorspar, anhydrite, gypsum, zeolites, and sulphur. Alunite is also commonly associated with clays in the northern part of the country and chlorite, montmorillonite, and minerals with mixed-layer structures are quite often found in hydrothermal deposits.

These deposits, because of their primary nature, are controlled by tectonics. Factors as faults, subsidence etc., affect their economy and exploitability.

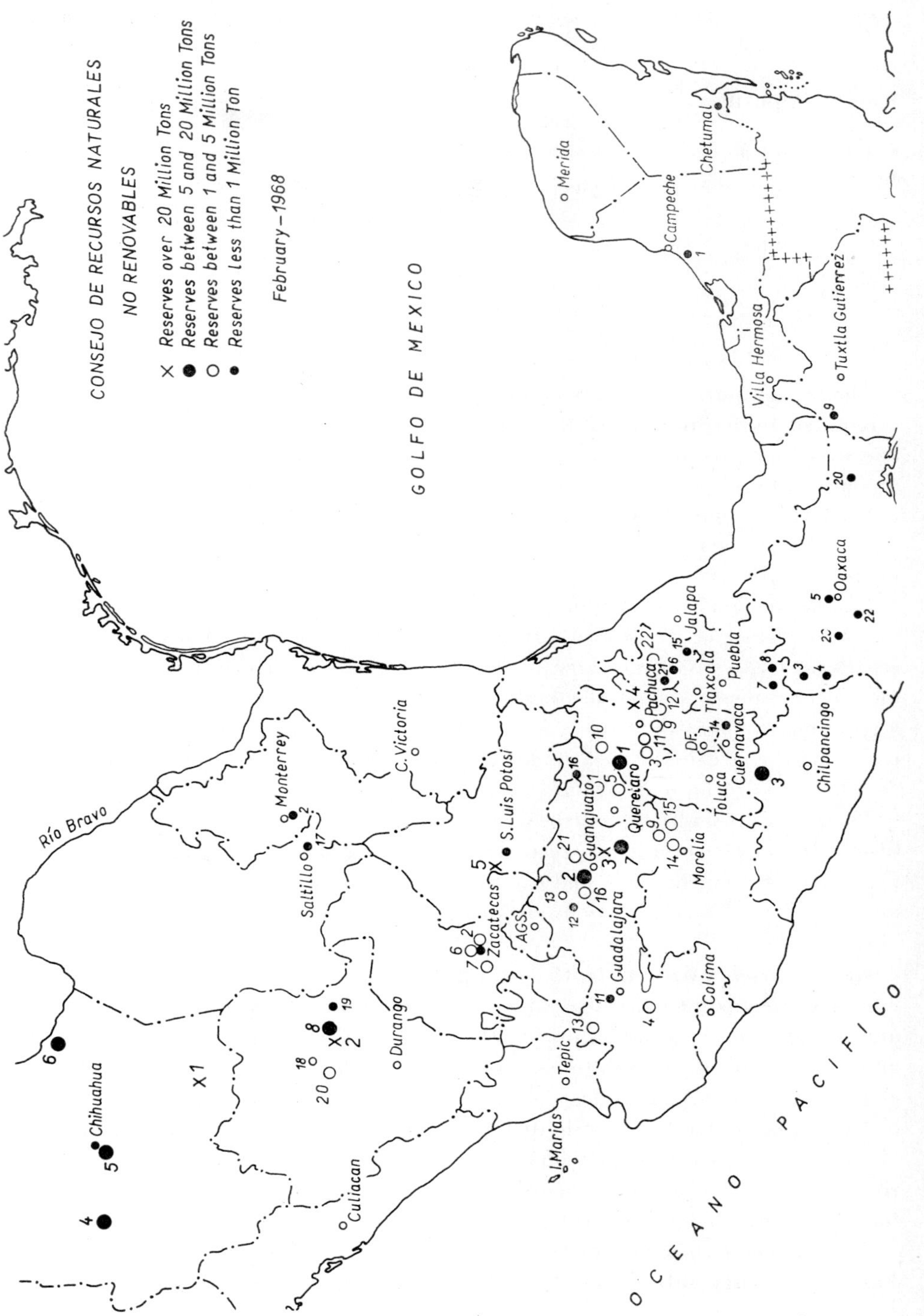
CONSEJO DE RECURSOS NATURALES
NO RENOVABLES
Reserves over 20 Million Tons
Reserves between 5 and 20 Million Tons
Reserves between 1 and 5 Million Tons
Reserves less than 1 Million Ton
February – 1968
GOLFO DE MEXICO
OCEANO PACIFICO
Río Bravo
Chihuahua
Monterrey
Saltillo
C. Victoria
Durango
Culiacan
Zacatecas
S. Luis Potosí
AGS.
Guanajuato
Querétaro
Guadalajara
Tepic
I. Marías
Colima
Morelia
Toluca
D.F.
Cuernavaca
Pachuca
Tlaxcala
Puebla
Jalapa
Chilpancingo
Oaxaca
Villa Hermosa
Tuxtla Gutiérrez
Campeche
Merida
Chetumal

The main mineral constituent is kaolinite, associated with quartz, chalcedony, alunite, gypsum, anhydrite, rutile, and others. Usually, clays utilizable in industry have over 90 % of clay substance, less than 15 microns in size, kaolinite from thermal deposits is usually of a disordered structure and often exhibits tubular form of crystals. The association with montmorillonite, endellite, and mineral with mixed-layer structure is common.

Local kaolins, generally being from primary, in situ, deposits, are coarse-grained, with a low content of clay substance less than 5 microns. Plastic kaolins or ballclays for ceramic use and the white fine kaolins for the paper industry are not common.

The standard methods of mineralogical analysis have been used in the study of clay minerals, such as X-ray diffraction and fluorescence, electron microscopy and diffraction, thermal analysis, sedimentation, emission spectrography, etc. Chemical analysis of the clay substance usually indicates 45—49 % SiO_2, 35 to 41.5 % Al_2O_3, less than 0.4 % TiO_2, less than 2.5 % Fe_2O_3, alkaline oxides less

---

Fig. 1—Kaolin deposits of Mexico

× 1 Jiménez, Chihuahua
2 Peñón Blanco, Durango
3 Juventino Rosas, Guanajuato
4 Guayacocotla, Veracruz
5 Ahualulco, San Luis Potosí

● 1 Pathé, Hidalgo
2 La Luz, Guanajuato
3 Iguala, Guerrero
4 Guerrero, Chihuahua
5 Chihuahua, Chihuahua
6 Ojinaga, Chihuahua
7 Celaya, Guanajuato
8 Yervanis, Durango

○ 1 Cadereyta, Querétaro
2 Guadalupe, Zacatecas
3 Tula de Allende, Hidalgo
4 Zacualco, Jalisco
5 Tequisquiapan, Querétaro
6 La Bufa, Zacatecas
7 Villa García, Zacatecas
8 San Felipe, Guanajuato
9 Acámbaro, Guanajuato
10 Zimapán, Hidalgo
11 San Agustín, Tlaxiaco, Hidalgo
12 Singuilucan, Hidalgo
13 Magdalena, Jalisco
14 Queréndaro, Michoacán
15 Hidalgo, Michoacán
16 León, Guanajuato
17 Salamanca, Guanajuato
18 Salvatierra, Guanajuato
19 Pachuca, Hidalgo
20 Sapiorís, Durango
21 Dolores Hidalgo, Guanajuato
22 Zapotitlán, Puebla

• 1 San José Champotón, Campeche
2 Región de Monterrey, Nuevo León
3 San Martín Zacatepec, Oaxaca
4 Mariscala de Iturbide, Oaxaca
5 Región de Oaxaca, Oaxaca
6 Ixtacamaxtitlán, Puebla
7 Acatlán, Puebla
8 Petlancingo, Puebla
9 La Razón-Cintalapa, Chiapas
10 Huichapan-Salitrera, Hidalgo
11 Zapopan, Jalisco
12 Lagos, Jalisco
13 Comanja, Jalisco
14 Cuautla, Morelos
15 Libres, Puebla
16 Río Blanco, Querétaro
17 Sierra de Saltillo, Coahuila
18 El Rodeo, Durango
19 Cuencamé, Durango
20 Santa María Chimalpa, Oaxaca
21 Chignahuapan, Puebla
22 San Mateo Mixtepec, Oaxaca
23 Magdalena Peñasco, Oaxaca

than 0.2 %, and loss on ignition 10—14 %. Trace elements are variable depending on the origin, Fe, Ti, Mn, Pb, Zn, Hg, Cu, Ca, S, F, are often detected. The chemical and mineralogical characteristics of the parent rock can be considered as those typical of rhyolitic rocks.

Kaolins are used in the paper, ceramic, refractories, rubber, chemical, and pharmaceutical industries, in accordance to their specific properties as well as to their locations. In Mexico, kaolin is a low priced commodity which does not usually permit processing of high transportation costs.

In a very general manner, it can be stated that kaolins from the central part of the country (states of Guanajuato, Querétaro, and Michoacan), white and with a low content of accessory minerals are preferred in the ceramic, refractories, and chemical industries. Finer clays often associated with metallic mineralizations, from Puebla and Hidalgo, are used in the paper industry. Those from San Luis Potosi, Zacatecas, and Guerrero, with some iron oxide, are used by the refractories and ceramic industries while those from Chihuahua, although of excellent quality, can be used only for special applications because of their high freight costs.

Washing plants for kaolin do not exist in Mexico. Some local kaolins contain 0 % residue on the 0.6 mm screen; compression strength after drying is high for some semiplastic kaolins from Michoacan and Puebla, and low for those from San Luis Potosi and Zacatecas; shrinkage varies from 8 to 12 %; porosity at 1410 °C is from 6 to 20 %, and refractoriness varies from 30 to 38 SC.

Grinding plants, usually equipped with hammer mills and air separators (pneumatic classification) are commonly distributed.

Kaolin reserves are difficult to estimate but a figure higher than 500,000,000 tons (including proven, probable, and possible reserves) can be stated. In a country where rhyolites are common, and volcanism and hydrothermal activity abundant, a wide distribution of kaolins may be expected. With new and faster developments in communications, new deposits are constantly reported.

Exploitation of kaolin deposits has been effected in open pits, except two deposits, in San Luis Potosi and Hidalgo. Thickness less than 1 meter, impurities higher than 15 % and remote consuming centers are the most common prohibitive factors.

The exploitation and mining is affected by factors such as freight, long distances, small and specialized markets and, frequently, variations in the mineralogy of the deposits.

	1962	1963	1964	1965	1966
Production (in tons)	56,000	46,000	64,300	81,200	96,600
Exportation (in tons)	5	4	10	3	29
Importation (in tons)	8,500	10,100	9,300	9,100	13,500
Consumption (in tons)	64,495	56,696	73,590	90,297	110,071

In Mexico, the exploitation of clays started long before the Spanish colonization and ceramic products of the various Indian Pre-Hispanic cultures are abundant. Kaolins and bentonitic clays from the vicinity of the Central Valley of Mexico were used by the Aztecs whereas those from Tula, Hidalgo and Michoacan were used respectively by Olmecs and Tarascos.

Black ceramic products, traditionally produced in Oaxaca, are made from local kaolinitic and bentonitic clays high in colouring oxides. In Jalisco—an important artistic ceramic center—kaolins are traditionally mixed with bentonites to increase their plasticity.

Deposits located on the Neo-volcanic axis, in the central part of the country, were developed in a more or less constant manner from the second half of the last century.

El Có and other mines in Hidalgo, Querétaro, and Guanajuato have supplied the local ceramic industry over a long period of time.

More recently, during the last 30 years, numerous mines have been opened in the states of San Luis Potosi, Zacatecas, Chihuahua, Durango, Guerrero, Puebla, Guanajuato, Querétaro, Hidalgo, and Michoacan which, with the exception of plastic ballclays and china clays for paper production, supply wholly the requirements of the national industry.

References

Aguilera J. (1898): Catálogo sistemático y geográfico de las especies mineralógicas.—Inst. Geol. Bol. II, pp. 104—105.

— (1907): Los Caolines de la Hacienda de Yextho, Hgo., Bol. Soc. Geol. Mex. III, pp. 25—33.

Aguilera, N. (1958): Arcillas en algunos sedimentos calcáreos de Campeche. — Publ. Cerám. I (1), pp. 31—34.

— (1959): Clays from some soils and calcareous sediments from the Yucatán Península, Cong. Geol. Int.—XX. Sesión, pp. 61—69.

Arellano, A. R. V., (1958): Las minas de Sta. Rosa y Anexas (Pathé, Tecozautla, Hgo.) Pub. Ceram. I (1), pp. 45—50.

— (1961): Distribución geográfica y geológica de ciertas materias primas para la industria cerámica. Pub. Ceram. I (2), pp. 105—118.

Barrera, T. (1930): Las arcillas y la fabricación de loza en Oaxaca. Inst. Geol. Anales IV, pp. 99—126.

Blasquez, L. y Lozano, R., (1946): Hidrología y minerales no metálicos del Edo. de Tlaxcala. Inst. Geol. Anales VIII, pp. 55—99.

— (1946): Hidrogeología y minerales no metálicos de la zona norte del Edo. de Michoacán.—Inst. Geol. Anales IX, pp. 59—76.

Cabrera A. et Caillère S. (1966): Etude Minéralogique de quelques argiles provenant du Mexique. Bulletin de la groupe française des argiles, Tome XVIII, Nouvelle Serie No. 14.

Castro, Carlos (1909): Análisis de una caolinita de una muestra de carbón de Villafuerte, Coah., Bol. Soc. Geol. Mex. V, 10.

Cummings, J. L. (1930): Arcillas, arenas, gravas y yeso en una Somarca septentrional del Edo. de Coahuila.—Inst. de Geol. Anales IV, pp. 81—84.

Delgado, J. (1962): Las arcillas refractarias utilizadas para la industria nacional.—Geol. y Met S. L. O. I (1), pp. 11—14.

De la Peña, L. (1959): Recursos Naturales No Renovables de México.—Min. y Met. 11, pp. 79—80.

De Pablo, L. (1958): Estudio de Arcillas.—Publ. Ceram. I (1), pp. 20—30.

— (1965): Caolinita Desordenada de Concepción de Buenos Aires, Jalisco. Inst. Geol. Bol. 76, pp. 39—69.

— (1962): Investigación de la Arcilla Pathé Min. y Met. 22, pp. 39—63.

— (1967): Estructura y Morfología de caolinitas desordenadas.—Int. Min. Cong. Canadá, 1967.

— (1965): Disordered kaolinite from Conceptión de Buenos Aires, Clays and clay minerals, pp. 143—150.

— (1956): Estudio de Mineralogía, parte 2a. Caolinita de Estructura desordenada de Concepción de Buenos Aires, Estado de Jalisco.—México. Universidad Nacional Autónoma de México.—Inst. de Geol.

— Las Arcillas (1964): Clasificación, uso y especificaciones industriales.—Soc. Geol. Mex. Tomo XXVII, No. 2.—pp. 449—42. Fig. 1, Tablas 1—9.—México.

Esquivel, J. y Zamora S. (1958): Informe sobre minerales no metálicos. Consejo de Recursos Naturales No Renovables. Bol. 44, pp. 1—152.

Fetter, H. y Velo, L. (1961): Análisis térmico de la arcilla Pathé.—Publ. Ceram. I (2), pp. 147—149.

Flores, T. (1954): Los Recursos Minerales No Metálicos en México.—Memorias de la Primera Convención Latinoamericana de Recursos Minerales, pp. 325—348.

— (1936): Algunos silicatos notables del Distrito Norte de la Península de Baja California.—Ingeniería 10, (1), pp. 5.

— (1951).—Memorias de la Primera Convención Interamericana de Recursos Minerales.—Los Recursos de Minerales No Metálicos en México.—Memoria del Instituto Nacional para la Investigación de recursos minerales, México.

Gonzáles, J. (1947): Riqueza Minera y Yacimientos Minerales de México.—Monografías del Banco de México, S. A.

Hernández, Alberto (1963): Minerales No Metálicos Mexicanos, Instituto Mexicano de Investigaciones Tecnológicas, 2, pp. 1—43.

— (1960): Minerales No Metálicos Mexicanos, Inst. Mex. de Inv. Tec. 1 (I—III).

— (1962): Minerales No Metálicos Mexicanos. Inst. Mex. de Inv. Tec. 60 (20), pp. 22—80.

Hernández, V. (1960): Ariel.— Minerales No Metálicos Mexicanos, Vol. I. Inst. Mex. de Inv. Tec. México 1960.

Martín del Campo, R. (1966): Estudio de una arcilla con alto contenido de alúmina, en el Mpio. de Guadalcazar, S. L. P. Geol. y Met. III (18), pp. 35—64.

Muñoz, G. M. (1961): Exploración de caolín en el municipio de Iguala, — Gro., Tesis. Fac. Ing., pp. 1—84.

Orozco, R. (1934): El Caolín en México.— Revista Industrial 2, (1), pp. 91—100.

Pesquera, R. (1959): Arcillas con alto contenido de alúmina de la región de Núñez, S. L. P. Min. y Met. 9, pp. 69—82.

Rodríquez, J. (1957): Estudios de Geología Económica. Inst. Geol. Anales XIII. 15.

Rubio, D. (1964): Plasticidad y moldeabilidad de algunas arcillas mexicanas. Bol. Asoc. Ing. Quim y Met. Inst. Pol. Nac. I (2), pp. 25—28.

Santillán, M. (1930): Arcillas y arenas de Cerro Blanco, Tlaxcala y sus alrededores. Inst. Geol. Anales IV, pp. 85—95 .

Schmitter, E. (1959): Clasificación basada en el análisis térmico diferencial de materiales arcillosos colectados en diferentes regiones del país. Inst. Geol. Bol. 59, pp. 105—111.

Schulze, C. (1960): Informe geológico sobre algunos yacimientos de caolín situados en las vecindades de Celaya y Guanajuato, Min. y Met. 14, pp. 39—51.

Ugalde, H. (1961): Observaciones geológicas al yacimiento alumínico de Nuñez, S. L. P. Soc. Geol. Mex. 24 (2), pp. 21—28.

Informacion acerca de Existencia de Caolin en Venezuela

MINISTERIO DE MINAS E HIDROCARBUROS, DIRECCION DE GEOLOGIA, DIVISION DE GEOLOGIA ECONOMICA
Venezuela

	Datos geográficos	Condic. geológicas	Condic. mineralógicas y geoquímicas	Eval. económica	Bibliografía
Cerro Copeyal	Upata, Edo. Bolívar. No desarrollado. Depósito pequeño.	Forma lenticular Area: 15.000 m^2. Espesor: ± 18 m. Otras condiciones prácticamente desconocidas.	Composición química promedio: SiO_2: 40.8 % Fe_2O_3: 2.9 % Al_2O_3: 36.6 %	Probable ± 500.000 T. M.	Candiales y Ascanio (1954); 17-1-54. 553.5V1 C. J.
Cerro La Bandera	Upata, Edo. Bolívar. No desarrollado. Condiciones favorables aparentes.		?	Probable ± 7.200.000 T. M.	Candiales y Ascanio (1954); 17-1-54. 553.5V1 C. J.
Isla La Orchila	Desconocidos	Desconocidas	Promedio: SiO_2: 59 % Al_2O_3: 29 % Fe_2O_3: 2 % CaO: Tr. MgO: Tr.		Schwarck A. (1950); 27-4-50. 553.612
Tinaquillo	Edo. Cojedes. No desarrollado. Desfavorable.	Desconocidas	Promedio: SiO_2: 71 % Al_2O_3: 19 % Fe_2O_3: 1 %		Schwarck A. (1950); 27-4-50. 553.612
Hacienda La Floresta	Tinaquillo Edo. Cojedes. No desarrollado. Muy poca importancia.	Forma irregular. Area: 6.000 m^2. Espesor: 1.0 m. Meteorización superficial de aplita en gneises hornabléndicos feldespáticos.	Promedio: SiO_2: 70 % Fe_2O_3: 1.9 % Al_2O_3: 19.7 % CaO: 0.6 % MgO: Tr. Análisis granulométrico granogrueso.	Parte central del depósito 2.900 T. M. (eliminado 40 % de porosidad absoluta).	Bellizzia A. (1956); 10-9-56. 553. 6 V 10
Hato La Margarita	Edo. Bolívar Dtto. Heres. No desarrollado. Condiciones favorables.	Depósito residual, alteración de rocas feldespáticas espesor aproximado: 2 m.	Caolín bruto: SiO_2: 65.7 % Al_2O_3: 23.8 % Fe_2O_3: 0.9 % Caolín lavado: SiO_2: 57.4 % Al_2O_3: 31.1 % Fe_2O_3: 0.4 %		Fernández de Caleya y Freile A. (1949); 25-1-49. 553.612 F. 363.
Rio Casiquiare	T. Amazonas. No desarrollado. Condiciones favorables aparentes.	Desconocidas.	Promedio: SiO_2: 48.15 % Al_2O_3: 33.12 % Fe_2O_3: 2.78 %		Sellier de Civrieux (1950); 18-4-50. 553.612 S. 467

Nota—Posibles depósitos de los cuales hay referendias someras en el informe sobre depósitos de Caolín existentes en el territorio nacional. Schwarck A. (1950) 27-4-50. Baruta, Edo. Miranda; Valera, Edo. Trujillo; Gran Sabana, Edo. Bolívar; Moitaco, Edo. Bolívar; Santa Rosa-Sur de Upata, Edo. Bolívar.

AUSTRALIA

1969 XXIII INTERNATIONAL GEOLOGICAL CONGRESS (1968) — VOL. 16 PAGE 115—150

Kaolin Deposits of Australia

A. J. GASKIN
Australia

Summary—Residual kaolins of sufficient purity to be of some industrial significance occur throughout many parts of Australia. Over 100 separate occurrences have been sampled but more than half of these have not been investigated in any detail, as there is little economic incentive to exploit deposits of average grade material situated at any distance from the local markets in the main coastal cities.

Geographical and geological data are given for all residual kaolin deposits known or recorded to date. A mode of origin for each is suggested, but conclusive evidence on this topic is available for only a few occurrences. Most deposits appear to be the result of leaching at the base of lateritic profiles or in deep weathering zones. Few can be linked with hydrothermal activity.

Occurrences are grouped according to the type of parent rock. The groups of kaolinised granitic and metamorphic rocks include some large bodies but the extent of many of the deposits of this type is not known with any precision. Some of the kaolinised sedimentary bodies are large enough and of sufficiently high grade to be industrially significant. The group of kaolinised dyke rocks includes most of the deposits that have been exploited as sources of high grade grit-free kaolinite for Australian consumption.

Local production has been mainly limited either to extraction of siliceous and aluminous fire-clays or to the preparation of filler clays and ceramic clays by dry-milling materials essentially in a condition as mined. Washed kaolins supply a small but increasing proportion of the local market and will become more important as reserves in accessible high-grade dyke deposits are exhausted.

The mineralogy of most of the deposits has not been studied in detail, except when specific matters of industrial or academic interest have been raised. Kaolinite is the dominant mineral, generally not very well crystallised and often cemented into compact masses by traces of material deposited under arid climatic conditions. In parts of certain deposits where kaolinisation has not been complete, illite is the most common minor component. In others traces of halloysite occur.

The relatively undeveloped state of most of the Australian deposits is largely a consequence of a combination of small local markets and large distances. Overland haulage costs are high and imported kaolins reach most points of consumption by sea transport as cheaply as local products brought from inland deposits.

Geographical data

Most of the known occurrences of residual kaolins are in the southern and eastern regions of Australia, but this may be only an effect of the greater degree

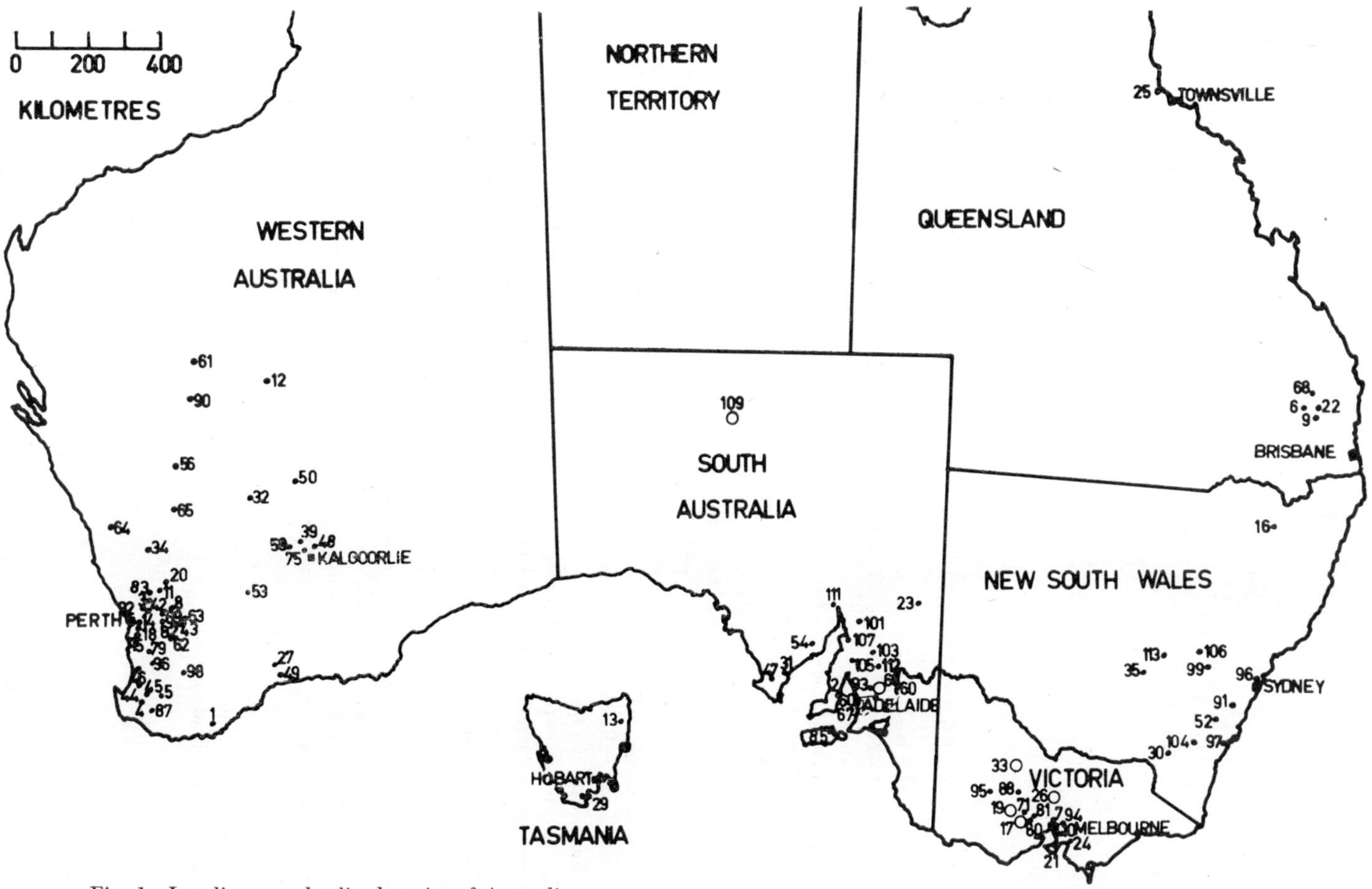

Fig. 1—Locality map kaolin deposits of Australia

of attention that has been given to exploration and sampling of deposits in these more populated areas.

The following list includes all recorded deposits which have or may have genetic or industrial significance. Australian literature on field geology and mining contains many other references to bodies of white and pale-coloured kaolin but insufficient data are given for any specific inferences to be drawn. A collection of typical references is given in the "Queensland Mineral Index" (Dunstan, Q. Geol. Surv. Publication, No. 241), published in 1913.

Deposits are listed, according to the type of parent rock, in four groups. This is not a precise system of classification because it is not always known, especially in the Pre-Cambrian systems of Western Australia and South Australia, whether parent rocks had been metamorphosed before kaolinisation, or whether the bodies were dykes, sills or flows. However, any other system of grouping the occurrences leads to an even less useful and reliable result. The parent rock classification at least gives an impression of the probable shape and extent of the deposits and generally indicates the type of mineral association to be expected.

The State in which the locality occurs is given by the notation:—W. A. (Western Australia); S. A. (South Australia); Vic. (Victoria); N. S. W. (New South Wales); Qld. (Queensland); Tas. (Tasmania). There are no kaolin deposits recorded from the Northern Territory.

Most of the deposits identified on the locality map are marked with small dots. This more often indicates unknown reserves than small reserves.

Group I: Kaolinised granites and related igneous rocks

(1) Albany (W. A.)
(2) Ardrossan (S. A.)
(3) Bolgart (W. A.)
(4) Bibilup (W. A.)
(5) Boyup Brook (W. A.)
(6) Brooklands (Qld.)
(7) Bulla (Vic.)
(8) Cunderdin (W. A.)
(9) Cooyar (Qld.)
(10) Dandenong (Voc.)
(11) Dowerin (W. A.)
(12) Duketon (W. A.)
(13) Fern Tree River (Tas.)
(14) Glen Forrest (W. A.)
(15) Hallam (Vic.)
(16) Inverell (N. S. W.)
(17) Lal Lal (Vic.)
(18) Lesmurdie (W. A.)
(19) Linton (Vic.)
(20) Minnievale (W. A.)
(21) Moorooduc (Vic.)
(22) Nanango (Qld.)
(23) Olary (S. A.)
(24) Pakenham (Vic.)
(25) Paluma (Qld.)
(26) Pyalong (Vic.)
(27) Ravensthorpe (W. A.)
(28) Roe (W. A.)
(29) Surges Bay (Tas.)
(30) Tumbarumba (N. S. W.)
(31) Tumby Bay (S. A.)
(32) Wagine Soak (W. A.)
(33) Wedderburn (Vic.)
(34) Wubin (W. A.)
(35) Wyalong (N. S. W.)
(36) Yarloop (W. A.)

Group II: Kaolinised metamorphic rocks

(37) Arthur River (W. A.)
(38) Baker's Hill (W. A.)
(39) Bardoc (W. A.)
(40) Boddington (W. A.)
(41) Calcarra (W. A.)
(42) Clackline (W. A.)
(43) Dangin (W. A.)
(44) Donnybrook (W. A.)
(45) Greenbushes (W. A.)
(46) Harvey (W. A.)
(47) Kapinnie (S. A.)
(48) Kanowna (W. A.)
(49) Kundip (W. A.)
(50) Leonora (W. A.)
(51) Lobethal (S. A.)
(52) Marulan (N. S. W.)
(53) Marvel Loch (W. A.)
(54) Miltalie (S. A.)
(55) Mount Helena (W. A.)
(56) Mount Magnet (W. A.)
(57) One Tree Hill (S. A.)
(58) Mount Maiden (W. A.)
(59) Ora Banda (W. A.)
(60) Paracombe (S. A.)
(61) Peak Hill (W. A.)
(62) Pingelly (W. A.)
(63) Quiarading (W. A.)
(64) Three Springs (W. A.)
(65) Westonia (W. A.)
(66) Williamstown (S. A.)
(67) Woodside (S. A.)
(68) Wondai (Qld.)

Group III: Kaolinised dyke rocks

(69) Balkuling (W. A.)
(70) Ballan (Vic.)
(71) Ballarat (Vic.)
(72) Birdwood (S. A.)
(73) Broad Arrow (W. A.)
(74) Burabadji (W. A.)
(75) Cardup (W. A.)
(76) Dromana (Vic.)
(77) Dunbible (N. S. W.)
(78) Duranillin (W. A.)
(79) Dwellingup (W. A.)
(80) Egerton (Vic.)
(81) Gordon (Vic.)
(82) Jacob's Well (W. A.)
(83) Jimperding (W. A.)
(84) Kalamunda (W. A.)
(85) Kangaroo Island (S. A.)
(86) Kunjin (W. A.)
(87) Manjimup (W. A.)
(88) Maryborough (Vic.)
(89) Meckering (W. A.)
(90) Meekatharra (W. A.)
(91) Mittagong (N. S. W.)
(92) Mokine (W. A.)
(93) Mount Crawford (S. A.)
(94) Ringwood (Vic.)
(95) Stawell (Vic.)
(96) Sydney (N. S. W.)
(97) Ulladulla (N. S. W.)
(98) Wagin (W. A.)

Group IV: Kaolinised sedimentary rocks

(99) Bathurst (N. S. W.)
(100) Birdwood (S. A.)
(101) Booleroo (S. A.)
(102) Buckaroo (N. S. W.)
(103) Burra (S. A.)
(104) Canberra District (Aust. Capital Territory and adjacent areas of N. S. W.)
(105) Clare (S. A.)

(106)	Clergate (N. S. W.)	(110)	Nairne (S. A.)
(107)	Crystal Brook (S. A.)	(111)	Port Augusta (S. A.)
(108)	Houghton (S. A.)	(112)	Robertstown (S. A.)
(109)	Imbitcha (S. A.)	(113)	Tichborne (N. S. W.)

Geological conditions

Few attempts have been made to study the general geological setting of kaolinised bodies in Australia. In the past, hydrothermal activity has often been suggested as the cause of kaolinisation of the igneous rocks and of some of the sedimentary masses. More recently, deep weathering has gained increasing prominence as the main mechanism of alteration.

The common weathering effects associated with kaolinisation appear to be: (a) leaching in pallid zones under laterites, (b) deep weathering in permeable rocks in elevated areas where the water table is far from the surface, (c) intense leaching by acidic ground waters derived from such features as brown coal deposits or oxidising sulphide bodies.

Kaolinised zones below lateritic cappings extend to depths of 30 metres in many parts of Western Australia and South Australia. Most of Australia has been subjected to some degree of lateritic weathering at various times since the beginning of the Tertiary and it is probable that kaolinisation by the action of this weathering process accounts for the origin of certain deposits in eastern and southern States where now only remnants of fossil laterites exist. In Western Australia, where this mode of origin of kaolin deposits is most extensively encountered, large areas of granitic and metamorphic rocks have been altered to depths depending on local factors such as the position of the water table, the topography, the amount of material eroded since kaolinisation and the susceptibility of the rocks to alteration. The expanse of altered granite in the Wubin area is typical. Throughout the southeast of the State there are many examples of the kaolinisation of complex Pre-Cambrian rock associations under the lateritic cappings of the uplifted block on the east side of the Darling Range fault. Typical schist-granitic gneiss assemblages cut by doleritic dykes have produced complex clay associations, the characteristics of the parent rocks being reflected in the grit content, colour and plasticity of the kaolins in the corresponding sections of each deposit. The altered complex in the Clackline district is an example.

In special circumstances of high rock permeability and low water table, kaolinisation caused by weathering processes other than lateritic weathering extends to depths of as much as 200 metres. Dykes injected through porous sedimentary formations are especially prone to such alteration. The kaolinisation of the dolerite dykes in the Sydney area is typical, alteration ceasing at the water table.

In more elevated areas where the water table has been far below the surface at some stage, alteration may extend to great depths, not necessarily correlated with the present day water table position. The kaolinised alkaline dyke at Egerton is one such example.

Examples of kaolinisation by acidic ground water action occur in Victoria, where granodiorites have been altered to depths of 20—30 metres over extensive areas once covered by swamps and brown coal deposits. The Lal Lal deposit is a typical situation. Kaolin bodies derived from the leaching of felspathic rocks by ground waters rendered acid through the oxidation of pyrite are found in various mineralised localities, such as the Kalgoorlie goldfield.

When the criteria of the various modes of origin of kaolins by weathering are applied to deposits throughout Australia, most of the occurrences fit one or other of the patterns mentioned. There is little need to consider hydrothermal activity as a major cause of kaolinisation, nor is there much direct or indirect evidence that it has been a factor in kaolin formation, except in a few instances. One well-defined example is the unusual deposit at Williamstown (S. A.), where a large body of coarse-grained kaolin occurs in the core of a zone of intense leaching and shearing. The kaolinised dykes and sediments in the same region, at Mt. Crawford and Birdwood, should also probably be related to local hydrothermal action, though there is far less supporting evidence available from the mineral associations in these bodies and the surrounding rocks. In New South Wales, deeply altered granitic bodies at Wyalong and Inverell could have been formed by hydrothermal action and there is some evidence of such a mode of origin for the kaolinised granites of the Nanango-Kingaroy district in Queensland.

Mineralogical and geochemical conditions

In all of the residual kaolins recorded in this article, kaolinite is the dominant clay mineral. Incompletely kaolinised sedimentary bodies and basic dykes contain more or less illite and some still show traces of montmorillonite and mixed-layer clay minerals. Small amounts of bauxitic minerals occur in the more intensely leached parts of deposits associated with laterites. Halloysite is rare, but minor amounts have been observed in certain sections of kaolinised alkaline dykes and granitic rocks, generally in fissured material saturated with ground water.

The proportions of clay in the deposits, as defined in terms of material less than 0.02 mm. and 0.002 mm. in effective diameter, can be calculated from the particle size distributions given in references [4], [5] and [6], for some localities, but not for all. Many of the samples used for the determinations in [4] and [5] were spot samples not collected in such a way as to be representative of the de-

posit. Those derived from granites and related acid rocks were on occasion freed from large quartz particles during, or shortly after sampling and the subsequent particle size determinations therefore refer only to arbitrarily separated clay fractions. A further factor complicating the determinations has been the incidence of compacted and lightly cemented kaolin in many of the deposits. Some material from the more arid regions contain enough amorphous silica, alunite or bauxitic material, in the form of cementing films between the kaolinite flakes to make it impractical to disperse these without grinding the sample and thereby altering the size distribution of the clay.

Detailed X-ray diffraction data are given for some New South Wales kaolins in reference [6], together with D. T. A. curves. In [4], [5] and [6], base exchange capacities are recorded. Chemical analyses are recorded in most of the references cited against individual clays.

Accessory minerals are described in [4], [5], [6], [24] and [25] in varying degrees of detail. Quartz is present in all the kaolins except those derived from certain Tertiary basic or alkaline-intermediate dyke rocks in Victoria and New South Wales. The much older basic dyke rocks in the Pre-Cambrian complexes of Western Australia and South Australia have generally been altered, with the development or introduction of free silica taking place, long before kaolinisation. The kaolins derived from these therefore contain at least minor amounts of fine quartz. The kaolinised pegmatites and felspathic sediments in the Mt. Crawford and Birdwood districts of South Australia are almost free of quartz.

Micas, in various stages of decomposition and subdivision form a common class of accessories especially in the kaolins derived from granitic rocks and in the less highly altered sediments. Sericitic mica is a major component of some leached sedimentary bodies in New South Wales and South Australia. Residual felspar is rarely encountered in Australian kaolin deposits, except in transition zones near the parent rocks.

Colouring accessories comprise mainly the various hydrated iron oxides, hematite and fine anatase stained with iron oxide. The pink tints in kaolins are caused by hematite, whilst yellow and brown colours indicate the presence of the hydrated iron oxides. Stained anatase gives a buff colour to kaolin.

Accessories strongly influence the utilisation of kaolin deposits in Australia. With very few washing plants operating, most attention has been given to the exploitation of deposits substantially free of quartz grit, except that clays suitable for siliceous refractories and ceramics manufacture are quarried and used without separation of the quartz content.

Apart from the quartz, iron oxides are the most important accessories controlling utilisation of kaolins. Hematite is very deleterious, as the staining power of this oxide is high and tends to be strongly developed during fine grinding, yielding pink-tinted products rated as particularly low-grade by the selective absorption photometry methods used in acceptance testing by the industries using filler

clays. Since these industries account for the bulk of the Australian kaolin market, there has been a tendency to exploit kaolin deposits containing only the hydrous iron oxides. A large proportion of the local market has in fact been generally satisfied from deposits of transported kaolin, usually having a grey or yellow colour component. This colour advantage has been considered more important than the contingent disadvantage of the proportion of fine quartz usually present in transported clays of the type not referred to elsewhere in this article.

Utilization of kaolins

As mentioned in the preceding sections, less than half of the Australian deposits recorded have been exploited and those that have been worked have produced mainly siliceous refractory and ceramic clays or substantially grit-free kaolins suitable for direct supply to filler-clay consumers after drying, hammer-milling and air-separating oversize particles.

In the production of low and medium duty refractories, kaolinised granitic rocks are the main sources of supply of residual kaolin to the manufacturers in capital cities other than Adelaide, where the local materials suitable for such applications are kaolinised sediments, metamorphic rocks and pegmatites.

For non-plastic components of ceramic bodies residual kaolins derived from basic dykes have been commonly used. The grit-free product from the large alkaline dyke at Egerton (Vic.) supplies a major proportion of this market.

The bulk of the kaolin produced, other than that used for refractories, is consumed by the paint, paper, plastics and rubber industries. Most of this material comes from dykes such as that at Egerton, and is dry-milled without any form of beneficiation. Filler clays for paper have been produced by levigation of kaolinised granitic rocks in Victoria and Tasmania and some coating clay has been made by the same technique. The Lal Lal and Linton deposits are the main sources of such clays, the Linton body having been only recently opened up as reserves of easily accessible clay in the Bulla deposits have declined during the past few years of extraction of material for the supply of levigated paper clay.

Some of the technological characteristics of South Australian and Western Australian kaolins, including reflectivities, plasticities, dry strengths, sorption capacities, drying and firing shrinkages, porosities and fired colours, are given in references [4], [5], [8] and [10]. Raw clay colours and reflectivities are recorded for New South Wales clays in [6]. Firing properties of Queensland clays are given in [7], and for Victorian clays in [9].

Acceptance of a local kaolin for marketing depends largely on whether it can meet the demands of the filler clay consumers to a sufficient extent. These specifications are stricter than those generally used in the ceramic industry, therefore

a filler clay which can be successfully worked and marketed will automatically gain a useful share of the ceramic market. Any kaolin with a reflectivity greater than 75 % of that of MgO and a grit content, as marketed, of 1—5 % by weight of particles larger than 0.05 mm. may gain some share of the total market, but a major share of the market would require a reflectivity of 80 % and less than 1 % of grit particles larger than 0.05 mm. Whilst some dyke deposits meet such requirements, reserves of such material that can be used without washing are limited and the preparation of levigated kaolin is beginning to be a significant factor in the main centres of industry.

Levigating plants of economic size could be sustained by the markets in at least Victoria and New South Wales at present and deposits such as Linton, Lal Lal, Inverell and Wyalong should provide suitable raw material for long-term production of kaolins at consistent grade. It is not at present clear whether such products could include the supply of the finest grades, such as coating clays, economically and at the rates required. Technically, however, it would be possible to produce an appreciable proportion of material to meet the requirements of 80—90 % less than 0.002 mm. and negligible quartz content. No local substitutes for the most specialised imported clays, such as those used in rubber reinforcing, are yet known.

Economics

The reserves of kaolin available in individual deposits in Australia have not been determined, except in a few special cases. Local markets have been small and attention has been directed more to the problem of sustaining minimum economic levels of production rather than to the question of whether reserves are adequate. Very little equipment for quarrying and beneficiating kaolin has been used and it has usually been possible to exploit a series of small deposits, or the best and most accessible portions of larger deposits, with little forward planning.

Material from small workings is generally routed through a few dry-milling and packaging plants scattered around the main industrial centres, but some consumers take kaolins, as mined, for direct use in industry. Production has tended to become more centralised over the years, with the bulk of the market being satisfied by the output of fewer and larger deposits, more consistent in product quality.

With the exhaustion of some of the smaller deposits close to the main cities, interest in the determination of reserves in new deposits has recently increased and drilling programmes have been undertaken in conjunction with the various State Mines Departments and Geological Surveys. Reserves are generally in-

vestigated only to the stage when it is certain that foreseeable local demands can be easily satisfied.

Dykes are commonly mined by underground methods. Shafts, tunnels and adits have produced much of the high-quality grit-free clay won from such deposits. Lower-grade siliceous kaolins are quarried by open-cast methods.

The limited size of local markets has been the main factor influencing kaolin production throughout Australia. Distances between centres of consumption are large and transport costs generally prevent the output of any deposit from reaching more than one industrial centre. Imported kaolins can reach all the main centres, which are in the coastal cities, at prices competitive with those of local products. Exceptions include low-cost bulk fireclays, quarried within a very short distance of the point of utilisation and the grit-free dyke kaolins, some of which are of high enough quality to stand the costs of shaft mining and land transport up to distances of several hundred kilometres.

In applications where kaolin can be used in the form of a thick suspension in water, as for instance in paper-making, it has been found economical to separate the clay fraction from kaolinised granite by levigation, thicken the slip in hydrocyclones and settling tanks, and then transport it directly to the point of consumption in tanker cars. Hauls of up to 200 km. are within the economic limits set by the prices of imported dry kaolins.

Statistical data

Few specific figures on kaolin production, import and consumption in Australia are available. Residual kaolins are not generally distinguished from transported clays in the annual records published by the State authorities and the "white clays" category to which imports are referred by Commonwealth authorities is similarly too broad to be used as a specific index for kaolins.

From the collected data given by Kalix et al. [57], it is evident that exports of kaolin from Australia are negligible. Imports of "white clays", mostly of the residual kaolin type, have been around 20,000 tons per annum. Annual local production of white clays known to be residual kaolins is also about 20,000 tons. The total Australian figures given [57] for annual kaolin production lie between 36,070 and 51,088 tons for the years 1960—64, but these include data for various deposits of transported clays, especially the deposits in New South Wales.

Imports of fireclays are small (4,311 tons in 1964) compared with local production, which is about 300,000 tons per annum. No complete data for fireclay production are available and it is in any case not possible to assess what proportion of the 300,000 tons represents residual kaolin.

Statistical data on the consumption of kaolins by various branches of Austra-

lian industry are not available. Most of the local production of siliceous kaolins is consumed in refractories manufacture. The ceramics industry uses about half the local output of kaolins from other deposits, such as dykes and sediments low in free silica. The rest is consumed as filler clay in rubber, paint, paper and plastics manufacture. Paper clays constitute the largest single item in the consumption of kaolins other than siliceous fireclays and the bulk of this item is imported.

Summarised data on deposits

Kaolinised granites and related igneous rocks

(1) Albany (W. A.)

At a locality 25 km. north of Albany on the south-west coast, Pre-Cambrian granite has been kaolinised, probably by deep lateritic weathering, yielding a clay which has low plasticity, although it contains up to 20 % of illite and has a wide particle size range [2] [5] [8].

(2) Ardrossan (S. A.)

At Pine Point, on the coast of Yorke Peninsula, 80 km. north-west of Adelaide, an aplite, intrusive into the Cambrian metamorphic basement, has been kaolinised to depths exceeding 30 metres, under a cover of further 20 metres of Tertiary sands and clays. The kaolin is relatively pure and shows good crystal outlines and has been considered as a potential source of paper clay. There is no direct evidence of hydrothermal activity. As the area is one which has been affected by deep lateritic weathering, this is the probable mode of origin. Reserves are large, exceeding 200,000 tons [4] [12] [13].

(3) Bolgart (W. A.)

A granitic rock at this locality 100 km. north-east of Perth has been kaolinised to a depth of at least 16 metres, yielding a non-refractory clay of moderate plasticity, 67 % of the material being in the sub-micron size range. Nodules of fine calcite are present and gibbsite has been recorded from some samples. The area has been affected by lateritic weathering and this has evidently been the mode of origin judging from the mineralogy of the clay [5] [10].

(4) Bibilup (W. A.)

Pre-Cambrian granite has been kaolinised at this locality 230 km. south of Perth, near Nannup. The clay is refractory and contains 30 % quartz grit. Deep lateritic weathering is the probable cause of the alteration [8].

(5) **Boyup Brook (W. A.)**

In this area, near the south-west corner of the State, 100 km. south-east of Bunbury, Pre-Cambrian granite has been altered to a quartzose vanadiferous kaolin, possibly by lateritic weathering [8].

(6) **Brooklands (Qld.)**

One of the kaolinised granites, probably of Permian age, 160 km. north-west of Brisbane in the Kingaroy-Nanango area. It outcrops in cliffs up to 10 metres in height but the full depth of alteration is not known. Though classed by Connah as a weathering product more recent studies suggest a connection with hydrothermal activity [18].

(7) **Bulla (Vic.)**

Palaeozoic granodiorite has been extensively kaolinised in a number of places in the Bulla and Broadmeadows districts 15 km. north of Melbourne. There is no clear evidence from the comparative mineralogy and trace element content of the fresh and altered rock in favour of the activity of hydrothermal solutions. The kaolinisation extends to a depth of 20 metres below the top surface of the granodiorite and has probably been caused by a weathering process. Kaolinite, mainly in the 1—5 micron particle size range, and coarse quartz are the dominant constituents of the fully altered zones, grading out through bleached biotite zones into fresh rock. Where an aplitic dyke occurs in the kaolinized area at Broadmeadows, the clay fraction is subordinate to fine quartz. Material from the Bulla deposit is levigated to provide filler clay for paper manufacture, the washed product being transported in slip form by tanker trucks to the paper mills [9] [20].

(8) **Cunderdin (W. A.)**

The granite-gneiss basement of this area 140 km. east of Perth has been kaolinised, probably by lateritic weathering, yielding a refractory non-plastic white clay containing 30 % of siliceous grit [8].

(9) **Cooyar (Qld.)**

Palaeozoic granite has been kaolinised at this locality near the southern extremity of the Kingaroy-Nanango district 120 km. north-west of Brisbane, in which similar deposits occur. The cause of the alteration has been attributed to hydrothermal activity but there is still a possibility that kaolinisation was caused by deep weathering throughout the region [7] [19].

(10) Dandenong (Vic.)

In this area, 30 km. south-east of Melbourne, bodies of Palaeozoic granodiorite have been kaolinised to depths of 10—30 metres, the general association being similar to that at many other localities throughout Victoria. Deep weathering is the probable cause of the alteration, though small pockets of remarkably coarse (0.5 mm) flaky kaolinite have been noticed in parts of the deposits suggesting mild hydrothermal activity. The clays are refractory and not very plastic but have been extensively quarried for firebrick manufacture [9].

(11) Dowerin (W. A.)

A kaolinised Pre-Cambrian granitic rock occurs at this locality 150 km. north-east of Perth. The clay contains 40 % of coarse quartz and is neither refractory nor plastic. It has probably been formed by deep weathering under laterite [8].

(12) Duketon (W. A.)

In this goldfield, 800 km. north-east of Perth, the Pre-Cambrian granite and greenstone basement complex has been kaolinised, probably in part by intensive leaching in the zone of oxidation of sulphides, though deep lateritic weathering would also have been operative [8].

(13) Fern Tree River (Tas.)

Kaolinised granite occurs in the workings of the Goshen Tin Mine at this locality 20 km. west of St. Helens. There are similar deposits, not necessarily associated with tin mineralisation, in other parts of north-east Tasmania such as Derby, Scottsdale and Gladstone. Most of the near-surface alteration of these granitic rocks may therefore have been caused more by weathering than by hydrothermal effects [1] [24].

(14) Glen Forrest (W. A.)

Pre-Cambrian granites and granitised sediments have been kaolinised at this locality and others nearby in the Darling Range 15 km. east of Perth. The Gooseberry Hill, Baker's Hill and Lesmurdie deposits are typical occurrences. All of these deposits appear to have formed by deep leaching in an elevated region that has been subjected to lateritic weathering since Tertiary times. The clays produced by this alteration are mainly siliceous kaolins, showing great variations in plasticity and refractoriness from one locality to another throughout the district. Some have been quarried for use in firebrick manufacture [1] [5] [8] [10] [21].

(15) **Hallam (Vic.)**

Kaolinised granodiorite similar to that at Dandenong, 5 km. to the west, has been extensively quarried at this locality, for refractory raw material. The clay is siliceous and not plastic. Weathering to depths of 10—20 metres appears to have been the cause of the alteration, but pockets of very coarse kaolin occur [1] [9].

(16) **Inverell (N. S. W.)**

In this area, 500 km. north of Sydney, a tin-bearing granite has been altered to a depth of 30 metres. Hydrothermal activity has been suggested as the cause of the kaolinisation, which has been complete, yielding a very white, well crystallised, fine kaolinite. The iron oxide content is 0.3 % and the clay disperses easily in water. There has been recent interest in the deposit as a source of paper and filler clay [58].

(17) **Lal Lal (Vic.)**

Palaeozoic granodiorite, 30 km. south-east of Ballarat, has been kaolinised over an area of several square miles at Lal Lal. The depth of alteration exceeds 40 metres but there are no indications of hydrothermal activity. The deposit forms the basement to Tertiary brown coal seams with a total thickness of 70 metres in places. Kaolinisation may have been caused by ground water activity related to the coal deposition but there is no clear evidence available. Large amounts of the clay have been used in firebrick and paper manufacture. Levigation produces a fair grade china clay, containing too much iron oxide (0.9 %) and titania (0.8 %) for the material to be generally used in ceramics. Some of the iron is contained in very fine leucoxene (anatase) and cannot be removed by sulphite bleaching. Small kaolinised felspathic dykes around the granodiorite have been worked for ceramic raw material, as these contain less iron and no residual quartz [1] [9] [11].

(18) **Lesmurdie (W. A.)**

Kaolinised granite occurs at this locality in the Darling Range 15 km. east of Perth. The clay is comparable with that at Glen Forrest but has a finer-grained kaolinite fraction. Probably derived under lateritic weathering conditions [1] [5] [8].

(19) **Linton (Vic.)**

A kaolinised Palaeozoic granodiorite, similar to that at Lal Lal, occurs in this area 40 km. south-west of Ballarat. Several square miles of the exposure show al-

teration to an unknown but considerable depth. Some of the material indicates that pegmatites were present in the parent rocks, as large masses of pure kaolinite can be found in the form of original felspar crystals. The kaolin is generally of high quality and is being considered as the main future source of paper clay for Victoria. Levigation can separate a fine fraction substantially free of fine quartz and mica and with good yields. Though the existence of pegmatitic phases suggests that hydrothermal alteration might have occurred, the extent of the deposit and its general similarity to that at Lal Lal, would indicate an origin by deep weathering [1] [9].

(20) **Minnievale (W. A.)**

A soft white clay derived from a Pre-Cambrian granite has been recorded in depth at this locality 170 km. north-east of Perth. It is a non-plastic kaolin associated with 45 % of quartz and is probably a product of deep lateritic weathering [8].

(21) **Mooroodue (Vic.)**

Palaeozoic granodiorite in this area 50 km. south of Melbourne has been kaolinised in places to a low-grade siliceous fireclay, probably by deep weathering [9].

(22) **Nanango (Qld.)**

In the Nanango-Kingaroy district 150 km. north-west of Brisbane, granites, probably of Permian age, have been altered to quartz-kaolinite assemblages. A deposit at Brooklands is typical and others occur at Tarong and Cooyar. Although hydrothermal action has been suggested as the mode of origin, the possibility of regional lateritic weathering remains as a cause of kaolinisation [7] [18] [19] [54].

(23) **Olary (S. A.)**

Pre-Cambrian granites and pegmatites in this area 350 km. north-east of Adelaide, have been kaolinised, yielding refractory clays containing variable amounts of fine quartz. Weathering is the probable mode of origin [4].

(24) **Pakenham (Vic.)**

Parts of a Palaeozoic granodiorite near the contact zone, and the adjacent sediments into which it has been intruded, have been kaolinised in this area east

of Hallam, probably by deep weathering. The material has been used as a low-grade fireclay [1] [9].

(25) **Paluma (Qld.)**

Upper Palaeozoic rhyo-dacite flows in the Paluma Range 40 km. north of Townsville, have been kaolinised, yielding a material of potential ceramic interest. The cause of the alteration is not yet established [36].

(26) **Pyalong (Vic.)**

Palaeozoic granodiorite has been kaolinised to depths exceeding 15 metres in this area, 90 km. north of Melbourne. Tertiary basalt covers the altered material as at Bulla. The alteration may have been caused by deep weathering. The granitic rock becomes fresher with increasing distance from the edge of the intrusion, a situation similar to that at Pakenham. Reserves are unknown but would be very large [9].

(27) **Ravensthorpe (W. A.)**

Pre-Cambrian granite has been kaolinised to an unknown depth at this locality 400 km. east of Bunbury. The clay is a non-plastic refractory type. No evidence as to the nature of the alteration is available [8].

(28) **Roe (W. A.)**

At Holland Soak in the Roe district, granitic rocks in the Pre-Cambrian basement have been kaolinised to a refractory non-plastic clay. Neither the cause of the alteration, nor the depth to which it extends, has been recorded [5] [8].

(29) **Surges Bay (Tas.)**

A syenite at this locality, some 30 miles south-west of Hobart, has been converted into a kaolin with an appreciable content of a mixed layer illite-montmorillonite. The depth of alteration is not known but the mineral assemblage suggests weathering as the mode of origin [24].

(30) **Tumbarumba (N. S. W.)**

Kaolinised granite occurs to the west of this town 400 km. south-west of Sydney. It is a coarse kaolinite associated with some illite. The depth of the alteration is unknown [6].

(31) **Tumby Bay (S. A.)**

Near this locality, on Eyre Peninsula, 240 km. east of Adelaide, granites and pegmatites of Pre-Cambrian age have been kaolinised, probably because of the lateritic weathering that has occurred in the region during the Tertiary, though fine tourmaline and talc occur in the district. The clay fractions produced by the alteration vary in plasticity and refractoriness according to locality, but one example, from a deposit 30 km. north-east of Tumby Bay is refractory and moderately plastic. The characteristics are reported under "Stokes" in ref. [4].

(32) **Wagine Soak (W. A.)**

The Pre-Cambrian granitic rocks of this district 200 km. north-west of Kalgoorlie have been kaolinised by deep lateritic weathering over an extensive area. Low cliffs of white, altered granite occur under a ferruginous duricrust [8].

(33) **Wedderburn (Vic.)**

In this district, 200 km. north-west of Melbourne, a large area of Palaeozoic granodiorite has been kaolinised to an unknown depth. The occurrence is similar to those at Bulla, Lal Lal ,Pyalong and other localities and there is no obvious evidence as to the mode of origin. The clay fraction is refractory but has an iron oxide content of 1.75 % and contains appreciable fine quartz grit [9].

(34) **Wubin (W. A.)**

Granitic rocks in the Pre-Cambrian basement of this district 230 km. north-east of Perth have been kaolinised to depths of 20—40 metres by deep weathering under laterite, the depth of altered rock being greater under hills than under valleys. This kaolinisation is general throughout the region extending to the north-west of Wubin, and is recorded in the goldfields of the Murchison district at Cue and Big Bell [47].

(35) **Wyalong (N. S. W.)**

Granitic rocks have been altered, probably hydrothermally, to a micaceous kaolin at this locality 380 km. east of Sydney. It is similar to that at Inverell but coarser. The depth of alteration is 50 metres, the material grading into a rock recorded as sheared diorite. Although the area has been subject to lateritic weathering, the depth of kaolinisation seems to indicate hydrothermal action, as does the coarseness of crystallisation [6].

(36) **Yarloop (W. A.)**

Granite in the Pre-Cambrian granite-gneiss complex of this area 120 km. south of Perth has been kaolinised, probably by weathering under laterite. Like the kaolin derived from metamorphic rocks nearby at Harvey, this is an unusually plastic clay with 50 % of a sub-micron fraction. It is not refractory, as the alteration has not been complete and residual microcline is commonly present [5].

Kaolinized metamorphic rocks

(37) **Arthur River (W. A.)**

A moderately fine-grained kaolin probably derived by weathering of a metamorphosed basalt or dolerite. The clay is not free from fine-grained quartz, as it would be if derived from an unmetamorphosed basic rock [2] [5].

(38) **Baker's Hill (W. A.)**

One of the kaolinized Pre-Cambrian granitic bodies of the Darling Range 30 km. east of Perth. Similar to the Glen Forest and Lesmurdie deposits, the kaolinisation extending to a depth of over 20 metres. Probably an example of deep lateritic weathering [1].

(39) **Bardoc (W. A.)**

At this locality 50 km. north of Kalgoorlie, a very fine-grained plastic kaolin has been produced by alteration, probably through weathering, of a metamorphosed Pre-Cambrian basic rock. The appreciable content of quartz (25 %) indicates that metamorphism preceded kaolinisation. The presence of some illite and montmorillonite and the transition from kaolin to nontronite at a depth of 20 metres suggests weathering as the cause of alteration [5] [8].

(40) **Boddington (W. A.)**

A non-plastic kaolin, probably formed by weathering of a metamorphosed Pre-Cambrian basalt or doleritic dyke, occurs in this area 120 km. south-east of Perth [8].

(41) **Calcarra (W. A.)**

A non-plastic refractory kaolin has been derived from Pre-Cambrian granitic gneiss, probably by weathering, in this area [8].

(42) **Clackline (W. A.)**

A Pre-Cambrian metamorphic complex in this area, 70 km. north-east of Perth has been extensively kaolinised, by weathering, down to a depth of 20 metres. The parent rocks include sillimanite and biotite schists, garnet schists and gneisses, altered basic rocks and dolerite dykes. The kaolin is fine-grained, only 20 % being above 5 microns in diameter and the proportion of clay to quartz and other residual minerals is variable. Sillimanite has been commonly kaolinised, but the clays have been extensively used in the manufacture of refractories [1] [5] [8] [21] [22] [23].

(43) **Dangin (W. A.)**

Pre-Cambrian granitic gneiss has been kaolinised in this area 140 km. east of Perth. The clay is physically similar to very fine mica but has some plasticity. It contains 30 % of fine quartz. By analogy with the deposits to the west, at Balkuling, an origin by weathering is presumed [8].

(44) **Donnybrook (W. A.)**

Granitic gneiss in this area 30 km. south-east of Bunbury has been locally weathered to a mixture of coarse quartz and white kaolinite of fine particle size, showing moderately high plasticity [8].

(45) **Greenbushes (W. A.)**

In this area, 70 km. south-east of Bunbury, a Pre-Cambrian complex of schistose amphibolite, granite and tin-bearing pegmatite has been extensively kaolinised. Hydrothermal activity associated with the mineralisation may have caused some of the kaolinisation but the area is one of deep lateritic weathering. White kaolin, in places containing fine tourmaline, occurs in the tin lodes. The deposits of clay are large in area but the depth of alteration is not recorded [8].

(46) **Harvey (W. A.)**

In this area, 30 km. north-east of Bunbury, a Pre-Cambrian granitic gneiss-greenstone complex has been kaolinised to a refractory clay-quartz mixture; the depth of alteration is unknown and there is no evidence of hydrothermal activity on record [2] [5] [8].

(47) **Kapinnie (S. A.)**

A smooth, lustrous kaolin encountered between 6 and 15 metres in a drill-hole in Pre-Cambrian rocks near Kapinnie railway station on Eyre Peninsula,

70 km. north-west of Port Lincoln. Reported under "Mitchell" in [4]. The parent rock could be one of the felspathic gneisses of the district, comparable to that at Miltalie. Kaolinisation of felspathic rocks by weathering, is general throughout much of Eyre Peninsula.

(48) **Kanowna (W. A.)**

In this gold field, near Kalgoorlie, a complex of schist, amphibolite and serpentine, formed by the alteration of basic lavas and intrusives of Pre-Cambrian age, has been leached throughout the zone of oxidation in the mineralised area. The resultant clays are variable mixtures of kaolinite, illite, halloysite, sericite and fine quartz and are generally non-refractory and of low plasticity. A further barrier to utilisation is the presence of salt and alunite, but the colour is impressive, the reflectivity being 94 % of that of magnesia [5] [8].

(49) **Kundip (W. A.)**

At this locality near Ravensthorpe, greenstone schists derived from Pre-Cambrian basalts have been kaolinised to depths of at least 30 metres. An early stage of hydrothermal activity is suggested by the presence of chalcedonic silica masses above the kaolin zone, where a thin seam of green clay rich in chromium and gold also occurs, but the main cause of kaolinisation is probably leaching by acid ground waters [8].

(50) **Leonora (W. A.)**

Greenstones derived from Pre-Cambrian basic rocks, have been altered to white kaolin in this goldfield 220 km. north of Kalgoorlie and in a similar situation at Lawlers, 120 km. north-west of Leonora. The depth of alteration is unknown but exceeds 20 metres. The origin may have been by ground water leaching in oxidized sulphide zones [8].

(51) **Lobethal (S. A.)**

A bed of refractory white kaolin occurs between two zones of mica schist at this locality in the Proterozoic sequence 16 km. east of Adelaide. The clay has low plasticity, contains fine residual quartz grains and has an appreciable vanadium content. It has been used in large quantities for local firebrick production. The material is a leached slate, the alteration probably having occurred during a period of lateritic weathering [4] [28].

(52) **Marulan (N. S. W.)**

Ordovician slates in this area 130 km. south-west of Sidney have been leached

by ground waters to depths of 20 metres, particularly in crush zones. In the more completely leached cores of these zones kaolinite is the dominant mineral mixed with fine quartz. The clays grade out into bentonitic zones away from the centres of activity. The best material has less than 1.0 % of iron oxide and has been used commercially as a filler clay [6].

(53) **Marvel Loch (W. A.)**

Chloritic schists of Pre-Cambrian age, derived from basic rocks, have been kaolinised in this goldfield 30 km. south-east of Southern Cross. The alteration appears to have been connected with sulphide oxidation as in other gold-fields and there is a typical development of green chromiferous kaolinite bands, as at Melville, in the Murchison region 100 miles east of Geraldton [8].

(54) **Miltalie (S. A.)**

In this area, at a locality 28 km. north of Cowell on the east coast of Eyre Peninsula, felspathic gneiss bands in Pre-Cambrian quartzite beds have been kaolinised to a depth which is unknown but exceeds 20 metres. The main bed of clay is up to 15 metres thick and dips steeply. It consists of quartz grit and kaolinite in approximately equal amounts, the kaolin being a non-refractory china clay. The alteration has probably been caused by ground waters penetrating the surrounding quartzite beds, the locality being in an area of deep lateritic weathering [29] [30] [31] [55].

(55) **Mount Helena (W. A.)**

The Pre-Cambrian schist-granite basement complex has been kaolinised in this area 30 km. east of Perth to an unknown depth, probably by lateritic weathering. The clays vary in grit content and plasticity and are not refractory [5] [8].

(56) **Mount Magnet (W. A.)**

Auriferous deep leads in this area 330 km. east of Geraldton overlie kaolinised greenstones derived from Pre-Cambrian basic rocks. The alteration has probably been the result of ground water leaching in the zone of oxidation of sulphides [8].

(57) **One Tree Hill (S. A.)**

Proterozoic slates and phyllites in this area, 10 km. north-east of Adelaide, have been leached to white kaolins to a depth of 20 metres, below which the unweathered rocks are brown in colour [35].

(58) **Mount Maiden (W. A.)**

In this goldfield 370 km. north of Kalgoorlie, greenstones derived from Pre-Cambrian basic rocks have been altered to white kaolin to a depth of 30 metres.

(59) **Ora Banda (W. A.)**

Greenstones and sheared porphyritic basic rocks of Pre-Cambrian age in this goldfield 50 km. north-west of Kalgoorlie have been altered to white kaolins to depths of 80 metres in the oxidised zone [8].

(60) **Paracombe (S. A.)**

A bed of felspathic gneiss in the Archaean basement of this area, some 30 km. north-east of Adelaide, has been altered to a mixture of white kaolinite and fine quartz by deep weathering under a Tertiary land surface. Reserves are of the order of 150,000 tons and the alteration is not recorded to extend below 10 metres depth [44]. The clay is variable in refractoriness containing up to 1.0 % of iron oxide and 2 % of alkalies, but has moderate plasticity [37] [38]. Another apparently similar deposit occurs to the south, near Milbrook [34].

(61) **Peak Hill (W. A.)**

In this goldfield in the Murchison region, a Pre-Cambrian chlorite schist has been kaolinised in the mineralised zone, down to a depth of 80 metres, probably by acidic ground water leaching in the zone of oxidation [8].

(62) **Pingelly (W. A.)**

In the Westbrook district 130 km. south-east of Perth, the granite-gneiss-dolerite basement rocks have been kaolinised, probably by lateritic weathering, to give a variety of clays, some plastic and rather refractory [8], others non-plastic and illitic [5].

(63) **Quairading (W. A.)**

Pre-Cambrian gneiss in this area 150 km. east of Perth has been kaolinised, probably by deep lateritic weathering. The clay contains a large proportion of fine quartz and is neither refractory nor plastic [8].

(64) **Three Springs (W. A.)**

Pre-Cambrian slates have been altered to white, siliceous kaolins in this area near Kadathinni 280 km. north of Perth. The kaolinisation extends to a depth

of at least 25 metres and has probably been associated with deep weathering as it is covered by a lateritic crust. The clay is moderately refractory but of rather low plasticity [5] [8].

(65) **Westonia (W. A.)**

Pre-Cambrian schists in this goldfield 350 km. north-east of Perth have been kaolinised to depths of at least 12 metres. Some of the clays are moderately plastic and refractory and some are rich in chromium [8].

(66) **Williamstown (S. A.)**

Pre-Cambrian schists and pegmatites, associated with quartz-sillimanite and kyanite rocks occur in a zone of intense metamorphism at this locality near Mount Crawford 40 km. north-east of Adelaide. Most of the rock in the core of the complex, between pronounced shear zones filled with coarsely crystallised damourite, has been kaolinised. Various neighbouring Proterozoic sediments have also been more or less altered to kaolinitic clays. The mineralogical characteristics of the kaolin and the geological setting suggest a hydrothermal mode of origin. The deposit is one of the most definite examples of hydrothermal kaolinisation in Australia.

Kaolinite is the main clay mineral in the bulk of the deposit, though there is evidence of the presence of some dickite. The flakes of kaolinite are poorly shaped, coarsely crystallised and of even particle size, mostly lying in the range 2—15 microns. In the unlevigated raw state the kaolinite is mixed with up to an equal amount of fine sillimanite, damourite, quartz and rutile, but most of this material can be separated by settling out the fraction larger than 20 microns.

The deposit has been worked on a large scale as a source of refractory clay and lump sillimanite, but the characteristics of the clay do not make it attractive to industry. It has high shrinkage, low plasticity and an appreciable grit content, but has been used as an aluminous fireclay component for many years [1] [4] [48] [49] [50] [51] [52].

(67) **Woodside (S. A.)**

Slates and intrusive albite-pegmatites in the Woodside-Balhannah area, 20 km. south-east of Adelaide, have been kaolinised by weathering under laterite during the Tertiary period. The alteration of a pegmatite in this Proterozoic basement complex near Balhannah has not been complete, as the clay still contains residual albite and has 3.4 % of soda, 1.0 % of magnesia present. The bleached and kaolinised slate, 3 km. north-east of Woodside, contains very little soda and potash but is not plastic and has 0.8 % iron oxide present [4].

(68) **Wondai (Qld.)**

Portions of a metamorphosed Ordovician volcanic sequence in this area north of Kingaroy and 180 km. north-west of Brisbane, have been kaolinised, yielding materials used in heavy clay products. The mode of alteration is not recorded and the clays are not of high purity [7] [36].

Kaolinised dyke rocks

(69) **Balkuling (W. A.)**

In this district 120 km. east of Perth Pre-Cambrian dolerite dykes have been altered by lateritic weathering to kaolins of low plasticity and low grit content. A deposit at Beverley, to the south of Balkuling, contains only trace amounts of grit. The clays are usually vanadiferous and not very refractory, containing minor amounts of illite [5] [8].

(70) **Ballan (Vic.)**

Felspathic dykes in this area, 60 km. north-west of Melbourne, have been kaolinised by leaching to depths greater than 30 metres. The clays can be free of grit in parts of the deposits. The plasticity is low, the colour is generally a clean white and the refractoriness is not high. The dykes are worked for clay only when the width is of the order of 1 metre [9].

(71) **Ballarat (Vic.)**

Kaolinised dykes similar to those in the Ballan area occur in the Ballarat goldfield 100 km. north-west of Melbourne. An example from a deposit at St. James's Hill was more plastic than is usual with these clays, of good colour and medium refractoriness. The dykes were probably injected early in the Tertiary and have been altered by ground waters penetrating porous country rocks to a deep water table [9].

(72) **Birdwood (S. A.)**

In the bleached and kaolinised basement complex of Proterozoic sediments in this district 30 km. east of Adelaide, felspathic dykes intersecting the beds have also been kaolinised yielding a different variety of clay. This is generally more coarsely crystalline, but contains no quartz grit. The area is auriferous and the alteration has been attributed to hydrothermal action [4] [55].

(73) **Broad Arrow (W. A.)**

A quartz keratophyre dyke at this locality 30 km. north of Kalgoorlie has been kaolinised to a refractory non-plastic clay. No record of the depth of alteration is available [8].

(74) **Burabadji (W. A.)**

An indurated kaolin with a wide particle size distribution has been derived from a Pre-Cambrian basic dyke at this locality in the south-west of the State. The clay is non-plastic and contains fine quartz veins. It is probably a product of lateritic weathering but the depth of alteration is unknown [5] [8].

(75) **Cardup (W. A.)**

In this area 20 km. south-east of Perth the Pre-Cambrian slates of the country rock have been extensively weathered and a dolerite dyke cutting through these beds has been converted into a fine-grained grit-free plastic kaolin, presumably by the same weathering process. A similar association of kaolinised sediments and dolerite occurs at Armadale, in the same district [8].

(76) **Dromana (Vic.)**

Fine-grained aplitic dykes through the Silurian sediments of this area, some 50 km. south of Melbourne, have been weathered to kaolins of low plasticity and low refractoriness. Some gold mineralisation is typical of the area but the kaolinisation is likely to have been caused by ground water leaching. The amount of fine residual quartz in the deposit detracts from the utility of the clay and it has not been worked commercially [9].

(77) **Dunbible (N. S. W.)**

Dolerite dykes in this area have been leached and kaolinised down to the water table by ground waters moving through the permeable sandstone country rock. Kaolinite is the dominant mineral, but illite is generally present in quantity where alteration has not been very complete [6] [25] [26].

(78) **Duranillin (W. A.)**

At this locality 200 km. south-east of Perth, clay has been formed by weathering of a doleritic rock probably a dyke of Pre-Cambrian age. Fine-grained kaolinite is the main component and the clay is moderately plastic but not refractory. The depth of alteration is unknown [2] [5] [8].

(79) **Dwellingup (W. A.)**

At this locality 80 km. south of Perth dolerite has been kaolinised to give a clay similar to that at Duranillin, but no details of depth or extent are known [8].

(80) **Egerton (Vic.)**

A high-grade kaolin derived from a tinguaitic dyke by deep ground water leaching. This deposit, 27 km. on the Melbourne side of Ballarat, is the main source of white grit-free kaolin in Victoria and has been used extensively in many industries both in the Melbourne area and interstate. The dyke has been worked for many years by underground mining. Good material can be obtained down to depths of 150 metres where the dyke cuts through a hill of permeable Palaeozoic sediments. In the valley, one mile north along the extension of the dyke, the kaolin has been worked only down to the present water table at about 30 metres. The width of the dyke is variable but generally less than 2 metres.

The clay, generally known as "Malone kaolin", is compact and has little plasticity until finely ground. Iron oxide ranges from 0.2 % to 0.6 % in the high grade material but more ferruginous clay is being encountered in recent workings. Kaolinite is practically the only mineral present and shows good crystal shape. In certain sheared zones a little halloysite occurs throughout the kaolin and this mineral also pseudomorphs nepheline in the rare instances where the original rock has not been completely altered. The clay is refractory and of good white colour and finds use throughout the ceramic industry and in paint and rubber manufacture as an extender and filler. The deposit is not large enough to supply the paper industry and the low degree of dispersion of the clay in water suspensions detracts from its use in this field. It is usually dry-milled and sized in air classifiers from the condition as mined [1] [9] [11] [56].

(81) **Gordon (Vic.)**

A kaolinised dyke of the same type as that at Egerton occurs 5 km. south of Gordon. Not as pure as Egerton kaolin but otherwise similar. The deposit has been worked at times but is more limited than Egerton and less easily mined and marketed [9].

(82) **Jacob's Well (W. A.)**

An aplite dyke and a dolerite dyke in the Pre-Cambrian basement of this locality near Balkuling have been kaolinised, probably by weathering. One of the deposits contains up to 30 % of quartz grit and is non-plastic. The other is more

plastic, less refractory, and has 13 % of grit. The clay fraction from the aplite body has only 0.11 % iron oxide [2] [5] [8].

(83) **Jimperding (W. A.)**

Two occurrences of non-plastic white kaolins are recorded from shafts in this locality 120 km. north-east of Perth. The clays are refractory and contain 14 to 17 % of quartz grit. The depth of alteration is unknown but it is probable that the deposits represent weathered dolerites [5] [8].

(84) **Kalamunda (W. A.)**

A kaolinised dolerite dyke occurs at the head of Woodlupine Brook 20 km. east of Perth. It is probably the product of lateritic weathering and contains a considerable amount of leucoxene. The plasticity is high and it is non-refractory [8]. Further to the north-east in Piesse Gully, a mixture of a similar clay with kaolinised granite has been used in tile and firebrick manufacture. This clay is not so plastic, but is more refractory and contains less leucoxene [5] [8].

(85) **Kangaroo Island (S. A.)**

In the Hog Bay area a pegmatite has been kaolinised, in part, to a non-plastic refractory clay. Alteration extends to a depth of 30 metres, but much of the body consists of fresh felspar. The abundance of tourmaline may suggest that hydrothermal activity caused the kaolinisation but there is no clear evidence available and exploitation of the deposit has long since ceased [3] [4].

(86) **Kunjin (W. A.)**

A kaolinised dolerite dyke occurs in the Pre-Cambrian complex of this area about 190 km. east of Perth. Although the clay contains 63 % of submicron particles it does not have good plastic properties. The deposit has been worked as a china clay but no records of extent or depth are available. It should be a useful filler clay, containing little grit, 0.4 % iron oxide and having a reflectivity of 94 % of that of magnesia. The mode of origin has probably been by lateritic weathering [5] [8].

(87) **Manjimup (W. A.)**

A kaolinised dolerite, associated with kaolinised graphitic schists occurs at this locality 220 km. south of Perth. The clay is moderately plastic, practically grit-free and non-refractory. Probably a weathering product, it extends to an unknown depth [8].

(88) **Maryborough (Vic.)**

Quartz porphyry dykes up to 7 metres in width have been kaolinised, probably by deep weathering, in this district 150 km. north-west of Melbourne. The clay is non-plastic and refractory and contains a large amount of free quartz [9].

(89) **Meckering (W. A.)**

A dolerite dyke in this area, some 120 km. north-east of Perth, has been kaolinised, probably by weathering, to an almost grit-free plastic clay, having moderate refractoriness [8].

(90) **Meekatharra (W. A.)**

In this goldfield 700 km. north-east of Perth, auriferous albite-porphyry dykes have been kaolinised to a depth of 100 metres, presumably by deep ground water leaching in the zone of oxidation of sulphides [8].

(91) **Mittagong (N. S. W.)**

Immediately above an anthracitic coal seam in mine workings in this area, 120 km. south-west of Sydney, a microsyenite dyke or sill has been kaolinised, probably by ground water leaching. Appreciable amounts of fine felspar still remain in the clay [6].

(92) **Mokine (W. A.)**

A non-plastic rather refractory clay low in grit content has been derived from dolerite at this locality near Clackline. The depth of alteration is unknown, but the area is one of deep lateritic weathering [8].

(93) **Mount Crawford (S. A.)**

The clay worked by shafts and tunnels at this locality, some 6 km. south of Williamstown, is generally known as Cromer "C" clay to distinguish it from the coarser kaolin in the sillimanitic deposits 4 km. to the north-east. The parent rocks appear to have been felspathic dykes and Proterozoic sediments. Only the kaolin derived from the dykes is grit-free. It contains a little halloysite but is otherwise pure and well crystallised, the bulk of the kaolinite lying in the 2—10 micron particle size range. The plasticity is low, the iron oxide content is around 0.4 % and the clay is refractory. The area is one of some hydrothermal activity, rutile having been commonly formed from the ilmenite and leucoxene of the original

sediments, so that it is possible hydrothermal activity caused the kaolinisation, though conclusive evidence is lacking [4] [32] [33].

(94) **Ringwood (Vic.)**

Small felspathic dykes, throughout the Melbourne region have been converted to kaolinite by the leaching action of ground waters penetrating the surrounding Silurian sediments, the main example at Ringwood 30 km. east of Melbourne being over a metre in width and free of quartz grit. The clays are non-plastic and refractory and have been used in ceramics, but reserves are very limited [9].

(95) **Stawell (Vic.)**

Quartz-porphyry dykes have been altered to semi-plastic mixtures of kaolinite and fine quartz in this district 200 km. north-west of Melbourne. The clay fraction is moderately refractory and is more plastic than the material derived from the less siliceous dykes of the Egerton-Ballan area. Deep weathering is probably the cause of the alteration, but no records of depths are available [9].

(96) **Sydney (N. S. W.)**

In the Sydney region, Tertiary dolerite dykes have been leached by ground waters penetrating the enclosing permeable sandstones, to depths dependent on the local water table. Alteration ceases at the water table, which may be at a depth of up to 30 metres. There are many such dykes, generally trending east to west, throughout the area between the coast and Asquith, a distance of 20 km., most of the commercially worked deposits being in the French's Forest district. The clays are not refractory and contain appreciable amounts of illite and leucoxene. The titania content of some examples approaches 6.0 % [6] [25].

(97) **Ulladulla (N. S. W.)**

A kaolinised dolerite, similar to those in the Sydney area, occurs at this locality on the south coast of N. S. W. By analogy, a weathering mode of origin is assumed [6] [45].

(98) **Wagin (W. A.)**

A weathered dolerite dyke, not appreciably silicified prior to kaolinisation, has been worked as a source of china clay in this area 200 km. south-east of Perth. Laterite occurs above the deposit but the depth of kaolinisation is not known. The clay is moderately plastic and rather refractory and has a good white colour [5] [8] [10] [46].

(99) **Bathurst (N. S. W.)**

Silurian slates have been bleached at several localities throughout the region extending from Bathurst towards Orange and Forbes, probably by hydrothermal activity, though the evidence does not rule out deep leaching by ground waters. Alteration extends to depths of 25 metres [6].

(100) **Birdwood (S. A.)**

Proterozoic sediments in this area, 30 km. east of Adelaide, have been hydrothermally altered to kaolins of different particle size distributions and fine quartz contents. The nature of the original beds is obscure, particularly those that have been converted to grit-free kaolins. The area is auriferous and the leaching solutions may have been associated with the mineralisation. In places the depth of alteration exceeds 30 metres. Different grades of clay have been worked as plastic and refractory raw materials or as fillers [4] [14] [15] [16].

(101) **Booleroo (S. A.)**

Large masses of Proterozoic sediments have been altered to white kaolin at this locality 230 km. north of Adelaide. The deposit is one of a sequence of similar occurrences on a line from Appila, in the south, through to Willowie in the north, a distance of some 60 km. The presence of numerous quartz veins suggests that the material has been affected by a stage of hydrothermal activity, but it is probable that lateritic weathering has been the main factor in the kaolinisation. The clay is plastic, with 38 % of the particles in the sub-micron range. In the early years of zinc refining it was used as a refractory bond clay in retort construction [4] [55].

(102) **Buckaroo (N. S. W.)**

One of the occurrences of leached Silurian slates in this region. Alteration extends to a depth of 25 metres. Quartz and fine sericitic mica occur in quantity and a hydrothermal origin is probable [6].

(103) **Burra (S. A.)**

As at Booleroo and related deposits to the west. Proterozoic sediments have been leached at Burra, a copper field 150 km. north of Adelaide. Some of the alteration, which has given rise to much quartz veining in the deposits, may have

been caused by hydrothermal solutions associated either with the copper mineralisation or the oxidation of sulphides. The clay is not very plastic and contains fine muscovite. With a potash content of 3.78 %, it fuses completely at 1300 °C [4] [55].

(104) **Canberra District** (Aust. Capital Territory and adjacent areas of N. S. W.)

Ordovician shales and mudstones, with rare andesitic dykes, have been bleached at various localities throughout the region. Examples are at Bungendore, Gungaderra, Gungahlin, Queanbeyan and Woodlands. It is thought that hydrothermal activity has been responsible but no clear evidence is available.

(105) **Clare (S. A.)**

A fine-grained plastic kaolin, similar to that at Booleroo, and also derived from the weathering of Proterozoic shales occurs at this locality 120 km. north of Adelaide. At 1400 °C., the material is not vitrified and is classed as a refractory bond clay. A considerable depth of similar kaolin occurs below 10 metres in a well, but no commercial workings are recorded [4].

(106) **Clergate (N. S. W.)**

In this area, 15 km. north of Orange, Silurian shales and slates have been leached to sericitic kaolins containing up to 35 % of fine quartz and less than 1 % iron oxide [6].

107) **Crystal Brook (S. A.)**

A large deposit of weathered Proterozoic sediments containing very little fine quartz has been worked at this locality 30 km. south-east of Port Pirie. Kaolinite is the only clay mineral present and shows a wide range of particle sizes. Alteration has not been as complete as in the similar deposits of the Appila-Booleroo-Willowie belt. Over 2 % of alkalies and 1.5 % of iron oxide remains in the clay reducing its whiteness and refractoriness. It has been used in large amounts as a bond clay for medium-temperature refractories in the Port Pirie lead-zinc smelting plant [4] [55].

(108) **Houghton (S. A.)**

Proterozoic shale has been locally bleached in this area east of Adelaide, probably by lateritic weathering. The clay contains fine silica, is non-plastic and not refractory. It has been used as a raw material for cream bricks [32]. A nearby

deposit of similar material at Anstey Hill contains more iron oxide and up to 5.4 % of alkalies.

(109) Imbitcha (S. A.)

Parts of an extensive flat-lying sequence of Cretaceous shales near Oodnadatta have been bleached, and cemented by alunite, to produce hard compact kaolins. The alteration, from a very plastic illitic clay of greenish grey colour, to brilliantly white masses of hard kaolin with a conchoidal fracture, may have occurred during the Tertiary, when a stage of intense weathering produced the siliceous duricrust of the area, though it is possible that the kaolinisation occurred during a later stage of laterisation. The depth to which alteration has extended is as great as 30 metres. Individual beds of very white kaolin, more or less contaminated with alunite, are 1—2 metres thick. Less altered clay, containing more iron oxide and less alunite, occurs in beds up to 10 metres thick for some miles throughout the district. Commercial exploitation of the kaolin has not been successful, because of the alunite content, the hardness of the material and the difficulties of mining and transportation associated with the remoteness of the area. Reflectance is 97 % of that of MgO [4] [27].

(110) Nairne (S. A.)

Proterozoic sediments in this area 30 km. east of Adelaide have been kaolinised, probably by weathering, to a fine, rather plastic material which has been extensively used as a refractory bond in firebrick manufacture. The clay contains up to 0.9 % iron oxide, 1.6 % titania and enough vanadium to cause green scumming in fired products [1] [3] [4].

(111) Port Augusta (S. A.)

Green Proterozoic shales, containing illite and chlorite, to the north and west of Port Augusta have been leached in the pallid zone of lateritic weathering to give moderately plastic kaolinite-quartz mixtures. The characteristics of the residual clays depend largely on the extent to which alteration has produced kaolinite in place of the original clay minerals. Generally these clays are more or less plastic but not refractory [4] [39] [40] [41] [55].

(112) Robertstown (S. A.)

Proterozoic tillites and other sediments have been altered to kaolinite-quartz mixtures in this area 100 km. north-east of Adelaide. The leached zones in places are covered by lateritic ironstone cappings and the cause of the kaolinisation is

probably deep lateritic weathering, though hydrothermal activity has been suggested [43].

(113) **Tichborne (N. S. W.)**

In this area, between Parkes and Forbes, 300 km. west of Sydney, Ordovician sediments have been leached to considerable depths during an era of intense weathering during the Tertiary period. The clays are typical of many such weathered deposits in the Palaeozoic sedimentary succession of the region. Moderately plastic, but non-refractory and containing around 2 % of iron oxide, the clays have found limited industrial application [6].

Table 1—Annual production of kaolins and fireclays in Australia (statistics from the Bureau of Mineral Resources)

I (in long tons)

Kaolin*	1963	1964	1965	1966
Queensland	168	401	373	45
New South Wales	27.161	22,089	27,608	21,336
Victoria	12.462	17,455	24,081	19,978
South Australia	3.833	4,542	7,383	7,589
Western Australia	920	631	238	1,162
Total	44,544	45,118	59,683	50,110

* includes some transported clays

II (in long tons)

Fireclay*	1963	1964	1965	1966
Queensland	13,998	16,188	17,642	16,668
New South Wales	104,578	102,298	94,177	116,015
Victoria	31,913	28,050	32,816	30,978
South Australia	29,544	27,170	27,445	24,036
Western Australia	25,002	47,251	61,102	98,487
Total	205,035	220,957	233,182	286,184

* includes some filler clays

Sources of information on Australian clays

Division of CSIRO (Commonwealth Scientific and Industrial Research Organisation).
(a) Division of Applied Mineralogy, Box 4331, G. P. O, Melbourne.
(Mineralogy, geochemistry and processing of clays).

(b) Division of Building Research, Graham Road, Highett, Vic.
(Mineralogy and utilisation of clays for ceramic building products).
(c) Division of Soils, Waite Road, Urrbrae, South Australia.
(Structural, chemical and physical properties of clay minerals).

Commonwealth Bureau of Mineral Resources. Geology & Geophysics, P. O. Box 378, Canberra City, A. C. T.
(Survey of production and economics.)

Australian Mineral Development Laboratories, Conyngham Street, Frewville, South Australia.
(Contract research on characteristics, beneficiation and utilisation of clays, recording of data in Clay Register).

TheNew South Wales Department of Mines
(a) Chemical Laboratory, 28 George Street, Sydney, N. S. W. (Clay analyses, physical properties).
(b) Geological Survey, Box 48, G. P. O., Sydney, N. S. W. (Deposit surveys).

The University of New South Wales, P. O. Box 1, Kensington, N. S. W.
(a) School of Applied Geology (Origin, nature and industrial utilisation of N. S. W. clays).
(b) Department of Ceramic Engineering (Ceramic clays)
(c) Unisearch Ltd. (Contract research on technical problems).

The Queensland Geological Survey, 2 Edward Street, Brisbane, Qld.
(Investigation of deposits in Queensland).

The University of Queensland, St. Lucia, S. W. 6., Qld. Department of Mineralogy and Geology.
(Origin and nature of clays in Queensland and northern N. S. W.)

The South Australian Department of Mines, Box 38, Rundle Street P. O. Adelaide, S. A.
(Survey of clay deposits, characteristics and reserves).

The University of Adelaide, Box 498D, G. P. O., Adelaide, S. A.
(a) School of Geology
(Clay mineralogy)
(b) Waite Agricultural Research Institute
(Structural and chemical studies on clays).

The Victorian Department of Mines. Treasury Place, Melbourne.
(Deposit surveys, analyses and properties of clays).

The Geological Survey of Western Australia. 26 Frances Street, Perth, W. A.
(Geology of deposits, reserves).

The Government Chemical Laboratories. Western Australia, 64 Adelaide Terrace, Perth, W. A.
(Characteristics, beneficiation and use of clay).

The Australian Ceramic Society.
Secretary, R. Hill, C/—CSIRO Division Building Research.

The Australian Clay Minerals Society.
Secretary, N. G. Buckley, C/—CSIRO Division of Applied Mineralogy.

References

[1] Aust. Bureau Mineral Resources, Bull. No. 72: Australian Mineral Resources: The Mineral Deposits pp. 113—126, Clays.

[2] Joplin, G. A. (1954): Mineral Resources of Australia, Summary Report No. 37, Clay (Bureau of Mineral Resources).

[3] Jack, R. L. (1926): Clay and Cement in South Australia. Geol. Surv. S. Aust. Bull. 12.

[4] Gaskin, A. J. and Samson, H. R. (1951): Ceramic and Refractory Clays of South Australia. Geol. Surv. S. Aust. Bull. 28.
[5] Cox, R. W., Frostick, A. C., Garrett, W. G. and Williamson, W. O. (1956): Ceramic and Refractory Clays of Western Australia. CSIRO Div. Indust. Chem. Tech. Paper No. 2.
[6] Loughnan, F. C. (1960): The Origin, Mineralogy and some Physical Properties of the Commercial Clays of New South Wales. Univ. N. S. W. Geol. Ser. No. 2.
[7] Robinson, E. V. (1960): Clays of Queensland. J. Geol. Soc. Aust. 7, Appendix 1.
[8] Simpson, E. S. (1952): Minerals of Western Australia. Vol. 3. pp. 1—41 (W. A. State Govt.).
[9] Keble, R. A. and Thomas, D. E. (1952): Clay and Shale Deposits of Victoria. Geol. Surv. Vic. Mem. 18.
[10] Bowley, H. (1940): The Ceramic Resources of Western Australia. J. Roy. Soc. W. A., 27, p. 181.
[11] Callister, R. C. (1924): Australian Clays in the Manufacture of White Pottery Ware. Aust. Inst. Sci. and Ind. Bull. 27.
[12] Ridgway, J. E. (1952): Kaolinised Aplite near Ardrossan. S. Aust. Min. Rev. 93, p. 93.
[13] Willington, C. M. and Jackson, N. (1952): Beneficiation of Pine Point Clay. S. Aust. Min. Rev. 93, pp. 139 and 149.
[14] Parkin, L. W. (1948): Clay Mine, Birdwood. S. Aust. Min. Rev. 86, p. 106.
[15] Ridgway, J. E. (1953): Birdwood Clay Deposit. S. Aust. Min. Rev. 95, p. 72.
[16] Wade, M. L. (1954): Birdwood White Clay Deposit. S. Aust. Min. Rev. 97, p. 34.
[17] Smith, A. D. (1956): Beneficiation of Birdwood Clay. S. Aust. Min. Rev. 101, p. 119.
[18] Connah, T. H. (1950): White Clays Goodger and Brooklands. Qld. Govt. Min. J. 51, p. 260.
[19] Denmead, A. K. (1929): Kaolin Deposits Suitable for Pottery Manufacture. Qld. Govt. Min. J. 30, p. 99.
[20] Gaskin, A. J. (1944): Kaolinised Granodiorite in the Bulla-Broadmeadows Area. Proc. Roy. Soc. Vic. 56, p. 1.
[21] Miles, K. R. and Stephens, H. S. (1950): Molding Sands. Dept. Mines W. A. Min. Res. Bull. 5, p. 74.
[22] Feldtmann, F. R. (1919): Clackline and Baker's Hill Clay Deposits. Dept. Mines W. A. Ann. Rep. p. 99.
[23] Matheson, R. S. (1937): Report on the Clackline Firebrick Clay Pits. Dept. Mines W. A. Ann. Rep. p. 68.
[24] Cole, W. F. and Carthew, A. R. (1953): The Mineralogical Composition of some Tasmanian Clays. Proc. Roy. Soc. Tas. 87, p. 1.
[25] Loughnan, F. C. and Golding, H. G. (1957): The Mineralogy of the Commercial Dyke Clays in the Sydney Area. J. Roy. Soc. N. S. W. 91, p. 85.
[26] Morrison, M. and Raggatt, H. G. (1928): The Mineral Industry of New South Wales. (Dept. Mines N. S. W.) p. 216.
[27] Heath, G. R. (1962): Imbitcha Clay Deposit. S. Aust. Min. Rev. 117, p. 38.
[28] Ridgway, J. E. and Johns, R. K. (1952): Lobethal Clay Deposits. S. Aust. Min. Rev. 92, p. 51.
[29] Johns, R. K. (1958): White Clay Deposit—Hundred of Miltalie. S. Aust. Min. Rev. 105, p. 80.
[30] Hiern, M. N. (1965): White Clay Deposit—Section 87, Hundred of Miltalie. S. Aust. Min. Rev. 119, p. 64.
[31] Ellerton, H. (1952): The Occurrence of China Clay on Eyre Peninsula and its Industrial Significance. S. Aust. Min. Rev. 93, p. 26.
[32] Broadhurst, E. (1954): Para Wirra Kaolin Mine. S. Aust. Min. Rev. 80, p. 103.
[33] Nixon, L. G. B. (1962): Cromer 'C' White Clay Deposit. S. Aust. Min. Rev. 113, p. 33.
[34] Clothier, E. A. (1959): Fireclay Deposit near Milbrook Reservoir. S. Aust. Min. Rev. 107, p. 35.
[35] Hiern, M. N. (1965): Clay Deposit near One Tree Hill. S. Aust. Min. Rev. 118, p. 32.
[36] Denmead, A. K. (1967): (Geol. Surv. Qld.) Personal communication.

[37] Wade, M. L. (1953): Paracombe White Clay Deposit. S. Aust. Min. Rev. 95, p. 77.
[38] Ellerton, H. (1953): Report on Paracombe Clay. S. Aust. Min. Rev. 95, p. 80.
[39] Betheras, F. N. (1952): High Silica Clay Deposit near Port Augusta. S. Aust. Min. Rev. 92, p. 162.
[40] Olliver, J. G. (1961): Sampling of White Clay Deposit near Port Augusta. S. Aust. Min. Rev. 114, p. 65.
[41] Johns, R. K. (1967): Geology and Mineral Resources of the Andamooka and Torrens Sheets. Geol. Surv. S. Aust. Bull. 41, (in press)
[42] Nixon, L. G. B. (1960): White Clay Deposit—Longwood. S. Aust. Min. Rev. 109, p. 95.
[43] Nixon, L. G. B. (1957): White Clay Deposit—Hundred of Stanley. S. Aust. Min. Rev. 104, p. 36.
[44] Loughnan, F. C., Grim, R. E. and Vernet J. (1962): Weathering of some Triassic Shales in the Sydney Area. J. Geol. Soc. Aust. 8, (2) p. 245.
[45] Harper, L. F. (1906): China Clay, Ulladulla. Ann. Rep. Dept. Mines N. S. W. p. 170.
[46] Bowley, H. (1923): Reports on China Clay-Wagin. Dept. Mines W. A. Ann. Rep. p. 166.
[47] Wiebenga, W. A. (1955): Geophysical Investigations of Water Deposits, Western Australia. Aust. Bur. Min. Res. Bull. 30.
[48] Armstrong, A. T. (1957): Williamstown (Mount Crawford) Clay Deposits. S. Aust. Min. Rev. 104, p. 31.
[49] Kaiser, C. F. (1957): Beneficiation of Williamstown Clay. S. Aust. Min. Rev. 104, p. 33.
[50] Cochrane, G. W. (1954): Clay Deposit—Williamstown. S. Aust. Min. Rev. 96, p. 51.
[51] Alderman, A. R. (1942): Sillimanite, Kyanite and Clay Deposits near Williamstown. Proc. Roy. Soc. S. Aust. Trans. 66, (1) p. 3.
[52] Jacskon, N. (1950): Preparation of Paper-Coating Clay from Mount Crawford Kaolin. S. Aust. Min. Rev. 90, p. 190.
[53] Loughnan, F. C. (1967): (Univ. N. S. W. Geol. Dept.) Personal communication.
[54] Croft, J. B. (1967): (Univ. Queensland, Geol. Dept.) Personal communication.
[55] Hiern, M. N. (1968): The Regional Setting of South Australian Ceramic Clays. S. Aust. Geol. Surv. Quart. Geol. Notes 25.
[56] Bain, A. D. N. and Spencer-Jones, D. (1953): Egerton Kaolin Mines. Min. Geol. J. (Vic.) 5:30.
[57] Kalix, Z., Fraser, L. M., and Rawson, R. I. (1966): Australian Mineral Industry: Production and Trade 1842—1964. C'wlth of Aust. Bur. Min. Resources. Bull. No. 81.

1969 XXIII INTERNATIONAL GEOLOGICAL CONGRESS (1968) — VOL. 16 PAGE 151—157

Kaolin Deposits of New Zealand

F. E. BOWEN
New Zealand

Summary—The major part of New Zealand kaolin reserves (18×10^6 plus tons) consist of the massive hydrothermal halloysites of Northland. Marginal hydrothermal kaolinites of Coromandel Peninsula are increasing in importance. Sub-coal measure kaolinites resulting from weathering processes are probably widespread but have not been extensively exploited because of mining difficulties.

Distribution and classification

New Zealand kaolins consist of the massive hydrothermal halloysites of Northland (Group 1), the marginal hydrothermal kaolinites of Coromandel Peninsula (Group 2), and the sub-coal measure kaolinites of the South Island (Group 3). The Northland halloysites are variable, depending on whether they were derived from dacite (1A), rhyodacite (1B) or rhyolite (1C). The distribution of individual deposits is shown in Fig. 1, in which the type of deposit is indicated in parentheses.

Geological conditions

The massive hydrothermal deposits of Northland (1A, 1B, 1C) are derived from rocks of the three formations of the Parahaki Volcanic Group. This group forms the middle part of a Cenozoic sequence which began with andesitic and concluded with basaltic volcanism. Hydrothermal halloysites are confined to the Parahaki rocks, except at Putahi where clays derived hydrothermally from both rhyolite and andesite are present. The andesite derived clays have no quartz or cristobalite fragments and, chemically, have less silica than the rhyolite derived clays. They have not been examined in detail but are unlikely to contain large reserves.

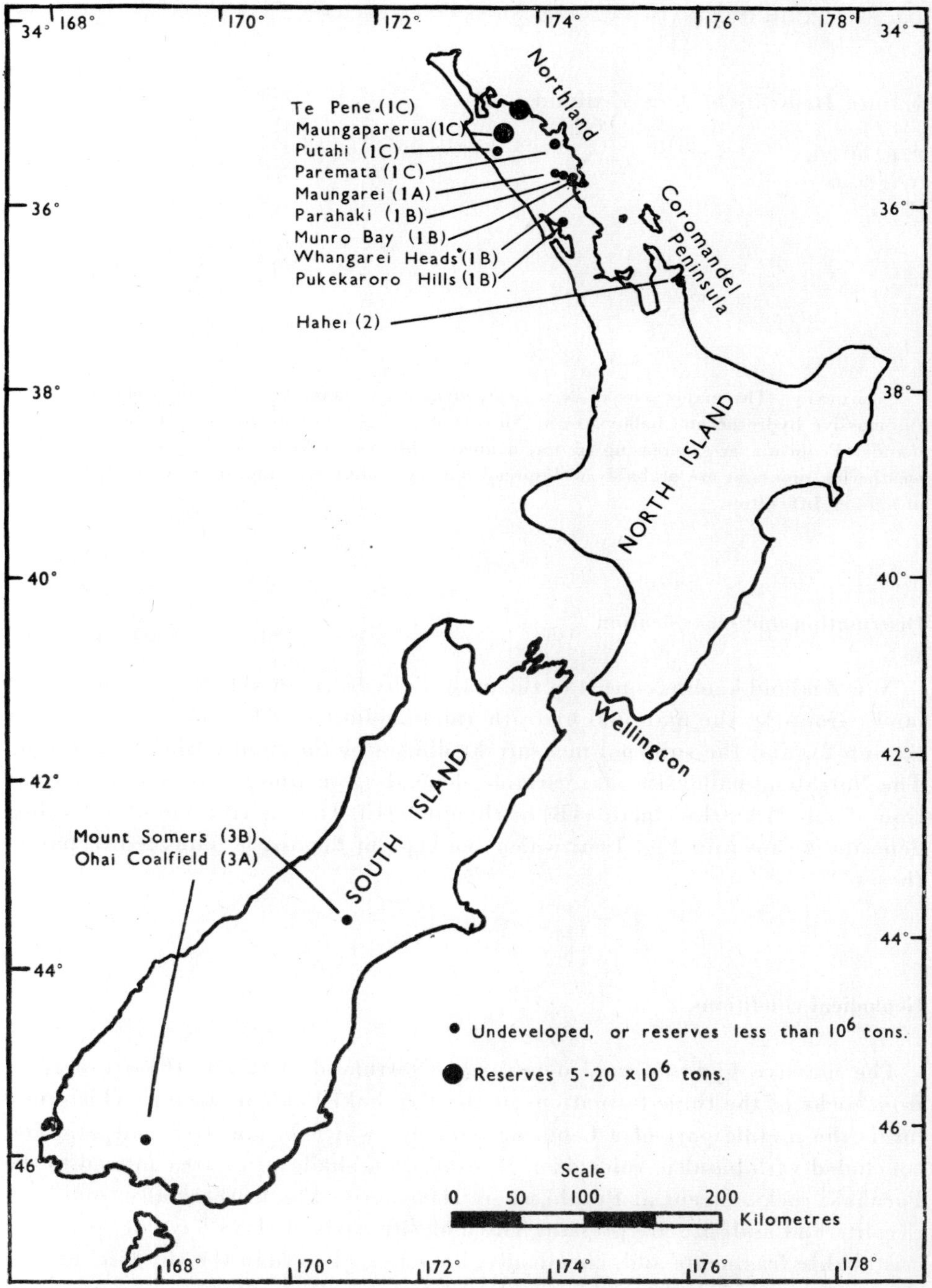

Fig 1—Kaolin deposits of New Zealand.

Table 1—Geological conditions of New Zealand kaolins

	Massive hydrothermal (1) halloysite	Marginal hydrothermal (2) kaolinite	Sub-coal measure (3) kaolinite
Region	Northland	Coromandel Peninsula	South and ?North Islands
Age	Plio-Pleistocene	Plio-Pleistocene	Cretaceous-Tertiary
Geological environment	Alteration of volcanic dome.	Zone around volcanic dome.	Redeposition in fresh-water basin.
Morphology and area	Massive, irregular. Individually up to at least 2.5 km^2.	Steep, irregular, pipe-like or marginal to dome. Individually less than 0.1 km^2.	Stratiform. Extent not known but could be widespread.
Depth of alteration	Up to at least 43 m.	Up to at least 50 m.	Up to at least 3 m.
Exploitable thickness	Up to at least 41 m.	Up to at least 12 m.	Up to at least 3 m.
Parent rock	Pliocene-Early Pleistocene dacite (1A), rhyodacite (1B), rhyolite (1C).	Plio-Pleistocene spherulitic and lithoidal rhyolite and rhyolite pipe breccia.	Triassic, indurated tuffaceous sediments, gabbro, diorite (3A). ?Cretaceous rhyolite (3B).
Overburden	Stained clays up to 6 m. thick. Local weathered basalt of variable thickness.	Locally sinter and terrace deposits of variable thickness.	Estuarine and marine mudstone and sandstone of variable thickness.
Genetic type	Acid hydrothermal.	Acid hydrothermal.	Peneplain weathering product.
Tectonic relations		Pene-contemporaneous faulting.	Post-depositional faulting and folding
Relations to other products		Pyrite, pyrophyllite and alunite may be present in adjacent zones.	
Conditions affecting exploitation	Water could be problem in deep mining.	Water seepage could be a problem in some places.	Local water problems. Deposits faulted and folded.
Total reserves	Visible 18×10^6 tons. Probable and possible, not known but could be many millions of tons.	Not known.	Not known, but could be large.

The kaolinites of Coromandel Peninsula are found in the pipes and around the margins of rhyolitic domes (Minden Rhyolites), and adjacent to associated faults and shears. Individually the deposits are not as extensive as the Northland halloysites, but only a few have yet been found.

Over large areas of New Zealand the closing stages of the late Mesozoic Rangitata Orogeny were marked by deep leaching of highly indurated, often tuffaceous, Mesozoic and older sedimentary rocks and associated intrusive and extrusive rocks. The succeeding Cretaceous and Tertiary coal measures are sometimes underlain by, and usually partly derived from, the products of this leaching. Contemporaneous and post depositional faulting and folding of the sequence has resulted in these clays occupying a variety of geological positions and varying greatly in form, thickness and extent. Most deposits are too deeply buried to be mined by opencast methods. Many of these clays contain illite so that, in the absence of precise mineralogical and physical data, it is not certain which of the deposits are true kaolins. The genetic differences between the clays of different deposits have not been investigated but may be related both to differences in the parent rocks and to differences in the relative effects of the original weathering and later leaching by circulating acid waters from the swamps in which the subsequent coal seams originated.

Mineralogical and chemical conditions

Massive hydrothermal clays derived from dacite are known at only one locality (Maungarei, 1A). The parent rock consists of large phenocrysts of plagioclase (andesine to labradorite), small phenocrysts of hypersthene and hornblende, and large grains of magnetite. Quartz is absent. The matrix is hyalopilitic. The clay, identified as halloysite by x-ray diffraction, is usually plastic, and has a linear shrinkage of 20—25 % at 1,250 °C. resulting in cracking of the fired product. The average chemical analysis of this clay is given in Table 2.

The rhyodacites are mainly holocrystalline, variably porphyritic rocks with phenocrysts of plagioclase (oligoclase to andesine), scarce hornblende, rare hypersthene and biotite. Quartz may be locally abundant but is usually absent. The finely crystalline groundmass usually contains orthoclase, plagioclase, quartz and magnetite, but locally may be glassy. The rhyodacite-derived clays are moderately coarse grained halloysites locally admixed with kaolinite. Shrinkage on firing is usually low (5.0—15.6 % at 1,250 °C.) but is high where the parent rock was glassy or had a high alkali content, as at Munro Bay (Table 2). Quartz fragments are only present where the parent rock contained quartz phenocrysts.

The Northland rhyolites contain phenocrysts of sanidine, albite (or anorthoclase), quartz, and scarce hornblende and muscovite, set in a microcrystalline

to rarely hyalophitic groundmass of quartz and feldspar. The clays are poorly plastic, and contain a high proportion of silica fragments. Linear shrinkage at 1,250 °C. varies from 3.2 % to 28.2 % but is usually low. A sample from Te Pene, identified by x-ray diffraction as highly hydrated halloysite, showed strong peaks at 10.1 A° (001) and 3.63 A° (003) with no second order reflection. Near the ground surface this clay contains 5 %—10 % quartz and 20 %—30 % cristobalite but there is a progressive downward decrease in the percentage of quartz and a complementary increase in that of cristobalite, although the relative percentage of halloysite remains fairly constant to at least 37 m. Of the clay substance content 86.3 % has a grain size less than 0.02 mm. and 57.6 % is less than 0.002 mm. Average chemical analyses for three deposits are given in Table 2.

Table 2—Chemical analyses of New Zealand kaolins and their parent rocks

		(1)	(2)	(3)	(4)	(5)	(6)	(7)	(8)	(9)	(10)	(11)	(12)
SiO_2	%	66.96	46.8	70.89	67.2	66.1	72.50	57.2	69.4	65.0	44.8	76.81	77.2
TiO_2	%	0.44	0.98	0.15	0.14	0.35	0.07	1.13	0.10	0.26	1.8	0.12	0.20
Al_2O_3	%	15.76	34.6	15.26	21.8	22.2	10.67	25.9	22.0	23.7	37.6	12.39	16.4
Fe_2O_3	%	1.15	2.40	0.49	0.88	1.46	4.16	3.40	0.15	0.71	1.7	0.63	0.35
FeO	%	1.77		0.60								0.39	
MnO	%	0.07		0.01								0.01	
MgO	%	0.64	0.05	0.84	0.10	0.44	0.52	0.58	—	0.18	0.1	0.07	0.13
CaO	%	2.50	0.10	1.37	0.34	0.23	1.58	0.17	0.12	0.08	0.1	0.88	0.06
Na_2O	%	4.04	0.05	2.62	0.51	0.15	3.38	1.33	0.18	0.10	0.1	2.82	0.02
K_2O	%	2.19	0.13	4.26	1.40	1.18	4.72		0.21	0.28	0.1	4.98	tr.
H_2O^+	%	3.23	14.8	1.40	7.83	7.96	2.40*	10.0	8.20	9.80	13.8	0.34	6.10
H_2O^-	%	0.73		1.93								0.57	
P_2O_5	%	0.14		0.02								0.06	
CO_2	%	nil		nil			nil					0.01	

* H_2O^+ org. matter

(1)—Dacite, Maungarei; (2)—Halloysite, Maungarei—11 samples; (3)—Rhyodacite, Munro Bay; (4)—Halloysite, Munro Bay—5 samples; (5)—Halloysite, Parahaki—5 samples; (6)—Rhyolite, Putahi; (7)—Halloysite, Putahi; 4 surface samples. (8)—Halloysite, Te Pene—6 samples. (9)—Halloysite, Maungaparerua—6 samples; (10)—Kaolinite, Ohai Coalfield; (11)—Rhyolite, Mount Somers; (12)—Kaolinite, Mount Somers.

The rhyolite domes of Coromandel Peninsula contain plentiful quartz and plagioclase and rare to scattered hypersthene, hornblende and biotite crystals, in a holocrystalline to spherulitic glassy groundmass. The clay is developed within the pipe leading to the dome or, where the dome has spread, around the margins. It consists of 5 %—45 % cristobalite, 5 %—10 % each of quartz and tridymite and 55 %—80 % kaolin, 64.8 % of which is less than 0.02 mm. and 44.1 % is less than 0.002 mm. in grain size. X-ray diffraction showed strong peaks at 7.1 A° (001) and 3.58 A° (002). An electron micrograph of Hahei clay shows plates

and some aggregates of kaolinite, together with cristobalite and amorphous quartz. No chemical analysis of this clay is available.

The parent rock of the sub-coal measure clays in the Ohai Coalfield, consists of Triassic indurated tuffaceous sandstone intruded by gabbro and diorite. At Mount Somers the parent rock is ?Cretaceous flow rhyolite, with some associated rhyolite breccias, consisting of phenocrysts of quartz, sanidine, plagioclase (oligoclase-andesine) and rare biotite set in a mesostasis of cryptocrystalline quartz and feldspar, or glass. The only other information recorded for the Mount Somers clays is the analysis given in Table 2. The Ohai Coalfield kaolinite showed good 7.12 A° (001) and 3.58 A° (002) x-ray diffraction peaks, while differential thermoanalysis showed a sharp endothermic reaction at 600 °C. and a similar exothermic reaction at 980 °C. An electron micrograph indicates the clay to be imperfectly crystalline but with a platy to hexagonal habit. Although other sub-coal measure clays, many of which may contain kaolin, are widespread in New Zealand, the cost of development is likely to be greater than that for the other types of clay.

Utilisation

New Zealand kaolins are mainly used for ceramics and as a filler in the paper, paint, adhesive and cosmetic industries. The only technological information available relates to the treatment of halloysite from Te Pene, which is dried in a rotary furnace and air separated into the required size fractions.

Economics

Data relating to reserves of kaolin are given in Table 1. Transport costs in New Zealand are high, so that underground mining is not feasible unless the deposit is close enough to the industry using the clay for the higher mining costs to be offset by savings resulting from short haulage. Extraction of the clay is thus normally by quarrying.

Statistical data

The following are the only statistical data available on the New Zealand kaolin industry.

	1961	1962	1962/3	1963/4	1964/5	1965/6
Imports (metric tons)	40,509	19,350	31,887	56,148	58,043	52,889

The tonnage for 1962 is for the first half of the year only; thereafter the tonnage refers to years commencing in June. Sources of imports were: the United Kingdom, Australia, the U.S.A., France, the German Federal Republic and the Netherlands.

	1964	1965	1966
Home output for ceramics (metric tons)	1,092	1,031	1,335
Home output for industry (metric tons)	2,304	3,344	1,982

History of investigation and exploitation

The small demand for kaolin by New Zealand industry has been traditionally met by imports and by the small scale use of local material that has often been inferior to the imported kaolins. In the past decade increasing interest in indigenous raw materials, together with technological advances in utilisation, has led to the mining of high grade halloysite, principally at Te Pene, but also in the Whangarei Heads area. In 1966, production of kaolinite commenced at Hahei, Coromandel Peninsula. Detailed prospecting is being carried out in other localities of Northland and Coromandel Peninsula in an attempt to meet the demands of local industry, and possibly to establish an export trade in treated clays.

References

Bowen, F. E. (1966): The Parahaki Volcanic Group and its Associated Clays. Report N. Z. G. S. 6 New Zealand Geological Survey, Department of Scientific and Industrial Research, Wellington.

Fieldes, M.; McDowall, I. C.; Claridge, G. G. C.; Williams, G. J. (in Williams, G. J. 1965): Clays. Economic Geology of New Zealand. Publ. 8th Commonwealth Min. Metal. Cong. Vol. 4. Austral. Inst. Min. Metal. Melbourne.

Harvey, C. C. (1967): Rock alteration in the south-east Whitianga Area. Univ. Auck. Thesis (Lodged in University of Auckland Library).

Henderson, J. (1943): Fireclay and Gannister in New Zealand. N. Z. Dep. Sci. Industr. Res. Bull. 8.

Unpublished reports and analyses are available from Chemistry Division, New Zealand Department of Scientific and Industrial Research, and research is being currently carried out by the N. Z. Pottery and Ceramic Research Association (Inc.). Papers on kaolins are most likely to appear in N. Z. J. Geol. Geophys. and N. Z. J. Sci.